# Interpretation of Biological and Environmental Changes across the Neoproterozoic-Cambrian Boundary

# Interpretation of
# Biological and Environmental Changes
# across the Neoproterozoic-Cambrian Boundary

Edited by

Loren E. Babcock
*Department of Geological Sciences*
*The Ohio State University*
*Columbus, Ohio 43210*
*USA*

2005

**ELSEVIER**

Amsterdam • Boston • Heidelberg • London • New York • Oxford • Paris
San Diego • San Francisco • Singapore • Sydney • Tokyo

**ELSEVIER B.V.**          ELSEVIER Inc              ELSEVIER Ltd              ELSEVIER Ltd
**Radarweg 29**           525 B Street, Suite 1900  The Boulevard, Langford Lane  84 Theobalds Road
**P.O. Box 211,**          San Diego,               Kidlington,               London
**1000 AE Amsterdam**     CA 92101-4495            Oxford OX5 1GB            WC1X 8RR
**The Netherlands**        USA                      UK                       UK

First edition 2005


Library of Congress Cataloging in Publication Data
A catalog record is available from the Library of Congress

British Library Cataloguing in Publication Data
A catalogue record is available from the British Library

Reprinted from Palaeogeography, Palaeoclimatology, Palaeoecology, Volume 220, Issues 1-2

ISBN    0 444 52065 1

Printed and bound by CPI Antony Rowe, Eastbourne
Transferred to digital print on demand, 2007

# Contents

Available online at www.sciencedirect.com

SCIENCE @ DIRECT°

ELSEVIER

Palaeogeography, Palaeoclimatology, Palaeoecology 220 (2005) 1–5

PALAEO

www.elsevier.com/locate/palaeo

Editorial

# Interpretation of biological and environmental changes across the Neoproterozoic–Cambrian boundary: developing a refined understanding of the radiation and preservational record of early multicellular organisms

## 1. Introduction

The Neoproterozoic–Cambrian transition was a time of fundamental change in the biosphere Between about 570 million years ago (the late Neoproterozoic) and 510 million years ago (the late early Cambrian, Grotzinger et al, 1995, Landing et al, 1998), marine organisms underwent considerable evolutionary innovation during a time of shifting ecological setting This dramatic activity culminated in the first stratigraphic appearances of many recognizable clades of animals, an "event" often referred to as the "Cambrian explosion" (e g, Cloud, 1968, Runnegar, 1982, Conway Morris, 1985; Erwin, 1991, Signor and Lipps, 1992, Bengtson, 1994, Valentine, 1994, Zhuravlev and Riding, 2001) In this rather protracted interval, predation became a significant factor in evolution (e.g, Babcock, 1993, 2003, Bengtson, 1994, Conway Morris and Bengtson, 1994, Conway Morris and Jenkins, 1995, Vannier and Chen, 2002, Huang et al, 2004, Zhu et al, 2004), and was likely one factor, combined with geochemical factors, involved in the appearance and later development of biomineralized skeletons In addition, there was a major change from a microbial mat-dominated sediment–water interface to a more extensively burrowed interface in shallow-marine settings (e g, Seilacher and Pfluger, 1994; Bottjer et al, 2000)

Biological changes during the Neoproterozoic–Cambrian transition occurred amidst physical and chemical changes of global importance (Knoll and Walter, 1992) Following the breakup of a Proterozoic supercontinent (Rodinia or Pannotia), high tectonic spreading rates and high volumes of mid-ocean ridges caused expansion of world oceans over many cratonic areas Melting of the extensive continental glaciers of the Neoproterozoic also played a role in marine flooding of the continents This flooding led to increased erosion rates and changes in seawater chemistry Most notably, increased oxygenation of the world ocean (Canfield and Teske, 1996) helped establish one of the most important prerequisites for the rise and later diversification of multicellular animals

The fossil record of biological evolution across the Neoproterozoic–Cambrian transition is related not only to the rise or ecological diversification of marine organisms, but also to the development of taphonomic and sedimentary conditions suitable for the preservation of mineralizing and non-mineralizing organisms The earliest multicellular organisms to construct mineralized skeletons appeared and diversified during

0031-0182/$ - see front matter © 2004 Elsevier B V All rights reserved
doi 10 1016/j palaeo 2004 09 013

the Neoproterozoic–Cambrian interval (Grant, 1990, Knoll and Walter, 1992, Knoll, 1996), but their diversity and abundance was greatly overshadowed by organisms lacking the capability to secrete biomineralized skeletons (e g , Conway Morris, 1985, 1986, Whittington, 1985, Briggs et al , 1994, Chen et al , 1997, Hou et al , 1999, Babcock et al , 2001) Our understanding that the vast majority of marine animals, plants and macroscopic bacteria lacked hard parts comes from deposits of exceptional preservation, including so-called Ediacaran biotas (e g , Glaessner, 1984, Seilacher, 1989, Conway Morris, 1990, Fedonkin, 1994, Narbonne, 1998, Gehling, 1999) and Burgess Shale-type deposits (e g , Conway Morris, 1985, 1990, Allison, 1988, Briggs et al , 1994, Butterfield, 1995, Babcock et al , 2001), which are also known as Konservat-Lagerstatten (Seilacher et al , 1985) Clearly, in order for non-biomineralized organisms to become fossilized, one of the most important taphonomic filters, biodegradation (including both carnivory and microbial decay), seems to have been suspended or reduced sufficiently under certain environmental circumstances to permit bodily remains to enter the sedimentary record Chemical conditions of the water or sediment columns (e g , anoxia or salinity extremes) are inferred to have played major roles in ensuring that bodily remains were preserved long enough for burial to occur (e g , Allison and Briggs, 1991, Babcock et al , 2001) Rapid burial, through sediment smothering (obrution) due to storms or turbidity currents, tidally influenced deposition or other means, also seems to have played a central role in many instances of exceptional preservation (e g , Conway Morris, 1986; Robison, 1991, Liddell et al , 1997, Babcock et al , 2001) Also important must have been the rapid onset of mineralization, particularly of non-biomineralized remains The action of microbial consortia, including fungi and autolithifying bacteria (see Borkow and Babcock, 2003), is inferred to be one pathway by which the remains of non-biomineralized metazoans and metaphytes were replicated as fossils

In addition to an improved body fossil record across the Neoproterozoic–Cambrian transition, the record of organism–sediment interactions, as demonstrated by trace fossils, underwent dramatic change (e g , Seilacher and Pfluger, 1994, Bottjer et al , 2000) Mat-dominated communities declined during this time, largely as the result of sediment disturbance by organisms (e g , Droser and Bottjer, 1988) Animals not only took refuge in sediment, but also sought prey in sediment (e g , Jensen, 1990, Babcock, 2003) Thus, predation, one of the major forcing factors involved in ecological and evolutionary change in the biosphere during the Neoproterozoic–Cambrian transition, also seems to have played a role in the change from mat-dominated communities of the late Neoproterozoic to increasingly fluidized substrates during the early Paleozoic. Although animals may have disturbed sediment at an increasing rate through the Phanerozoic, some stabilization of sediment grains must have occurred in order for traces and sedimentary structures to have been preserved (see Schieber, 1999) Microbes within the sediment may have played a role in sediment binding even during and following the decline of mat-dominated communities

Escalation in predator–prey systems through the late Neoproterozoic and early Paleozoic (Vermeij, 1995, Babcock, 2003) evidently resulted in changing morphologies among predators and prey, and some changes in burrowing behavior may have accompanied those changes Together, ecological and evolutionary changes in the biosphere, and animal–sediment interactions that resulted from evolutionary changes, have been linked together under the larger concept of an Early Paleozoic Marine Revolution (Babcock, 2003)

The study of Neoproterozoic and Cambrian deposits of exceptional preservation is of crucial importance for unraveling the complex circumstances leading to and resulting from ecological and evolutionary diversification of multicellular organisms Study of such deposits is also of crucial importance for decoding the complex factors involved in the preservation of fossils, both body fossils and trace fossils Refining our understanding of life's history from the Proterozoic through the Early Paleozoic Marine Revolution is one of the most exciting challenges in paleontology today The lessons we learn from this interval of Earth history are likely to have an important impact on our interpretation of the subsequent stratigraphic and evolutionary records

The symposium "The Precambrian–Cambrian Biotic Transition Interplay of Biological and Environmental Changes" held at the North American

Paleontological Convention in Berkeley, California (NAPC 2001), was devoted to an exploration of some of the emerging concepts and techniques used to develop greater insight into the early record of organismic diversification and the preservational record of that diversification The meeting brought together researchers from eight countries, and represented various research interests in paleontology, stratigraphy, sedimentology and geochemistry This volume includes a selection of papers developed as an outgrowth of the NAPC symposium

## 2. Summary of contents

This volume is organized with papers emphasizing two major areas of interest 1, Proterozoic organisms and environments, followed by 2, early and middle Cambrian organisms and environments The contribution by Babcock et al offers new insight into the phylogenetic affinities of an Ediacaran-type animal from Brazil Interpretation of the animal as a cnidarian emphasizes the role of predation in guiding evolutionary processes during the Neoproterozoic (and prior to the Cambrian radiation) Trace fossil preservation and its relationship to evolution of early animals are examined in the paper by Jensen et al Based on a lack of a significant trace fossil record prior to the beginning of the Cambrian, large benthic bilaterian animals are inferred to have arisen relatively late in the Neoproterozoic–Cambrian transition interval

Interpretation of Cambrian deposits of exceptional preservation comprises the central theme for most of the papers in this issue. The paper by Zhu et al examines the diagenetic history of fossils from the Chengjiang deposit of China using scanning electron microscopy (SEM) and energy dispersive X-ray spectrographic analysis (EDS) techniques Early mineralization, particularly by apatite, pyrite and iron-rich aluminosilicates, is inferred to have a played a key role in the preservation of non-biomineralized organisms An association also seems to be present between tissue type and method of mineralization The "Cambrian substrate revolution" and some of the major organisms that played roles in it are the subject of a contribution by Dornbos et al The authors argue that adaptive radiation of benthic organisms during

the "Cambrian explosion" was driven in part by the "Cambrian substrate revolution" as benthic animals were forced to adapt to development of a mixed layer in the sediment Paleoecology of the Sinsk Biota of Siberia is summarized and interpreted in a paper by Ivantsov et al. Inferred to have been deposited under anoxic conditions, Sinsk organisms primarily include sessile filter-feeders, vagrant detritophages, and diverse carnivores or scavengers adapted to life under dysaerobic conditions A remarkable occurrence of articulated sponges from the Hetang Biota of South China is described by Xiao et al Sponges are shown to have been taxonomically diverse and morphologically complex by the early Cambrian Peng et al describe some early enigmatic fossils now interpreted as sessile cnidarians and further emphasize the importance of predation in driving evolution during the Cambrian Work by Steiner et al describes exceptionally preserved animal fossils from new localities in South China Early Cambrian taxa from these sites include a variety of arthropods, worms, sponges, chancelloriids, cnidarians and cap-shaped fossils The contribution by Hollingsworth summarizes information on the earliest occurrences of trilobites and brachiopods in Laurentia The first occurrence of trilobites in Laurentia is inferred to approximate the first appearances of trilobites both in Siberia and Gondwana The paper by Skinner describes the taphonomy of exceptionally preserved fossils from the Kinzers Biota of Pennsylvania Articulated remains of non-biomineralized animals are rare in the Kinzers Formation, but fragments evidently left behind as the result of successful predation and scavenging are common and provide strong evidence that predation was a major taphonomic filter during the Cambrian Similar to work on the Chengjiang deposit (Zhu et al ), Skinner also describes the diagenetic history of the Kinzers fossils as deduced through SEM and EDS techniques Skinner links autolithifying microbes responsible in part for biodegradation to early diagenesis of non-biomineralized skeletons Gaines et al outline a new hypothesis for the preservation of organic carbon-based tissues in the Wheeler Formation of Utah The authors argue that a combination of factors reduced permeability in sediment and thus reduced oxidation of carbon, which in turn restricted microbial decomposition of organic tissues

One paper in this volume addresses the development of early Phanerozoic reef communities Zhuravlev and Naimark characterize the faunal composition of early Cambrian reefs through qualitative and quantitative means and conclude that the early Phanerozoic evolution of reef communities was expansive, although diversity was limited

## Acknowledgements

Many people have played a role in producing this special issue the authors who transformed talks into papers, colleagues who served as referees and those who helped assemble the volume into printable form Referees for submitted papers include R B Blodgett, G Budd, L E Holmer, S Conway Morris, R R Gaines, G Geyer, J W Hagadorn, L Holmer, S A Leslie, B S Lieberman, J H Lipps, C G Maples, B R Pratt, A L Rode, S M Rowland, M.R Saltzman, J Schieber, E S Skinner, B Waggoner, M D. Wegweiser and M Y Zhu K Polak assisted with editing this special issue Finally, I would like to thank the Elsevier staff, including Femke Wallien, for encouragement and advice during the editing of this volume This work was supported in part by grants from the National Science Foundation (EAR 9405990, 9526709, 0073089, 0106883, EAR-OPP 9614989, 0229757, 0326709)

## References

Allison, P A, 1988 Konservat-Lagerstatten cause and classification Paleobiology 14, 331–344

Allison, P A, Briggs, D E G, 1991 Taphonomy of nonmineralized tissues In Allison, P A, Briggs, D E G (Eds), 2001 Taphonomy Releasing the Data Locked in the Fossil Record Plenus Press, New York, pp 26–70

Babcock, L E, 1993 Trilobite malformations and the fossil record of behavioral asymmetry J Paleontol 67, 217–229

Babcock, L E, 2003 Trilobites in Paleozoic predator–prey systems, and their role in reorganization of early Paleozoic ecosystems In Kelley, P H, Kowalewski, M, Hansen, T A (Eds), 2001 Predator–Prey Interactions in the Fossil Record Kluwer Academic/Plenum Publishers, New York, pp 55–92

Babcock, L E, Zhang W T, Leslie, S A, 2001 The Chengjiang Biota record of the early Cambrian diversification of life and clues to the exceptional preservation of fossils GSA Today 11 (2), 4–9

Bengtson, S, 1994 The advent of animal skeletons In Bengtson, S (Ed), Early Life on Earth Columbia University Press, New York, pp 412–425

Borkow, P S, Babcock, L E, 2003 Turning pyrite concretions outside-in role of biofilms in pyritization of fossils Sed Rec 1 (3), 4–7

Bottjer, D J, Hagadorn, J W, Dornbos, S Q, 2000 The Cambrian substrate revolution GSA Today 10 (9), 1–7

Briggs, D E G, Erwin, D H, Collier, F J, 1994 The Fossils of the Burgess Shale Smithsonian Institution Press, Washington 239 pp

Butterfield, N J, 1995 Secular distribution of Burgess Shale-type preservation Lethaia 28, 1–13

Canfield, D E, Teske, A, 1996 Late Proterozoic rise in atmospheric oxygen concentration inferred from phylogenetic and sulphur-isotope studies Nature 382, 127–132

Chen, J Y, Zhou, G Q, Zhu, M Y, Yeh, K Y, 1997 The Chengjiang Biota A Unique Window of the Cambrian Explosion National Museum of Natural Science, Taichung, Taiwan 222 pp

Cloud, P E, 1968 Pre-metazoan evolution and the origins of the Metazoa In Drake, E T (Ed), Evolution and Environment Yale University Press, New Haven, pp 1–72

Conway Morris, S, 1985 Cambrian Lagerstatten their distribution and significance Philos Trans R Soc Lond, B 311, 49–65

Conway Morris, S, 1986 The community structure of the middle Cambrian phyllopod bed (Burgess Shale) Palaeontology 29, 423–467

Conway Morris, S, 1990 Late Precambrian and Cambrian soft-bodied faunas Annu Rev Earth Planet Sci 18, 101–122

Conway Morris, S, Bengtson, S, 1994 Cambrian predators possible evidence from boreholes J Paleontol 68, 1–23

Conway Morris, S, Jenkins, R J F, 1995 Healed injuries in early Cambrian trilobites from South Australia Alcheringa 9, 167–177

Droser, M L, Bottjer, D J, 1988 Trends in depth and extent of bioturbation in Cambrian carbonate marine environments, western United States Geology 16, 233–236

Erwin, D H, 1991 Metazoan phylogeny and the Cambrian radiation Trends Ecol Evol 6, 131–134

Fedonkin, M A, 1994 Early multicellular fossils In Earth Bengtson, S (Ed), Early Life on Earth Columbia University Press, New York, pp 370–388

Gehling, J G, 1999 Microbial mats in Proterozoic siliciclastics Ediacaran death masks Palaios 14, 40–57

Glaessner, M F, 1984 The Dawn of Animal Life A Biohistorical Study Cambridge University Press, Cambridge 244 pp

Grant, S W F, 1990 Shell structure and distribution of *Cloudina*, a potential index fossil for the terminal Proterozoic Am J Sci 290-A, 261–294

Grotzinger, J P, Bowring, S A, Saylor, B Z, Kaufman, A J, 1995 Biostratigraphic and geochronologic constraints on early animal evolution Science 270, 598–604

Hou, X G, Bergstrom, J, Wang, H F, Feng, X H, Chen, A L, 1999 The Chengjiang Fauna Exceptionally Well-Preserved Animals from 530 Million Years Ago Science and Technology Press, Yunnan 170 pp

Huang, D Y, Vannier, J, Chen, J Y, 2004 Anatomy and lifestyles of early Cambrian priapulid worms exemplified by *Corynetis* and *Anningvermis* from the Maotianshan Shale (SW China) Lethaia 37, 21–33

Jensen, S , 1990 Predation by early Cambrian trilobites on infaunal worms—evidence from the Swedish Mickwitzia Sandstone Lethaia 23, 29–42

Knoll, A H , 1996 Daughter of time Paleobiology 22, 1–7

Knoll, A H , Walter, M , 1992 Latest Proterozoic stratigraphy and earth history Nature 356, 673–678

Landing, E , Bowring, S A , Davidek, K L , Westrop, S R , Geyer, G , Heldmaier, W , 1998 Duration of the early Cambrian U–Pb ages of volcanic ashes from Avalon and Gondwana Can J Earth Sci 35, 329–338

Liddell, W D , Wright, S H , Brett, S E , 1997 Sequence stratigraphy and paleoecology of the Middle Cambrian Spence Shale in northern Utah and southern Idaho Brigham Young Univ Geol Stud 42, 59–78

Narbonne, G M , 1998 The Ediacara Biota a terminal Neoproterozoic experiment in the evolution of life GSA Today 8 (2), 1–6

Robison, R A , 1991 Middle Cambrian biotic diversity examples from four Utah Lagerstatten In Simonetta, A M , Conway Morris, S (Eds ), 2001 The Early Evolution of Metazoa and the Significance of Problematic Taxa Cambridge University Press, Cambridge, pp 77–98

Runnegar, B , 1982 The Cambrian explosion animals or fossils? J Geol Soc Aust 29, 395–411

Schieber, J , 1999 Microbial mats in terrigenous clastics the challenge of identification in the rock record Palaios 14, 3–12

Seilacher, A , 1989 Vendozoa organismic construction in the Proterozoic biosphere Lethaia 22, 229–239

Seilacher, A , Pfluger, F , 1994 From biomats to benthic agriculture a biohistoric revolution In Krumbein, W S , et al , (Eds ), 2001 Biostabilization of Sediments Biblioteks und Informationsystem der Universitat Oldenberg, Oldenberg, Germany, pp 97–105

Seilacher, A , Reif, W -E , Westphal, F , 1985 Sedimentological, ecological, and temporal patterns of fossil-Lagerstatten Philos Trans R Soc Lond , B 311, 5–24

Signor, P W , Lipps, J H , 1992 Origin and early radiation of the Metazoa In Lipps, J H , Signor, P W (Eds ), 2001 Origin and Early Evolution of the Metazoa Plenum Press, New York, pp 3–23

Valentine, J W , 1994 The Cambrian explosion In Bengtson, S (Ed ), Early Life on Earth Columbia University Press, New York, pp 401–411

Vannier, J , Chen, J Y , 2002 Digestive system and feeding mode in Cambrian naraoiid arthropods Lethaia 35, 107–120

Vermeij, G J , 1995 Economics, volcanoes, and Phanerozoic revolutions Paleobiology 21, 125–152

Whittington, H B , 1985 The Burgess Shale Yale University Press, New Haven 151 pp

Zhu, M Y , Vannier, J , Van Iten, H , Zhao, Y L , 2004 Direct evidence for predation on trilobites in the Cambrian Proc R Soc Lond , B (Suppl , online)

Zhuravlev, A Y , Riding, R (Eds ), 2001 The Ecology of the Cambrian Radiation Columbia University Press, New York 525 pp

Loren E Babcock
*Department of Geological Sciences,*
*The Ohio State University,*
*Columbus, OH 43210, USA*
*E-mail address* babcock 5@osu edu

27 June 2004

Available online at www.sciencedirect.com

Palaeogeography, Palaeoclimatology, Palaeoecology 220 (2005) 7–18

ELSEVIER

www elsevier com/locate/palaeo

# *Corumbella*, an Ediacaran-grade organism from the Late Neoproterozoic of Brazil

Loren E. Babcock[a,b,*], Anne M. Grunow[b], Georg Robert Sadowski[c], Stephen A. Leslie[d]

[a]*Department of Geological Sciences, The Ohio State University, Columbus, OH 43210, USA*
[b]*Byrd Polar Research Center, The Ohio State University, Columbus, OH 43210, USA*
[c]*Instituto de Geociências, Universidade de São Paulo, São Paulo, Brazil*
[d]*Department of Earth Science, University of Arkansas at Little Rock, Little Rock, AR 72204, USA*

Received 15 October 2002, accepted 31 January 2003

## Abstract

The morphology and phylogenetic affinities of *Corumbella werneri* [Hahn, G, Hahn, R, Leonardos, O H, Pflug, H D, and Walde, D H G, 1982 Korperlich erhaltene Scyphozoen-Reste aus dem Jungprakambrium Brasiliens Geologica et Paleontologica, 16 1–18 ], type species of the genus, is reinterpreted based on new material from the Tamengo Formation (Corumbá Group, Upper Neoproterozoic) of western Mato Grosso do Sul, Brazil *Corumbella* secreted a narrow, elongate, tetraradially symmetrical tube Reinterpretation of tube morphology and new evidence of reproduction by means of budding indicate close affinities with the present-day coronate scyphozoan cnidarian *Stephanoscyphus*, and possibly the conulariids *Corumbella* species were distributed between the Amazon craton and Laurentia *Corumbella* is interpreted as a sessile predator, and provides evidence of the importance of predation in animal evolution during the Late Neoproterozoic
© 2004 Elsevier B V All rights reserved

*Keywords* Brazil, Neoproterozoic, Precambrian, Scyphozoa, Ediacaran. Predation

## 1. Introduction

The Pantanal region of western Brazil yields the largest assemblage of Neoproterozoic life forms recorded from South America The Neoproterozoic Corumbá Group, which is sporadically exposed in the

---
* Corresponding author Department of Geological Sciences, 125 South Oval Mall, The Ohio State University, Columbus, OH 43210, USA
*E-mail address* babcock 5@osu edu (L E Babcock)

Pantanal, contains a moderately diverse assemblage of fossils and biologically mediated sedimentary structures including stromatolites (Jones, 1985, O'Connor and Walde, 1986, Zaine, 1991), organic-walled microfossils (Zaine, 1991), putative eukaryotic algae (Zaine, 1991), putative metazoans (Hahn et al, 1982; Glaessner, 1984, Zaine, 1991), and trace fossils (Zaine, 1991) The assemblage provides further documentation, developed from localities globally, for a relatively diverse biota following the Varanger (Marinoan) glaciation, and further evidence

0031-0182/$ - see front matter © 2004 Elsevier B V All rights reserved
doi 10 1016/j palaeo 2003 01 001

for a rather protracted increase in biotic diversity through the Late Neoproterozoic

*Corumbella werneri*, which has been interpreted as a scyphozoan cnidarian, is the most abundant macroscopic body fossil in the Tamengo Formation (Corumbá Group, Upper Neoproterozoic) of Mato Grosso do Sul, Brazil (Hahn et al, 1982) Its presence there, together with the small conical shell *Cloudina*, has been used to link the Tamengo Formation fossils to the Ediacaran biota (Hahn et al, 1982, Glaessner, 1984, Hahn and Pflug, 1985, Trompette et al, 1998) Recent collecting in Mato Grosso du Sol has yielded new material that sheds considerable new light on the morphology of this organism Based on this new information, we provide a new interpretation of the morphology of *Corumbella*, and a revised interpretation of its phylogenetic affinities Comparison with an undescribed species of *Corumbella* from the western United States (Hagadorn and Waggoner, 2000) provides important support for a Late Neoproterozoic age for the Tamengo

Formation of Brazil, and for a relatively close juxtaposition of the Amazon craton and the Laurentian craton during the latest Neoproterozoic

## 2. Geologic setting

General geology, stratigraphy, and paleontology of western Mato Grosso do Sul, Brazil, and adjacent areas of Bolivia (the Corumbá, Brazil-Puerto Suárez, Bolivia, region) has been summarized in a number of publications and theses, including those of Almeida (1944, 1945, 1957, 1958a,b, 1964, 1965, 1968), Dorr (1945, 1973), Weiss and Sweet (1959), Walde et al (1981), Jones (1985), O'Connor and Walde (1986), Hoefs et al (1987), Zaine (1991), Pimentel and Fuck (1992), Pimentel et al (1996), Boggiani (1998), and Trompette et al (1998) Neoproterozoic sediments in this area accumulated in or adjacent to a triple junction tectonic setting according to some authors (e g, Jones, 1985, Trompette et al, 1998). Corumbá, Brazil, is

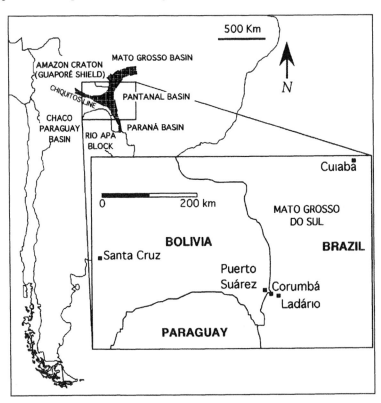

Fig 1 Map of part of South America showing Corumbá region, western Brazil, and adjacent area of eastern Bolivia The Itaú quarry (Fig 2), from which specimens of *Corumbella werneri* discussed here were collected, is located in Ladário, Brazil

located near the southeastern tip of the Amazon craton and north of the Rio Apa block (Rio de La Plata craton?), the folded portion of the Proterozoic (Brasiliano-age) Paraguay belt occurs about 50 km to the east (Fig 1) Two sets of extensional structures are present in the Corumbá region WNW–ESE-trending faults associated with the Chiquitos–Tucavaca aulocogen, and NE–SW-trending faults (Tromppette et al , 1998). The NE–SW-trending faults are thought to have been active until the Late Neoproterozoic in developing the putative Corumbá graben system (Trompette et al , 1998), the system extends no more than 20 km in width

The general stratigraphic succession in the Corumbá graben system includes a high standing, thick detrital and Fe–Mg-rich interval (the Jacadigo Group, including the Urucum and Santa Cruz Formations), apparently overlain by a carbonate-fine siliciclastic interval (the Corumbá Group, including the Cadiueus, Cerradihno, Bocaina, Tamengo, and Guaicurus Formations, Almeida, 1965, Boggiani, 1998, Fig 2) The Jacadigo Group is a thick succession of basin fill containing glacial sediments associated with late Varanger (Mar-

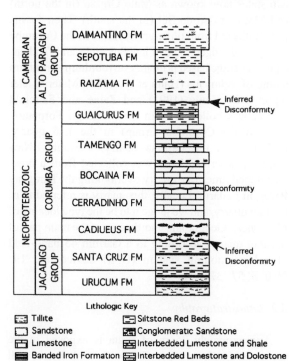

Lithologic Key

| | |
|---|---|
| ▦ Tillite | ▦ Siltstone Red Beds |
| ☐ Sandstone | ▦ Conglomeratic Sandstone |
| ▤ Limestone | ▦ Interbedded Limestone and Shale |
| ▬ Banded Iron Formation | ▦ Interbedded Limestone and Dolostone |

Fig 2 Generalized section of Proterozoic–Cambrian strata in the vicinity of Mato Grosso do Sul, Brazil

inoan) glaciation, and intercalated beds of banded iron and manganese Trompette et al (1998) inferred a hydrothermal origin for the iron and manganese, which may have been leached from mafic plutons associated with graben formation Alternatively, the origin of iron, including iron–ooid layers, and rare manganese beds in the Jacadigo Group could be related to the precipitation of cryptocrystalline iron oxyhydroxides from seawater enriched in Fe, Al, and Si in a manner analogous to that described for the origin of iron ooids from the Holocene of Indonesia and the Ordovician of northern Europe (Sturesson et al , 2000) In the Indonesian and European examples, and perhaps also in the case of the Brazilian iron–ooid-bearing layers, sedimentary iron is associated with the presence of hydrothermal fluids, volcanic ash falls into a shallow basin, or rapid weathering of fresh volcanic rocks

The Corumbá Group (Fig 2), which represents a spectrum of related environments ranging from rift basin fill (mostly fine siliciclastics) through rift margin sedimentation (mixed siliciclastics and carbonates) and adjacent shallow platform sedimentation (carbonates and fine siliciclastics), overlies the Jacadigo Group General structural and stratigraphic evidence indicates that the Corumbá Group-Jacadigo Group contact is unconformable, although we have not directly observed an unconformable relationship on outcrop

The Cadiueus Formation (Fig 2), which is the lowermost formation of the Corumbá Group, ranges up to 300 m in thickness (Almeida, 1965) and consists of red shale, arkose, and conglomerate The unit is of inferred lacustrine and alluvial fan origin (Boggiani, 1998)

The Cerradinho Formation (Fig. 2) ranges from 15 to 20 m in thickness, and consists of carbonate grainstone, shale, lithic arenite, and arkose The unit is of inferred shallow marine shelf to distal alluvial plain origin (Boggiani, 1998) A disconformity separates the Cerradinho Formation from the overlying Bocaina Formation

The Bocaina Formation (Fig 2) ranges from 10 to 15 m in thickness, and consists of light colored carbonates, mostly lime mudstone, ooidal rudstone, and carbonate breccia, with thin intercalated shale beds and phosphatic layers in places Stromatolites, ranging in shape from laminar to domal, are abundant in the formation Much of the carbonate in the unit is in the form of dolostone The Bocaina Formation is of inferred shallow marine to intertidal origin (Boggiani, 1998)

The Tamengo Formation (Figs 2 and 3) ranges from 80 to 100 m in thickness, and consists of rudstone, oncoidal rudstone, ooidal grainstone (sometimes dolomitized), thin-bedded shales with microphosphorite layers, dark gray lime mudstone, intraformational breccia, and quartz arenite The unit is of inferred rift-margin (eolian through shallow marine platform, basin edge, and deep basin) origin (Boggiani, 1998) The presence of hummocky cross-stratified beds indicates occasional storm impingement on the depositional site

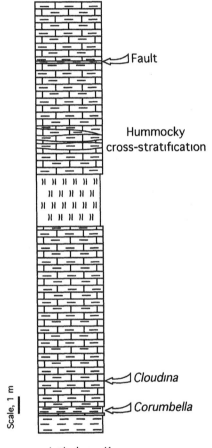

Fig 3 Stratigraphic section through part of the Tamengo Formation (Neoproterozoic), in the Itaú quarry, Ladário, Brazil, showing positions from which *Corumbella werneri* and *Cloudina lucianoi* were collected

Some material in the Tamengo Formation from the platform and basin edge has been resedimented by turbidites or other sediment-gravity flows (Boggiani, 1998) *Corumbella*, described here, is from a rhythmically thin-bedded shale lithofacies

Lastly, the Guaicurus Formation (Fig 2), which comprises the upper part of the Corumbá Group, ranges up to a few meters in thickness, and consists of olive-gray shale and lime mudstone It is of inferred shallow marine-shelf origin (Boggiani, 1998)

## 3. Location and stratigraphy

### 3 1 Location

*Corumbella werneri* Hahn et al (1982) was first reported from the Claudia limestone quarry, owned by the Itaú Cement Company, in Corumbá–Ladário, Mato Grosso do Sul, southwestern Brazil (Fig 3) Hahn et al. (1982) reported that their specimens were collected from Mato Grosso, Brazil Subsequent to the time of the report, the state of Mato Grosso was divided into two states now known as Mato Grosso (in the north) and Mato Grosso do Sul (in the south) Their fossils were collected from olive-green shales in the Corumbá Group, and judged to be of late Precambrian (Vendian-equivalent) age based primarily on rubidium–strontium dating of sediments (Hahn et al , 1982). Material later reported by Zaine (1991) and Zaine and Fairchild (1987) was collected from the Tamengo Formation (part of the Corumbá Group) in the Itaú quarry, Ladário, and the Lajinha quarry, Corumbá New material, described here, was collected from a now-abandoned limestone quarry overlooking the Paraguay River, in Ladário Locally, this quarry is referred to as the Itaú quarry, and we judge it to be the type locality of *C werneri* Geographic coordinates of the main part of the Itaú quarry, according to a Garmin 45 hand-held global positioning system (GPS) device, are 19° 00 0′ S, 57° 37 2′ W

### 3 2 Lithostratigraphy

Approximately 25 m of strata is exposed in the quarry at Ladário, Mato Grosso do Sul, Brazil All rocks in the Ladário quarry are assigned to the Tamengo Formation (Figs 2 and 3) The rocks consist predominantly of dark gray lime mudstones (weathering tan) that are medium- to thick-bedded and have thin shaly partings (Fig 3) The limestone is interrupted near the base of the quarry section by gray, greenish-gray, tan, or reddish shale interbeds ranging up to about 1 m in thickness A tan bed, 2 4 m thick, near the middle of the quarry section, interpreted as a K-bentonite, contains small zircons Determination of radiometric age dates based on the zircons has not yet been completed Approximately 2 m above the top of the bentonite, hummocky cross-stratified beds are present Still higher in section, asymmetrically and symmetrically rippled beds, carbonate flaser beds, and lenticular beds are present Evidence of bioturbation is lacking throughout the exposed section Thrust and tear faulting in the quarry has twice resulted in partial repetition of the section The *Corumbella*-bearing shale interval now appears in three locations in the vertical profile of the quarry Strata in the quarry are distorted by drag folding along a prominent fault

Organic remains in the Tamengo Formation at Ladário include *Corumbella werneri* Hahn et al (1982), the conical fossil *Cloudina lucianoi* (Beurlen and Sommer, 1957, reassigned from *Aulophycus* by Zaine and Fairchild, 1987), a putative metaphyte, *Tyrasotaenia* sp ; and the organic-walled microfossil *Sphaerocongregus variabilis* (see Zaine, 1991) *Corumbella* material occurs in abundance in a 0 4-m-thick shale interval near the floor of the Ladário quarry (Fig 3) Fragments of *Tyrasotaenia* occur in the same interval *Cloudina* remains occur sporadically through the section beginning 1 5 m above the top of the shale containing *Corumbella* (Fig 3)

## 3 3 Age

Biostratigraphic evidence suggests a Late Neoproterozoic age for the Corumbá Group, at least up through the Tamengo Formation The vase-shaped microfossil *Melanocyrillium* sp occurs in subjacent strata of the Urucum Formation (Jacadigo Group) at Morro do Urucum, Bolivia (Fairchild et al , 1978, Zaine, 1991) The presence of this form suggests a Neoproterozoic age for the Urucum Formation *Cloudina, Corumbella, Sphaerocongregus*, and *Tyrasotaenia* occur in the Tamengo Formation (Corumbá Group) at Ladário (Zaine, 1991; Fairchild, 1984, Zaine and Fairchild, 1987) and nearby localities in

Brazil Individually and collectively these fossils indicate a Late Neoproterozoic age for this part of the succession Elsewhere in the world, *Cloudina* is present in strata spanning most of the interval yielding taxa characteristic of the Ediacaran biota (Knoll, 1996) Recent recalibration of the Neoproterozoic time scale (Grotzinger et al , 1995, Knoll, 2000) constrains this interval to between 570 and 544 Ma.

A report of *Corumbella* from the western United States (Hagadorn and Waggoner, 2000) tends to substantiate the inferred latest Neoproterozoic (Ediacaran-equivalent) age for the Tamengo Formation At present, the United States material constitutes the only known occurrence of *Corumbella* outside of Brazil Specimens from the lower part of the Wood Canyon Formation in the Montgomery Mountains, Nopah Range, Nye County, Nevada, occur in association with cloudinid tubes and *Swartpuntia* Although *Swartpuntia* ranges upward into the Lower Cambrian (Hagadorn et al , 2000), the co-occurrence of cloudinids with *Corumbella* in Nevada implies a Late Neoproterozoic age for *Corumbella* in Brazil

The Raizama Formation (Alto Paraguay Group), which is superjacent to the Corumbá Group, contains a trace fossil assemblage that is most likely Early Cambrian in age, although a Late Neoproterozoic age cannot be ruled out at present Trace fossils reported from the lower Raizama Formation near Cáceres, Brazil (Zaine, 1991), are *Cochlichnus*, cf *Lockeia*, *Palaeophycus*, and *Planolites*

Radiometric estimates of the age of the Corumbá Group are debatable but tend to support a Neoproterozoic age assignment The age estimates are based mainly on Rb–Sr age determinations of the Diamantino Formation (569±20 Ma, Cordani et al , 1978, 1985) and the Araras Group (595±4 Ma; Kawashita et al , 1997) The Diamantino Formation occurs well above the top of the Corumbá Group, and the Araras Group, which crops out to north of the outcrop range of the Corumbá Group, is thought to be a deeper-water unit correlative of the Corumbá Group An age estimate of approximately 600 Ma for sediments of the Corumbá Group, based on Rb–Sr isotopic techniques, has been reported (Cordani in Hahn et al , 1982) Another age estimate of 489±29 Ma for the Tamengo Formation, based on Rb–Sr isotopic techniques (Fugita, 1979, Cordani et al , 1985), most likely reflects the time of most recent metamorphism

## 4. Biogeography and paleoecology

*4 1 Biogeography and paleogeographic implications*

As discussed previously (Hagadorn and Waggoner, 2000), *Corumbella* may have some value for interpreting paleogeographic relationships between terranes rifted from the most recent Late Neoproterozoic supercontinent (Rodinia or Pannotia, Dalziel, 1997, Unrug, 1997) *Corumbella* species are known from only two Neoproterozoic localities (western Brazil and the western United States; these areas represent parts of the Amazon craton and the Laurentian craton, respectively) Although it is possible that the limited evidence available on the distribution of *Corumbella* renders a biogeographic interpretation premature, it is worth reiterating the point that the shared presence of *Corumbella* on both the Amazonian and Laurentian cratonic areas is consistent with the hypothesis that these areas were in relatively close juxtaposition at the close of the Neoproterozoic (Hagadorn and Waggoner, 2000, Leslie et al, 2001, compare Dalziel, 1997) Likewise, the shared presence of the small, shelly, tubular fossil *Cloudina* in western Brazil and the western United States (Hagadorn and Waggoner, 2000), further supports the hypothesis of a relatively close paleogeographic position between the Amazonian and Laurentian cratons

*4 2 Paleoecological implications*

Interpretation of cnidarians, including *Corumbella*, in Neoproterozoic strata is important from a paleoecological perspective, as it provides indirect evidence of trophic relationships during that interval of time Considerable evidence now exists (e g., reviews in Conway Morris and Bengtson, 1994, Babcock, 2003) that predation was a forcing factor in animal evolution by the early part of the Cambrian, but much less predation evidence (Bengtson and Zhao, 1992, Lipps, 2002) has been identified in the Neoproterozoic The presence of animals interpreted as cnidarians in Neoproterozoic strata provides strong, but indirect, support for the view that predation was already important in marine ecosystems by the Late Neoproterozoic (Lipps, 2002) *Corumbella* is interpreted as a sessile predator, and its food source is unknown Given the relatively small size of the apertural area, it

is possible that *Corumbella* fed upon small nektonic animals, perhaps microplankton

## 5. Morphologic and phylogenetic interpretation

*5 1 Previous interpretation*

Originally, *Corumbella werneri* was interpreted (Hahn et al , 1982) as having a frondlike morphology superficially resembling such characteristic Ediacaran organisms as *Charnia* and *Charniodiscus* (see Glaessner, 1984), and also resembling present-day pennatulate cnidarians The external covering, termed periderm, was interpreted as chitinous in composition. Two rather distinct body regions were reconstructed (1) a proximal region consisting of an elongate tube or stalk, termed the primary polypar, and composed of ring-shaped elements containing four sclerosepta internally, and (2) a distal region consisting of biserially arranged secondary polypars, each having a distinct, small peridermal tube The primary and secondary polypars were reconstructed as rounded in cross-sectional shape Structure of the stalk was interpreted as resembling the recent scyphozoan *Stephanoscyphus*, and for this reason, Hahn et al (1982) referred *Corumbella* to the class Scyphozoa (phylum Cnidaria) The inferred presence of biserially arranged secondary polypars, however, is a character lacking in scyphozoans, and for this reason, Hahn et al (1982) erected the monogeneric family Corumbellidae embracing *Corumbella*, the monofamilial order Corumbellida embracing the Corumbellidae, and the monoordinal subclass Corumbellata embracing the Corumbellida

Until recently, *Corumbella* was known only from the Brazilian species *Corumbella werneri* In a report of a variety of Ediacaran fossils from the Great Basin of the United States, however, Hagadorn and Waggoner (2000) illustrated and described another form attributed to *Corumbella*, which they identified as *Corumbella* new species A This species was interpreted as having a thin annulated tube that was nearly square in cross section but with rounded corners, and a single longitudinal groove along the midline of each face Members of the species are helically twisted No evidence of secondary polypars was reported Based on the two known species, Hagadorn and Waggoner

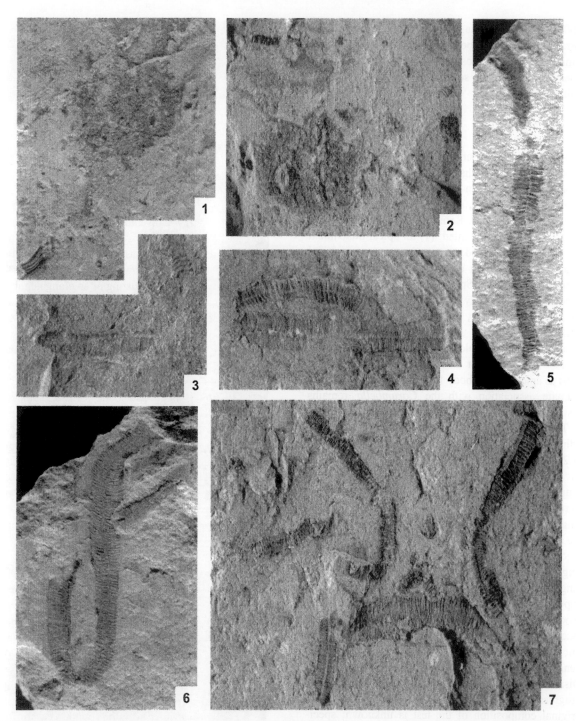

Fig. 4. *Corumbella werneri* Hahn et al., 1982, from the Itaú quarry, Ladário, Brazil. 1, 2: part–counterpart slabs of organic mass containing small pieces, including pieces inferred to be of the apical attachment structures, ×3, OSU 46397A (1), OSU 46397B (2); 3: piece of tube from apical region (lower) and small fragment of larger tube (upper), from same slab as in panel 7, ×6, OSU 46398; 4: long section of detached tube folded over on itself, ×3, OSU 46399; 5: isolated tube, ×3, OSU 46400; 6: large segment of twisted and folded tube, ×3, OSU 46401; 7: cluster of tubes preserved in close association, ×3, OSU 46398.

(2000) emended the diagnosis of the genus to include narrow annulated tubes having a fourfold radial symmetry. The diagnosis was expanded to include tubes having a cross sectional shape that is square with rounded corners.

## 5.2. Reinterpretation

New material, preserved as limonite-coated molds in siltstone and silty shale (Figs. 4 and 5), affords an improved interpretation of the morphology of *Corumbella werneri* (Fig. 6), and provides stronger support for a cnidarian affinity. Remains of two separate parts of the body are represented: (1) an apical attachment region (here referred to as the apex or apical structure); and (2) an elongate tubular region (here referred to as a tube). Specimens showing larger tubes joined to the previously unknown apical structures are not known. However, the apical structures can be confidently linked to the larger tubes because they have the same morphological structure externally, and because parts of larger tubes are present on the same slabs as apices. Other body fossils are not preserved in intimate association with the *Corumbella* remains described here.

Tubes of *Corumbella werneri* range up to an estimated 80 mm in length and have a diameter of 20 to 25 mm. Systematic expansion along the length of an individual tube is imperceptible. However, all available specimens are compacted in shale and show some differences in width, as preserved, along the lengths of the specimens. Such differences in width are interpreted to be the result of compaction of a tube whose original cross section was squarish. Transverse ringlet structures, approximately 0.5 mm in sagittal length, provide the tube with an accordian-like appearance. At the midlines of the faces, longitudinal grooves extending the length of the tube are present externally. Internally, carinae are present at the midline. The tube was quite flexible, and capable of bending through 360 degrees (Fig. 4.4 and 4.6). Disarticulation of the tube involves breakage into progressively shorter lengths of attached ringlets. The original composition of the test is unknown; specimens are now preserved by iron oxide and as impressions in shale.

Apices of *Corumbella werneri* show the same external structure as the tubes, but their width has not

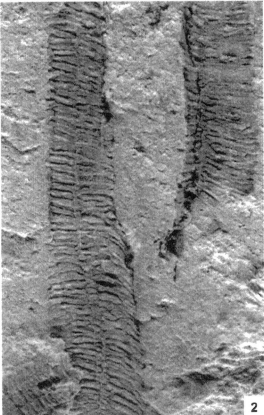

Fig. 5. *Corumbella werneri* Hahn et al., 1982, from the Itaú quarry, Ladário, Brazil. 1: short segments of tubes clustered in close association, ×3, OSU 46402; 2: detail of specimen in Fig. 3.6, apical end to base of photograph, ×9, OSU 46401.

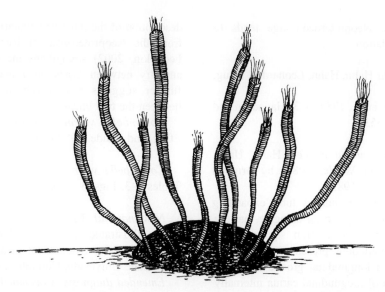

Fig 6 New reconstruction of *Corumbella werneri* The species is reconstructed as a colonial cnidarian attached to an organic substrate Tentacles are conjectural

been observed to exceed 2 mm Expansion in the apical region is subtle The apex is simple, and in a few specimens, associated with an indistinct, rounded, organic mass The round organic masses, up to 10 mm in diameter, consist of fine, intertwined, netlike structures (Fig 4 1 and 4 2); from these masses arise the apices of *C werneri* The round organic masses represent either intertwined connections between individuals budded from a single colonial progenitor, or individuals attached to another organic structure (perhaps bacteria or colonial algae) Even if the organic masses represent organisms other than *Corumbella*, the assemblages of *Corumbella* together with the organic masses make it difficult to rule out the possibility of budding in *Corumbella*

As newly reconstructed, *Corumbella werneri* (Fig 6) bears striking similarity to the living *Stephanoscyphus*, a coronate scyphozoan (see Werner, 1967, 1969) Further, the likely capability of asexual reproduction through budding in *C werneri* tends to support a cnidarian affinity

## 6. Systematic paleontology

Phylum CNIDARIA Hatschek
Class SCYPHOZOA Gotte

Order Uncertain
Family CORUMBELLIDAE Hahn, Hahn, Leonardos, Pflug, and Walde, 1982

*Type genus Corumbella* Hahn, Hahn, Leonardos, Pflug, and Walde, 1982, by original designation (Hahn et al , 1982)

*Emended diagnosis* Narrow, annulated, mostly parallel-sided tube having fourfold radial symmetry, annulations meet at longitudinal grooves marking the midlines of faces, internally, the midlines are marked by carinae

*Discussion* On the basis of available evidence, the monogeneric family Corumbellidae is interpreted as having closest affinities with representatives of either the order Coronatae or the order Stauromedusae Because we have no knowledge of the medusa phase, or even if *Corumbella* possessed a medusa phase, we cannot discern the order with confidence Secondary polypars (*sensu* Hahn et al , 1982), which were used as the primary morphologic criterion for recognizing both the subclass Corumbellata and the order Corumbellida, are evidently absent, thus the continued use of these names is unsupported Collins (2002) interpreted both the Coronatae and Stauromedusae as basal clades within Cnidaria Recognition of *Corumbella* as either a coronate or stauromedusid scyphozoan of Late Neoproterozoic age tends to support this interpretation

Genus *Corumbella* Hahn, Hahn, Leonardos, Pflug, and Walde, 1982

*Corumbella* Hahn et al , 1982, p 3, Hagadorn and Waggoner, 2000, p 353

*Type species Corumbella werneri* Hahn, Hahn, Leonardos, Pflug, and Walde, 1982, by original designation (Hahn et al , 1982)

*Emended diagnosis* Annulated tube expanding slightly from a narrow apex to an elongate, narrow, parallel-sided section; cross section squarish with rounded corners, annulations present on each face, midline consists of a longitudinal groove externally that reflects position of longitudinal carina internally

*Discussion* Several features of *Corumbella* suggest a possible sister-group affinity with the conulariids, which themselves commonly have been inferred to have cnidarian affinities (e g., Van Iten, 1992a,b, Jerre, 1994; Hughes et al , 2000, Ivantsov and Fedonkin, 2002) Corumbellids apparently have a radial symmetry similar to conulariids, although in conulariids a bilateral pattern on the four faces (perhaps secondary to development of an underlying radial symmetry) exists Both *Corumbella werneri* and the species reported as *Corumbella* new species A (Hagadorn and Waggoner, 2000) even have a squarish cross section, which is similar to many conulariids External longitudinal grooves at the midlines of the four faces are present in both conulariids and corumbellids Lastly, in corumbellids, annulations of the tube resemble, and may be homologous with, ridges of the conulariid exoskeleton In conulariids, these ridges are produced by thickenings of the phosphatic layers that comprise the exoskeleton Although some conulariids have been shown to disarticulate into short segments of these thickened areas (termed rods, Babcock and Feldmann, 1986), they are not independent of the integument In both corumbellids and conulariids, the ridges are slightly offset at the midlines Corumbellids and conulariids differ significantly in general shape of the tube. corumbellids have long, narrow, minimally expanding tubes, whereas conulariids have rapidly expanding tubes This character may not, however, have significance for ascertaining the phylogenetic affinities of the two animals Recent

description of the putative conulariid *Vendoconularia* from the Neoproterozoic of Russia (Ivantsov and Fedonkin, 2002) strengthens the case for a shared ancestry between *Corumbella* and conulariids, and further suggests a Precambrian divergence time between the two lineages

*Corumbella werneri* Hahn, Hahn, Leonardos, Pflug, and Walde, 1982 (Figs 4–6)

*Corumbella werneri* Hahn et al , 1982, p 4–9, Tables 1–3, Figs 3–5, 9, 11

*New material* More than 200 fragmentary specimens. Illustrated specimens (Figs 5 and 6) are reposited in the Orton Geological Museum of The Ohio State University, Columbus, Ohio

*Emended diagnosis Corumbella* expanding up to 5 mm in diameter, annulations across each of four faces closely spaced (approximately 40–45/cm), well-defined, annulations meet and are offset slightly at midline of each face

*Discussion* To date, only two species, *Corumbella werneri* from the Tamengo Formation of Mato Grosso do Sul, Brazil, and *Corumbella* new species A (Hagadorn and Waggoner, 2000) from the Wood Canyon Formation of the Nopah Range, Nye County, Nevada, have been included in *Corumbella* The illustrated specimen of *Corumbella* new species A (Hagadorn and Waggoner, 2000, Fig 5 4–5 6) is not well preserved, but appears to be distinct from the type species of the genus, *C werneri*, primarily in having fewer and less distinct annulations *Corumbella* new species A shows a helical twist of long amplitude along the tube Similar helical twists are present on some specimens of *C werneri* (Fig 4 7), and it is uncertain whether the twist has any taxonomic value

## Acknowledgments

We would like to thank J C Miller for much helpful advice concerning the geology of Brazil; and A G Collins, L Gershwin, J H Lipps, and B. Waggoner for helpful discussion of *Corumbella* This work was supported in part by grants from the National Science Foundation (EAR 9526709 to Grunow and Babcock, and EAR 0106883 to Babcock)

# References

Almeida, F F M de, 1944 O diastrofismo Taconico no Brasil An Acad Bras Ciênc 16 (2), 125–135

Almeida, F F M de, 1945 Geologia do Sudoesta mato-grossense Bol -Div Geol Mineral, Dep Nac Prod Miner 116, 1–118

Almeida, F F M de, 1957 Novas ocorrências de fósseis no Pré-Cambiano brasileiro An Acad Bras Ciênc 29 (1), 63–72

Almeida, F F M de, 1958a Ocorrência de fósseis nos Dolomitos Bocaino, em Corumbá, Mato Grosso Relatório Annual da Divisão de Geologia e Mineralogia Departamento Nacional de Produção Mineral, pp 87–88

Almeida, F F M de, 1958b Ocorrência de Collenia em dolomitos da Série Corumbá Notas Prelim Estud -Div Geol Mineral 106, 1–11

Almeida, F F M de, 1964 Geologia do centro-oeste mato-grossense Bol -Div Geol Mineral, Dep Nac Prod Miner 215, 1–137

Almeida, F F M de, 1965 Geologia da Serra Bodoquena (Mato Grosso), Brasil Bol -Div Geol Mineral, Dep Nac Prod Miner 219, 1–96

Almeida, F F M de, 1968 Evolução tectônica do Centro-Oeste Brasileiro no Proterozóico superior An Acad Bras Ciênc (Suplemento Simpósio de Manto Superior) 40, 285–296

Babcock, L E, 2003 Trilobites in Paleozoic predator–prey systems, and their role in reorganization of early Paleozoic ecosystems In Kelley, P H, Kowalewski, M, Hansen, T A (Eds), Predator–prey interactions in the fossil record Kluwer Academic/Plenum Publishers

Babcock, L E, Feldmann, R M, 1986 Devonian and Mississippian conulariids of North America Part A General description and Conularia Ann Carnegie Mus 55, 348–410

Bengtson, S, Zhao, Y, 1992 Predatorial borings in late Precambrian mineralized exoskeletons Science 257, 367–369

Beurlen, K, Sommer, F W, 1957 Observaçãoes estratigráficas e paleontológicas sobre o Calcário Corumbá Div Geol Mineral, Dep Nac Prod Miner 168, 1–35

Boggiani, P C, 1998 Análise Estratigráfica da Bacia Corumbá (Neoproterozóico)—Mato Grosso do Sul PhD Thesis Universidade de São Paulo, São Paulo 181 pp

Collins, A G, 2002 Phylogeny of Medusozoa and the evolution of cnidarian life cycles J Evol Biol 15, 418–432

Conway Morris, S, Bengtson, S, 1994 Cambrian predators possible evidence from boreholes J Paleontol 68, 1–23

Cordani, U G, Kawashita, K, Thomaz-Filho, A, 1978 Applicability of the rubidium–strontium method to shales and related rocks In Contributions to the geologic time scale, 6 AAPG, Stud Geol pp 93–117

Cordani, U G, Thomaz-Filho, A, Brito-Neve, B B, Kawashita, K, 1985 On the applicability of the Rb–Sr method to argillaceous sedimentary rocks some examples from Precambrian sequences of Brazil J Geol 47, 253–280

Dalziel, I W D, 1997 Neoproterozoic–Paleozoic geography and tectonics review, hypothesis, environmental speculation Geol Soc Amer Bull 109, 16–42

Dorr II, J V N, 1945 Manganese and iron deposits of Morro do Urucum, Mato Grosso, Brazil U S Geol Surv Bull 946A 47 pp

Dorr II, J V N, 1973 Iron-formation in South America Econ Geol 68, 1005–1022

Fairchild, T R, 1984 Caution an "Ediacaran" or Early Cambrian age for the Corumbá and Jacadigo Groups (SW Brazil) still requires definitive proof Abstracts 27th International Geological Congress, Moscow, vol 1, pp 38–39

Fairchild, T R, Barbour, A P, Haralyi, L E, 1978 Microfossils in the "Eopaleozoic" Jacadigo Group at Urucum, Mato Grosso, southwest Brazil Boletim Instituto de Geociências, vol 9 Universidade de São Paulo, pp 74–79

Fugita, H H, 1979 Geochronologia da região de Corumbá (MT) Relatório de estágio no Centro de Pesquisas Geocronológicas da Universidade de São Paulo (22 pp, not seen, cited in Cordani et al, 1985)

Glaessner, M F, 1984 The Dawn of Animal Life Cambridge University Press, Cambridge 244 pp

Grotzinger, J P, Bowring, S A, Saylor, B Z, Kaufman, A J, 1995 Biostratigraphic and geochronology constraints on early animal evolution Science 270, 598–604

Hagadorn, J W, Waggoner, B M, 2000 Ediacaran fossils from the southwestern Great Basin, United States J Paleontol 74, 349–359

Hagadorn, J W, Fedo, C M, Waggoner, B M, 2000 Early Cambrian Ediacaran-type fossils from California J Paleontol 74, 731–740

Hahn, G, Pflug, H D, 1985 Die Cloudinidae n fam, Kalk-Rohren aus dem Vendium und Unter-Kambrium Senckenb Lethaea 65, 413–431

Hahn, G, Hahn, R, Leonardos, O H, Pflug, H D, Walde, D H G, 1982 Korperlich erhaltene Scyphozoen-Reste aus dem Jungprakambrium Brasiliens Geol Palaeontol 16, 1–18

Hoefs, J, Muller, G, Schuster, K A, Walde, D H G, 1987 The Fe–Mn ore deposits of Urucum, Brazil an oxygen isotope study Chem Geol 65, 311–319

Hughes, N C, Gunderson, G O, Weedon, M, 2000 Late Cambrian conulariids from Wisconsin and Minnesota J Paleontol 74, 828–838

Ivantsov, A Y, Fedonkin, M A, 2002 Conulariid-like fossil from the Vendian of Russia a metazoan clade across the Proterozoic/Palaeozoic boundary Palaeontology 45, 1219–1229

Jerre, F, 1994 Anatomy and phylogenetic significance of Eoconularia loculata, a conulariid from the Silurian of Gotland Lethaia 27, 97–109

Jones, J P, 1985 The southern border of the Guaporé Shield in western Brazil and Bolivia an interpretation of its geologic evolution Precambrian Res 28, 111–135

Kawashita, K, Mizusaki, A M P, Kiang, C H, 1997 Razões 87Sr/86Sr em sedimentos carbonáticos do Grupo Bambuí (MG), vol 1 Congreso Brasileiro Geoquímica, Anais, Porto Alegre, pp 133–137

Knoll, A H, 1996 Daughter of time Paleobiology 22, 1–7

Knoll, A H, 2000 Learning to tell Neoproterozoic time Precambrian Res 100, 3–20

Leslie, S A, Babcock, L E, Grunow, A M, Sadowski, G R, 2001 Paleobiology and paleobiogeography of Corumbella, a late Neoproterozoic Ediacaran-grade organism Paleobios 21 (2, Suppl), 83–84

Lipps, J H , 2002 Early evolution of marine trophic structures Neoproterozoic to Cambrian Abstr Programs-Geol Soc Am 34 (6), 170

O'Connor, E A , Walde, D H G . 1986 Recognition of an Eocambrian orogenic cycle in SW Brazil and SE Bolivia Zentrallblat Geol Palaontol 1, 1441–1456

Pimentel, M M . Fuck. R A , 1992 Neoproterozoic crustal accretion in central Brazil Geology 20, 375–379

Pimentel, M M , Fuck, R A , De Alvarenga. C J S , 1996 Post-Brasiliano (Pan-African) high-K granitic magmatism in central Brazil the role of late Precambrian–Early Paleozoic extension Precambrian Res 80, 217–238

Sturesson, U . Heikoop, J M , Risk, M J , 2000 Modern and Palaeozoic iron ooids—a similar volcanic origin Sediment Geol 136. 137–146

Trompette, R , De Alvarenga. C J S , Walde. D , 1998 Geological evolution of the Neoproterozoic Corumbá graben system (Brazil) Depositional context of the stratified Fe and Mn ores of the Jacadigo Group J South Am Earth Sci 11. 587–597

Unrug, R , 1997 Rodinia to Gondwana the geodynamic map of Gondwana supercontinent assembly GSA Today 7 (1), 1–6

Van Iten, H , 1992a Microstructure and growth of the conularid test implications for conularid affinities Palaeontology 35, 359–372

Van Iten, H , 1992b Morphology and phylogenetic significance of the conularid test Palaeontology 35, 335–358

Walde. D H G , Gierth, G , Leonardos, O H , 1981 Stratigraphy and mineralogy of the manganese ores of Urucum. Mato Grosso, Brazil Geol Rundsch 70, 1077–1085

Weiss, M P , Sweet, W C , 1959 Stratigraphy and Structure of the Mutun Mountains, Department of Santa Cruz, Bolivia Congreso Geológico Internacional, XXᵃ Sesión, Ciudad de México, Sección XIII Geología Aplicada a la Ingeniería y a la Minería, pp 399–413

Werner, B , 1967 *Stephanoscyphus* Allman (Scyphozoa Coronatae), ein rezenter Vertreter der Conulata? Palaontol Z 41, 137–153

Werner, B , 1969 Neue beitraege zur evolution der Scyphozoa und Cnidaria I Simposio International de Zoofilogenia Universidad de Salamanca, Salamanca, Spain, pp 223–244

Zaine. M F , 1991 Analise dos fosseis de Parte da Faixa Paraguai (MS, MT) e Seu Contexto Temporal e Paleoambiental Unpublished PhD Thesis Universidade de São Paulo, São Paulo, 218 pp

Zaine, M F . Fairchild, T R , 1987 Novas considerações sobre os fósseis da Formação Tamengo, Grupo Corumbá. SW do Brasil Anais Xo Congreso do Brasil, Paleontología, Rio de Janeiro, Brazil, pp 797–806

Available online at www.sciencedirect.com

Palaeogeography, Palaeoclimatology, Palaeoecology 220 (2005) 19–29

www.elsevier.com/locate/palaeo

# Trace fossil preservation and the early evolution of animals

Sören Jensen[a,*], Mary L. Droser[a], James G. Gehling[b]

[a]Department of Earth Sciences, University of California, Riverside, CA 92521, USA
[b]South Australian Museum, North Terrace, Adelaide, South Australia 5000, Australia

Received 18 March 2002, accepted 1 September 2003

## Abstract

The trace fossil record is an important element in discussions of the timing of appearance of bilaterian animals A conservative approach does not extend this record beyond about 560–555 Ma Crucial to the utility of trace fossils in detecting early benthic activity is the preservational potential of traces made close to the sediment–water interface Our studies on the earliest Cambrian sediments suggest that shallow tiers were preserved to a greater extent than typical for most of the Phanerozoic This can be attributed both directly and indirectly to the low levels of sediment mixing The low levels of sediment mixing meant that thin event beds were preserved The shallow depth of sediment mixing also meant that muddy sediments were firm close to the sediment–water interface, increasing the likelihood of recording shallow tier trace fossils in muddy sediments

Preservation of surficial trace fossils in this type of setting remains problematic but the above factors suggest that also these can be expected to have left a reasonable record Overall, the trace fossil record can be expected to provide a sound record of the onset of bilaterian benthic activity The lack of convincing trace fossils significantly before the Cambrian supports models of late appearance of macroscopic benthic bilaterians
© 2004 Elsevier B V All rights reserved

*Keywords* Trace fossils, Taphonomy, Cambrian, Proterozoic

## 1. Introduction

The appearance and subsequent diversification of bilaterian animals are topics of current controversy (e g., Wray et al , 1996, Fortey et al , 1996, 1997,

* Corresponding author Current address Area de Paleontologia, Facultad de Ciencias, Universidad de Extremadura. Badajoz 06071, Spain
*E-mail address* soren@unex es (S Jensen)

Knoll and Carroll, 1999, Budd and Jensen, 2000, in press, Conway Morris, 1998, 2000, Collins and Valentine, 2001, Erwin and Davidson, 2002) There are three principal sources of evidence body fossils, trace fossils (trails, tracks, and burrows of animal activity recorded in the sedimentary record), and molecular data in the form of divergence times calculated by means of the molecular "clock" theory A literal reading of the body fossil record suggests that the diversification of bilaterian animals did not

0031-0182/$ - see front matter © 2004 Elsevier B V All rights reserved
doi 10 1016/j palaeo 2003 09 035

significantly precede the Neoproterozoic–Cambrian boundary (ca 545 Ma) This is in line with a conservative evaluation of the trace fossil record, the oldest certain traces likely to have been made by bilaterians have been dated at 555 3 Ma (Martin et al, 2000) Even taking into consideration uncertainties in the precise correlation of trace fossil-bearing Neoproterozoic strata, this date is likely close to the age of the oldest trace fossils. On the other hand, there are reports of bilaterian trace fossils (Seilacher et al., 1998), as well as molecular clock data (Wray et al, 1996), which suggest that diversification of bilaterian groups may had commenced more than 1000 Ma There is, however, a considerable spread in the results from molecular clock studies, with some coming close (Ayala et al, 1998), or even very close (Peterson and Takacs, 2002), to the pattern seen from body fossils Furthermore, the reports of deep Proterozoic trace fossils are doubtful, as briefly discussed below The trace fossil record is of particular interest in that it should, compared to the body fossil record, be less prone to taphonomic biases towards animals with mineralized hard parts Several authors have made a case for bilaterians being primitively benthic, arguing that many of the morphological features of bilaterians could only have evolved in a moderately large animal with a benthic lifestyle (e g, Valentine, 1994, Budd and Jensen, 2000, Collins and Valentine, 2001) This is by no means universally received: others favor an extended history of small meiofaunal or planktic larva-like bilaterians (e g, Davidson et al, 1995, Fortey et al, 1996, 1997) Nevertheless, if a benthic cradle is accepted, trace fossils should give a useful estimate to the timing of appearance of moderately large mobile bilaterian animals

Trace fossils, however, do come with their own set of problems Distinguishing simple trace fossils from sedimentary structures of inorganic origin is far from trivial Furthermore, trace fossils have their own set of taphonomic limitations With respect to early bilaterian trace fossils, it is particularly important to recall that in most modern marine settings, traces made on or near the sediment surface have a low preservational potential (e g, Seilacher, 1978)

The purpose of this paper, therefore, is to discuss trace fossil preservation, with particular reference to the preservational potential of surface and near-surface trace fossils in the Proterozoic and earliest Cambrian

## 2. Trace fossil preservation—introductory considerations

It is widely thought that surface traces have a low preservational potential in marine settings (e g, Seilacher, 1978, Hallam, 1975). One contributing factor is that surface sediment is generally rather soupy and therefore prone to reworking by weak currents In most shallow marine settings, the upper portion of the sediment is extensively mixed, resulting in a diffuse mottled texture and rapid destruction of shallow discrete trace fossils Preservation of discrete trace fossils generally requires that the trace represents a deep tier, sometimes referred to as elite trace fossils (cf Bromley, 1996) However, there are situations where relatively shallow-tiered burrows—not necessarily surface traces—are preserved A particularly important type of trace fossil preservation occurs along an interface, such as a sand sharply overlying mud (Fig 1A and B) (e g, Seilacher, 1978) This type of sediment interface generally results from event bedding, such as tempestites These events generate sharp boundaries of contrasting lithologies that are ideally suited to preserving trace fossils In addition, some of the loose muddy surface sediments are brought into suspension, which further contributes to a sharp interface and a relatively firm muddy surface Animals that burrow into the sediment to the mud–sand interface may leave trace fossils that are preserved on the base of the sandstone (Fig 1A) Much of the descriptive terminology for trace fossil preservation and stratinomy reflects this relationship (e g, Seilacher, 1964, Martinsson, 1965); burrows are described with respect to their relationship to a particular sand bed Providing that the sand layer is thin, this can preserve shallow infaunal traces (Fig 1B) Open burrows made within the mud may also be exhumed by currents and filled by sand (Fig 1C) This is particularly characteristic of turbidite settings, but also can occur in storm-influenced settings. In this situation, the upper part of the burrow commonly is eroded, and there often is a tell-tale sign of the trace having been washed out (e g, Seilacher, 1982) Indeed, trace fossils have been used to measure stratigraphic completeness (Wetzel and Aigner, 1986)

Much of the pattern of trace fossil changes across the Proterozoic–Cambrian boundary is based on discrete trace fossils that were preserved according

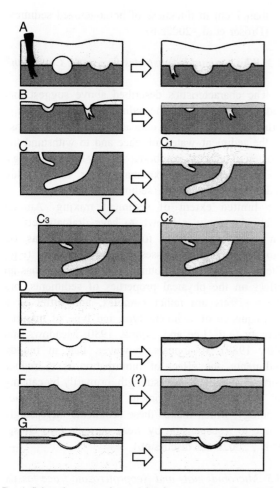

Fig 1 Selected scenarios for trace fossil preservation with particular emphasis on preservation of shallow tiers in a sub-tidal marine setting (A) Preservation along a sand–mud interface resulting in instantaneous casting of sand into mud, or collapse of sand into depression in the mud (B) A thin sand layer may become removed by winnowing leaving sand only where trapped in burrows These casts may subsequently become attached to the base of new layer of sand (C) Open burrows constructed in firm mud may be preserved by casting of sand, to which it remains attached (C1) Winnowing and subsequent deposition lead to attached burrows (C2) as in (B) In a setting with sediment bypass, burrows may serve as traps Depending on subsequent sedimentation, these casts may "float" in muddy sediment (C3) or become attached (C2) (see Droser et al, 2002a for a more exhaustive discourse of this topic) (D) Preservation potential of surface burrows in sand is enhanced if there is a thin blanket of mud at time of burrowing (E) Preservation potential of surface burrows in sand may also be enhanced blanketing layer of mud introduced shortly after trace formation (F) The preservation potential of surface burrows in mud is a matter of contention (see text) (G) Preservation of a trace formed by an animal that burrowed along a buried microbial mat Sand is eventually displaced into the open tunnel

to the principles outlined above The preservation of true surface traces remains problematic, as will be discussed below

## 3. The record of Neoproterozoic and Cambrian trace fossils

### 3 1 Characteristics of Lower Cambrian trace fossils in muddy sediments

The greatest diversity of Cambrian trace fossils has been documented from shelf settings that have a heterolithic bedding characterized by moderately thin, generally centimeter-scale, often sharp-based sand and siltstone beds, separated by layers dominated by mudrock We have previously demonstrated that firmground conditions relatively close to the sediment–water interface were widespread in these types of settings in the Early Cambrian (Droser et al, 2002a,b) This was based on the examination of over 10 Cambrian units on four continents

Of the units that we examined, the Lower Cambrian formations exhibited a number of shared ichnological and sedimentological characteristics These characteristics include the following

1  Preservation of shallow tiers and scarcity of deep tiers. Earliest Cambrian sediments preserve a range of trace fossils that represent shallow tiers The geometry and style of preservation of these trace fossils suggest that they formed less than a few centimeters below the sediment–water surface (Droser et al, 2002a,b) These trace fossils include *Treptichnus pedum* and other treptichnids, as well as *Gyrolithes*, all of which are common in Lower Cambrian shallow marine terrigenous rocks

2  Quality of preservation. Even though treptichnids were constructed close to the sediment–water interface, they have sharp walls, and in certain cases, they preserve delicate surface ornamentation There is no sign of actively reinforced burrow margins The extent of compaction of the burrows is also relatively minor Several other trace fossils of shallow emplacement show sharp preservation of detail This includes the vertical spiral burrow *Gyrolithes* and shallow *Rusophycus*

This quality of preservation is ubiquitous in the Lower Cambrian units examined

3 Styles of preservation In most shallow marine settings, burrows preserved on the base of sandstone beds are created by animals that burrow through the sand to the interface with the underlying finer-grained sediment (e.g, Seilacher, 1970) In Lower Cambrian strata, particularly, but also commonly in Middle and Upper Cambrian strata, a fundamentally different style of preservation appears to be particularly common Cambrian sand-filled burrows are generally preserved in one of two manners The burrow may be cast by a source bed to which it remains attached, or the burrow may be cast by a sand source which bypasses that environment and thus the cast is attached to the base of a different bed, or may even be preserved as a sand-filled burrow completely surrounded by muddy sediment (Fig 1C) (Droser et al, 2002a) This is a common style of preservation of treptichnids, *Gyrolithes*-type burrows, and *Palaeophycus/Planolites*-type burrows, and was found to be the most common style of preservation in all of the Lower Cambrian units examined This type of preservation requires that the burrows be open and, given the preservation of shallow tier trace fossils in this manner, that the muddy sediment be rather resistant to erosion

4 Sedimentary structures These units contain sedimentary structures that must have been formed close to the sediment–water interface. This includes *Kullingia*-type scratch circles, which form when a tethered organism is rotated by currents to imprint delicate concentric grooves (Jensen et al., 2002) These structures formed in mud or fine sands, and were cast by a coarser material

5 Lack of homogenized sediment· Some animals that burrow do not leave well-defined, discrete trace fossils Instead, the record produced is one of some degree of homogenization That is, primary sedimentary structures are not preserved and the final texture is one of a mottled appearance (e g, Bromley, 1996) This is direct evidence of a mixed layer In sedimentary rocks of the earliest Cambrian, there is virtually no evidence of such homogenization, with rare isolated examples less

than 1 cm in thickness of homogenized sediment (Droser et al, 2002a,b)

## 3 2 Why are muddy substrates firm in the Cambrian?

The characteristics described above suggest that sediment just beneath the sediment–water interface is firm enough to maintain delicate features formed close to the sediment–water interface and to withstand the erosion of currents in subtidal shallow marine settings

The most likely explanation for firm muddy sediments close to the sediment–water interface is the limited extent of sediment mixing Animal infaunal activity is known to have dramatic effects on sediment surface topology, as well as on biogeochemical processes in the sediment (e g, Ziebis et al, 1998) Bioturbation clearly also has an effect on the physical properties of sediments, but these effects are rather complex, and differ as a consequence of sediment type and type of infaunal activity (e g., Lee and Swartz, 1980; Meadows and Tait, 1989). It appears, however, that in muddy sediments, the effect of bioturbation is to reduce surface–sediment shear strength, primarily by causing an increased water content and irregular surface topology, which makes these sediments increasingly prone to resuspension by weak currents (see Droser et al, 2002b for discussion)

## 3 3 Microbial mats and Neoproterozoic trace fossils

The Neoproterozoic trace fossil record is dominated by essentially horizontal unbranched traces (e g., Gehling, 1999, Seilacher, 1999, Jensen, in press) It has been suggested that the orientation of these trace fossils was controlled by exploitation of microbial mats buried by a thin veneer of sand (Seilacher and Pfluger, 1994; Seilacher, 1999, Gehling, 1999) Indeed, the trace fossils are often preserved along the interface of two sandstone beds with no evidence of a muddy parting, here the interface was sand/microbial mat rather than sand/mud (Fig 1G) There is convincing (and growing) indirect evidence for these Proterozoic microbial mats, which were spread to a greater extent than commonly seen in normal marine settings (Gehling, 1999).The microbial mats would have lowered sand erodability (e g, Paterson, 1994), and also created taphonomic conditions favorable to the preservation of the non-mineralized Ediacaran biota (Gehling, 1999) These microbial mats may have been an important influence on early bilaterians (Seilacher, 1999, Bottjer et al, 2000), adaptations to matground conditions may be seen also in some Cambrian trace fossils (Hagadorn and Bottjer, 1999)

## 4. Preservation of surface (superficial) traces

Based on the criteria discussed above, there is good evidence that Neoproterozoic–Cambrian burrows formed within the sediment—even at very shallow depths—would have had a reasonable chance of entering the stratigraphic record

The preservational potential of true surface traces in a marine setting is, however, more problematic It should be noted that there is no absolute definition for what constitutes a surface trace There is a gradation from surface movement in which no sediment is displaced to that in which the animal is partly submerged but remains in more or less continuous contact with the sediment–water interface. This depends not only on the animal's activity, but also the nature of the sediment, especially water content, grain size, and within-bed heterogeneity, all of which are of great importance for the quality of preservation as well as the resulting trace morphology (e g, Graf, 1956, Knox and Miller, 1985) In subaerially exposed sediments, the depth of the trace increases with water

content (e g, Graf, 1956) The effect will be less pronounced in subaqueous settings but nevertheless be a factor

The surface traces discussed here all involve various degrees of sediment displacement A possible alternative mode of surface trace formation is of interest Collins et al. (2000) studied experimentally produced mucociliary creeping trails of animals, such as cerianthanan anemones and flatworms, and compared these with certain Neoproterozoic trace fossils The movement of these animals above a soft substrate resulted in sediment trapped in the mucus film It would, however, appear that relatively little sediment was actually trapped by the mucus, and that most of the relief was provided by mucus It is therefore questionable how resistant such a trace would be to sediment loading Mucus containing little sediments also would be prone to disruption by gentle currents Most of the Neoproterozoic ichnotaxa with which Collins et al (2000) compared their experimental traces clearly owe their preservation and morphology to displacement/excavation of sediment to a much greater degree than can be envisaged for such mucus creeping trails Further experiments on the preservation potential of mucus trails would be of great interest

### 4 1 Surface traces in mud

The surfaces of muddy sediments are highly unlikely sources for finding the earliest bilaterian benthic epistratal activity (Fig 1F) A muddy sub-

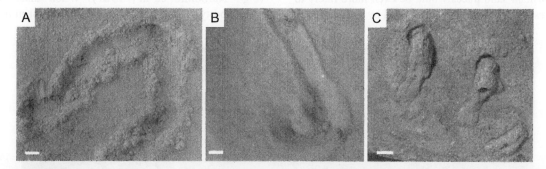

Fig 2 Trace made by the gastropod *Hinia reticulata* in experiments using natural sediments in marine tanks Scale bars are 10 mm (A) Surface trace from crawling over a muddy surface resulting in poorly defined levees (B) Surface traces made in sand Surficial movement resulted in a smooth flat base and narrow distinct lateral levees Near lower margin of picture, the animal has dug deeper into the sediment and was by this stage seen only from its "snorkel" (C) Experimentally produced undertraces (following the protocol of Jensen and Atkinson, 2001, with a sand layer about 3 cm thick) resulting from the behavior seen in the deeper portion of the trace in (B) The flat surface corresponds to the base of a storm bed, with the gastropod traces forming pronounced raised ridges Note that the pronounced depression surrounding the traces formed as mud moved up into the sand

strate with high water content will have low potential of recording the traces as well as a low potential of preservation (Fig. 2A) A completely firm muddy substrate, on the other hand, will not record any such surficial movement in the first place A possible exception is sediments that largely consist of silt, which will be discussed in Section 4 2

*4 2 Surface traces in sand and silt*

Traces formed close to the sediment water interface in mobile sand clearly have negligible preservational potential Movement over sheets of sand and silt that had been deposited by storms offshore, could, however, provide a more favorable scenario because these are unlikely to be rapidly reworked by additional currents A likely range of traces that may form in this setting was provided by experiments with the netted dog whelk *Hinia reticulata* This animal exhibited a range of behaviors from "ice skating" on top of the sediment to burrowing slightly beneath the sediment with a "snorkel," providing contact with the sediment–water interface (Fig. 2) In these experiments, movement on the surface resulted in lateral levees of displaced sand (Fig 2B) By and large, this type of trace represents involuntary sediment displacement Upon digging deeper in the sediment, a pronounced V-shaped furrow formed At this stage, the animal impinged across a mud–sand interface and produced what would amount to interface trace fossils if preserved at the base of sandstone (Fig. 2C)

Surficial sand traces would be even more likely to be preserved if the sand was covered by a thin blanket of muddy sediment at the time of trace formation (Fig 1D), or became covered with muddy sediment shortly afterwards In a sense, the mud serves to "protect" the trace (Fig 1E) Neoproterozoic trace fossils that can be assigned to *Archaeonassa* appear to represent movement over sandy substrates, and may be analogous to the situation of the creeping gastropods (Fig. 3) While these traces show displacement of sediment to various depths, they clearly were made by a producer that kept in contact with sediment–water interface

## 5. Trace fossil producers

As is the case for the majority of Phanerozoic trace fossils, it is difficult to assign producers to Neoproterozoic trace fossils. Stem group molluscs and priapulids are likely candidates for some of the early trace fossils (Valentine, 1994), but specific evidence is lacking The flat central area in specimens of *Archaeonassa* certainly would be in agreement with a mollusc-type producer Priapulids (sensu lato) may have been responsible for some of the earliest penetrative burrows Unfortunately, little is known of the types of burrows constructed by Recent priapulids (Powilleit et al , 1994) Typical arthropod-type trace fossils are not known from the Neoproterozoic, and bilobed forms with characteristic scratch patterns (*Rusophycus*) first appear somewhat above the currently defined base of the Cambrian (e g , Narbonne et al , 1987) When discussing arthropod-type trace fossils, it should be kept in mind

Fig 3 Neoproterozoic trace fossils representing movement close to the sediment–water interface in fine sand (A) *Archaeonassa* sp from the UstPinega Formation. Winter Coast of the White Sea, north–west Russia (SM 27518) Scale bar is 10 mm (B) Several morphologic varieties of *Archaeonassa* sp in the Ediacara Member, Flinders Ranges. South Australia, reflecting depth of animal movement within the sediment (specimen in collection of J G Gehling, Adelaide) Scale bar is 10 mm

that preservation of such features that would identify a trace as produced by an arthropod is under strong taphonomic control (Fig 4) Furthermore, the earliest arthropods may not have had the equipment with which typical arthropod-type trace fossils are identified For example, Budd (1996) derived arthropods from lobopodians with the development of lateral lobes, and segmentation, occurring prior to the development of sclerotized limbs

Although most Neoproterozoic traces can be reasonably assumed to have been produced by bilaterian animals, it is worth considering alternatives There are burrowing cnidarians, but these generally form simply vertical burrows Simple vertical plugs such as *Bergaueria* generally are interpreted as formed by actinians or even by pennatulacean holdfasts (Alpert, 1973, Seilacher-Drexler and Seilacher, 1999) More complex, short, branching, cnidarian-made burrows have been described (e g., Bradley, 1981, Jensen, 1992), which could be confused with bilaterian traces Collins et al (2000) observed surface creeping in cerianthid cnidarians, which trapped sediment in a mucus string or lateral levees As discussed above, the preservational potential of this type of mucus-bound surface trace is unclear

Protists rarely figure in the discussion of producers of trace fossils Numerous observations, however, suggest that foraminiferans are likely producers of trace fossils Buchanan and Hedley (1960) showed that *Astrorhiza limicola* will create a trace identical to *Archaeonassa* by the leading edge of the test pushing a wave of sand that disperses to the sides as two ridges Severin et al (1982) observed that burrowing of the

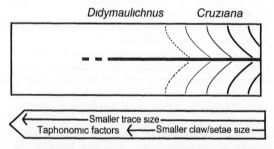

Fig 4 Schematic drawing to illustrate preservational control on trace fossil morphology The morphology of a trace produced by an animal with paired appendages will depend on the animals size as well as sediment properties Loss of distinct scratches yields a smooth bilobed trace (e g, *Didymaulichnus*) Further loss of preservational detail renders the trace essentially smooth

benthic foraminiferan *Quinqueloculina impressa* produced vertical traces thought to represent escape, and a horizontal and vertical maze-like system The ability to burrow apparently is present also in forms with a mostly planktic habit Hilbrecht and Thierstein (1996) observed foraminifera burrow and produce pits by removing sediment with rhizopodia Hilbrecht and Thierstein (1996) suggested that there was evidence for active lining of the burrow These lined burrows would potentially preserve as small dimples

The above discussion suggests that producers other than bilaterians must be considered for simple trace fossils, and perhaps also for short-branched forms That some of the nonbilaterian trace makers discussed above represent groups that on phylogenetic grounds are unlikely to have been present in the Neoproterozoic is besides the point Nevertheless, the relatively gradual increase in diversity and complexity of trace fossils at the Neoproterozoic–Cambrian boundary suggests that it reflects bilaterian radiation The biology of a number of Ediacaran organisms has long been debated The lack of evidence of mobility has been one important factor for nonmetazoan interpretations of the Ediacaran fossils Recent finds of associations of Ediacarans with trace fossils appear to provide evidence of mobility and thus support for metazoan affinities In the White Sea area of northern Russia, Ivantsov and Malakhovskaya (2002) found series of imprints of *Dickinsonia* preserved in positive relief on the base of a sandstone slab, terminated by a *Dickinsonia* preserved in negative relief, with the latter being the common mode of preservation for this organism This suggests that the *Dickinsonia* had made repeated depressions in the microbial mat that remained open until filled by the smothering sand (Ivantsov and Malakhovskaya, 2002) Although these traces bear little resemblance to the traditional terminal Proterozoic trace fossils, this should rekindle interest in Ediacarans as trace makers (cf Gehling, 1991, Seilacher, 1997)

## 6. How far back does the trace fossil record extend?

The low levels of bioturbation in the Cambrian, and even more so in the Proterozoic, mean that these were times in which the preservation of shallow tier trace fossils in marine settings would have been particularly favorable The most obvious consequence is the reduced destruction of thin event beds (e g , Sepkoski, 1982) As discussed above, the low levels of sediment mixing would have led to a more rapid dewatering of muddy sediments with resulting firm sediments close to the sediment–water interface The early development of infaunal activity should therefore be readily observable in burrows preserved by sediment filling open mud burrows, where an animal burrowed through a sand–mud interface, and probably also from surficial traces The widespread microbial mats also would have been beneficial to the preservation of shallow tier trace fossils In either case, the trace fossil record should be relatively sensitive to the onset of infaunal activity in storm-influenced settings

There are several reasons why particular care should be taken in examining Proterozoic sedimentary structures of suspected biogenic origin The Neoproterozoic contains a range of sedimentary structures of problematic origin For many of these structures, such as *Arumberia*, opinions are divided about whether their origin is biogenic or physical, or a combination of the two In our opinion, *Arumberia* is an obvious physical sedimentary structure, as outlined by McIlroy and Walter (1997) This conundrum probably can be explained by unusual preservational conditions High cohesiveness of the sediment may explain preservation of these aberrant sedimentary structures, the physical processes causing these structures were not necessarily unusual but the preservational conditions were Additionally, in the absence of bioturbation, the sedimentological record would be dominated by physical sedimentary structures The common presence of microbial mats also would have led to unusual preservational conditions (e g , Seilacher, 1997, Gehling, 1999) It therefore may not be an exaggeration to consider much of the Proterozoic as being a sedimentary structure preservation Lagerstatte It follows that it is a dubious approach to favor a biogenic origin of a problematic structure merely because there is no obvious explanation for it as an inorganic sedimentary structure

The distinction between a trace fossil and a pseudofossil is far from trivial (see discussion in Ekdale et al , 1984, pp 29–36) Particularly in the Neoproterozoic, there is the additional problem of confounding trace fossils with metaphytes It is

impossible to find definitive criteria that will give the true origin of a sedimentary structure but some of the features to look for include the following

– Gradual tapering in dimensions, where this is not stratinomically controlled, argues against a trace fossil origin
– Angular terminations may be expected from breakage of an organism with a modular growth, but not from a trace fossil
– Displacement of sediment Active displacement of sediment, such as levees and cross-cutting of other sedimentary structures, is a strong indicator of a trace fossil, although care must be taken with compactional deformation.
– Meniscate fill, if accompanied by fecal pellets or sorting yielding different properties to different parts of the trace, suggests a trace fossil origin
– Carbonized remains strongly suggest a body fossil origin

There have been numerous reports of pre-Neoproterozoic trace fossils, but upon critical evaluation, these traces turn out to be misidentified inorganic structures or metaphytes, or to have been misdated (e g , Hofmann, 1992, Fedonkin and Runnegar, 1992) Nevertheless, there has been a steady flow of new reports, of which Table 1 presents a partial listing

Table 1
A selection of recent reports of purported trace fossils 600 Ma and other with brief evaluations

*Neonereites uniserialis*, ca 600 Ma, Scotland, Brasier and McIlroy (1998), dubiofossil (also Brasier and Shields, 2000)
*Lockeia* sp , ca 600 Ma, Mexico, McMenamin (1996) undiagnostic, probably pseudofossil
*Palaeophycus tubularis*, ca 600 Ma, Mexico, McMenamin (1996), undiagnostic, probably pseudofossil
*Cochlichnus anguineus*, Riphean, India, Kulkarni and Borkar (1996b), pseudofossil, shrinkage cracks (Chakrabarti, 2001)
*Vermiforma antiqua* gen et sp nov , 620 Ma, USA, Cloud et al (1976), pseudofossil (Seilacher et al , 2000)
Metazoan burrows, ca 1 6 Ga, India, Seilacher et al (1998), dubiofossil, shrinkage cracks?
Grazing traces, Mesoprot , USA, Breyer et al (1995), dubiofossil
*Changchengia dahongyuensis* igen et isp nov , Mesoprot , P R China, Gao et al (1993), pseudofossil
*Dahongyuichnus dahongyuensis* igen et isp nov , Mesoprot P R China, Gao et al (1993), pseudofossil
Trace-like fossils, Mesoprot , western Australia, Rasmussen et al (2002a,b), dubiofossil

Below we will briefly consider some of the reports of trace fossil older than 600 Ma

Breyer et al (1995) reported grazing traces more than 1 Ga old However, the evidence for a continuous meander is unconvincing Possibly this structure has an origin similar to that of the pseudofossil *Arumberia* (cf Bekker, 1980, Fig 2) Kulkarni and Borkar (1996a) reported the vertical trace fossil *Skolithos linearis* from Riphean beds, which appears similar to Phanerozoic examples This needs further study In addition to the possibility that these are water escape pipes, more information on age constraint on this report would be welcome

Perhaps the most widely known recent report of ancient trace fossils is that of Seilacher et al (1998) from India Now known to be close to 1 6 Ga in age (Rasmussen et al, 2002a), these were described as trace fossils of triploblastic animals The irregular crinkly development is unlike that of trace fossils and better explained by the release of tensile strength in a compacting sediment Similarly, the presence of numerous pointed terminations is better explained by an origin as shrinkage cracks It should also be noted these structures occur together with a range of other sedimentary structures (e g, Sarkar et al, 1996)

Another noteworthy recent report is of more than 1 5 Ga hair pin-like ridges on the base of sandstone beds in Western Australia (Rasmussen et al, 2002b) A major concern with these purported traces is in reconciling the preservation on the base of sandstone beds with the proposed mode of preservation as casts of surface traces consisting of mucus and displaced sediment (Budd and Jensen, in press)

## 7. Conclusions

1 Cambrian (particularly earliest Cambrian) muddy sediments in shelfal settings were firm close to the sediment–water interface to a much higher degree than what is typical of the Phanerozoic The most likely interpretation for this is the low level and intensity of bioturbation of surface sediments

2 Firm sediments and low levels of disruptive bioturbation increased the likelihood of preservation of trace fossils made close to the sediment–water interface as well as sedimentary structures There also should have been an increased likelihood in the preservation of superficial trace fossils Proterozoic sediments therefore should have been particularly sensitive in recording the onset of infaunal activity

3 Despite reports to the contrary, there is no widely accepted trace fossil record from sediments older than about 560–555 Ma

4 The above conclusions place serious constraints on the time of appearance of bilaterian animals For example, assuming that key bilaterian features could only have been acquired in moderately large benthic animals, the absence of an ancient trace fossil record suggests that the Cambrian "explosions" are a reality in terms of the relatively rapid appearance and diversification of macroscopic bilaterians

## Acknowledgements

This work was supported, in part, by grants from National Geographic and the National Science Foundation (grant EAR-0074021 to M L D ) S J gratefully acknowledges funding for experimental production of trace fossils from the Leverhulme Trust and NERC (grant GR3/10713 to Simon Conway Morris) This paper benefited greatly from the constructive criticism of two anonymous reviewers and the editorial assistance of Loren Babcock

## References

Alpert, S P, 1973 *Bergaueria* Prantl (Cambrian and Ordovician), a probable actinian trace fossil J Paleontol 47, 919–924

Ayala, F J, Rzhetsky, A, Ayala, F J, 1998 Origin of the metazoan phyla molecular clocks confirm paleontological estimates Proc Natl Acad Sci U S A 95, 606–611

Bekker, Yu R, 1980 Novoe mestonakhozhdenie fauny ediakarskogo tipa na urale Dokl Akad Nauk SSSR 254, 480–482

Bottjer, D J, Hagadorn, J W, Dornbos, S Q, 2000 The Cambrian substrate revolution GSA Today 10 (9), 1–7

Bradley, J, 1981 *Radionereites, Chondrites*, and *Phycodes* trace fossils of anthoptilod sea pens Pac Geol 15, 1–16

Brasier, M D, McIlroy, D, 1998 *Neonereites uniserialis* from c 600 year old rocks in western Scotland and the emergence of animals J Geol Soc (Lond ) 155, 5–12

Biasiei M D , Shields, G , 2000 Neoproterozoic chemostratigraphy and correlation of the Port Askaig glaciation. Dalradian Supergroup of Scotland J Geol Soc (Lond ) 157, 909–914

Breyer, J A , Busbey, A B , Hanson, R E , Roy III, E C , 1995 Possible new evidence for the origin of metazoans prior to 1 Ga sediment-filled tubes from the Mesoproterozoic Allamoore Formation Trans-Pecos Texas Geology 23, 269–272

Bromley, R G , 1996 Trace fossils, biology, taphonomy and applications, 2nd ed Chapman and Hall, London 361 pp

Buchanan, J B , Hedley, R H . 1960 A contribution to the biology of Astrorhiza limicola (Foraminifera) J Mar Biol Assoc U K 39, 549–560

Budd G E 1996 The morphology of Opabinia and the reconstruction of the arthropod stem-group Lethaia 29, 1–14

Budd, G E , Jensen, S , 2000 A critical reappraisal of the fossil record of the bilaterian phyla Biol Rev 75, 253–295

Budd, G E , Jensen, S , 2003 The limitations of the fossil record and the dating of the origin of the Bilateria Syst Assoc Spec Publ (in press)

Chakrabarti, A , 2001 Are meandering structures found in Proterozoic rocks of different ages of the Vindhyan Supergroup of central India biogenic a scrutiny Ichnos 8, 131–139

Cloud, P , Wright, J , Glover III, J , 1976 Traces of animal life from 620 million-year-old rocks in North Carolina Am Sci 64, 396–406

Collins, A G , Valentine, J W , 2001 Defining phyla evolutionary pathways to metazoan body plans Evolut Develop 3 432–442

Collins, A G , Lipps, J H , Valentine, J W , 2000 Modern mucociliary creeping trails and the bodyplans of Neoproterozoic trace-makers Paleobiology 26, 47–55

Conway Morris, S , 1998 Early metazoan evolution reconciling paleontology and molecular biology Am Zool 38, 867–877

Conway Morris, S , 2000 Evolution bringing molecules in the fold Cell 100, 1–11

Davidson, E H , Peterson K J , Cameron, R A , 1995 Origin of bilaterian body plans evolution of developmental regulatory mechanisms Science 270, 1319–1325

Droser, M L , Jensen, S , Gehling, J G , Myrow, P M , Narbonne, G M , 2002a Lowermost Cambrian ichnofabrics from the Chapel Island Formation, Newfoundland implications for Cambrian substrates Palaios 17, 3–15

Droser, M L , Jensen, S , Gehling J G , 2002b Trace fossils and substrates of the terminal Proterozoic–Cambrian transition implications for the record of early bilaterians and sediment mixing Proc Natl Acad Sci U S 99, 12572–12576

Ekdale, A A , Bromley, R G , Pemberton, S G , 1984 Ichnology the use of trace fossils in sedimentology and stratigraphy SEPM Short Course 15 317 pp

Erwin, D H , Davidson, E H , 2002 The last common bilaterian ancestor Development 129, 3021–3032

Fedonkin, M A , Runnegar, B N , 1992 Proterozoic metazoan trace fossils In Schopf, J W , Klein, C (Eds ), The Proterozoic Biosphere A Multidisciplinary Study Cambridge University Press, pp 389–395

Fortey, R A , Briggs, D E G , Wills, M A , 1996 The Cambrian evolutionary 'explosion' decoupling cladogenesis from morphological disparity Biol J Linn Soc 57, 13–33

Fortey, R A , Briggs, D E G , Wills, M A , 1997 The Cambrian evolutionary 'explosion' recalibrated BioEssays 19, 429–434

Gao, Jian-hua, Cai, Keqing, Yang, Shipu, Fu, Heping, 1993 Discovery of the oldest trace fossils in the Changchengian of Jixian Chin Sci Bull 83, 1891–1895

Gehling, J G , 1991 The case for Ediacaran roots to the metazoan tree Mem Geol Soc India 20, 181–223

Gehling, J G , 1999 Microbial mats in terminal Proterozoic siliciclastics ediacaran death masks Palaios 14, 40–57

Graf, I , 1956 Die Fahrten von Littorina littorea Linne (Gastr ) in verschiedenen sedimenten Senckenb Lethaea 37, 305–318

Hagadorn, J W , Bottjer, D J , 1999 Restriction of a late Neoproterozoic biotope suspect-microbial structures and trace fossils at the Vendian–Cambrian transition Palaios 14, 73–85

Hallam, A , 1975 Preservation of trace fossils In Frey, R W (Ed ), The Study of Trace Fossils Springer-Verlag, New York, pp 55–63

Hilbrecht, H , Thierstein, H R , 1996 Benthic behavior of planktic foraminifera Geology 24, 200–202

Hofmann, H J , 1992 Proterozoic and selected Cambrian megascopic dubiofossils and pseudofossils In Schopf, J W , Klein, C (Eds ), The Proterozoic Biosphere A Multidisciplinary Study Cambridge University Press, pp 1035–1053

Ivantsov, A Yu , Malakhovskaya, Ya E , 2002 Giant traces of Vendian animals Dokl Earth Sci 385A, 618–622

Jensen, P , 1992 Cerianthus vogti Danielssen, 1890 (Anthozoa Ceriantharia) A species inhabiting an extended tube system deeply buried in deep-sea sediments off Norway Sarsia 77, 75–80

Jensen, S , 2003 The Proterozoic and earliest Cambrian trace fossil record, patterns, problems and perspectives Integr Comp Biol 43 (in press)

Jensen, S , Atkinson, R J A , 2001 Experimental production of trace fossils, with a discussion of allochthonous trace fossil producers Neues Jahrb Geol Palaontol , Monatsh 10, 594–606

Jensen, S , Gehling, J G , Droser, M L , Grant, S W F , 2002 A scratch circle origin for the medusoid fossils Kullingia Lethaia 35, 291–299

Knoll, A H , Carroll, S B , 1999 Early animal evolution emerging views from comparative biology and geology Science 284, 2129–2137

Knox, L W , Miller, M F , 1985 Environmental control of trace fossil morphology In Allen Curran, H (Ed ), Biogenic Structures Their Use in Interpreting Depositional Environments Society of economic petrologists and mineralogist's special publication, vol 35, pp 167–176

Kulkarni, K G , Borkar, V D , 1996a A significant stage of metazoan evolution from the Proterozoic rocks of the Vindhyan Supergroup Curr Sci 70, 1096–1099

Kulkarni, K G , Borkar, V D , 1996b Occurrence of Cochlichnus hitchcock in the Vindhyan Supergroup (Proterozoic) of Madhya Pradesh J Geol Soc India 47, 725–729

Lee, H , Swartz, C , 1980 Biological processes affecting the distribution of pollutants in marine sediments Part II Biodeposition and bioturbation In Baker, R A (Ed ), The Contaminants and Sediments, vol 2 Ann Arbor Science, pp 555–606

Martin, M W, Grazhdankin, D V, Bowring, S A, Evans, D A D, Fedonkin, M A, Kirschvink, J L, 2000 Age of Neoproterozoic bilaterian body and trace fossils, White Sea, Russia implications for metazoan evolution Science 288, 841–845

Martinsson, A, 1965 Aspects of a Middle Cambrian thanatotope on Oland Geol Foren Stockh Forh 87, 181–230

McIlroy, D, Walter, M R, 1997 A reconsideration of the biogenicity of *Arumbera banksi* (Glaessner and Walter) Alcheringa 21, 79–80

McMenamin, M, 1996 Ediacaran biota from Sonora, Mexico Proc Natl Acad Sci U S A 93, 4990–4993

Meadows, P S, Tait, J, 1989 Modification of sediment permeability and shear strength by two burrowing invertebrates Mar Biol 101, 75–82

Narbonne, G M, Myrow, P, Landing, E, Anderson, M A, 1987 A candidate stratotype for the Precambrian–Cambrian boundary, Fortune Head, Burin Peninsula, southeastern Newfoundland Can J Earth Sci 24, 1277–1293

Paterson, D M, 1994 Microbiological mediation of sediment structures and behaviour In Stal, L J, Caumette, L N (Eds), Microbial Mats Springer-Verlag, Berlin, pp 97–109

Peterson, K J, Takacs, C M, 2002 Molecular clocks, snowball earth, and the Cambrian explosion Society for Integrative and Comparative Biology, Annual Meeting, Final Program and Abstracts, p 341

Powilleit, M, Kitlar, J, Graf, G, 1994 Particle and fluid bioturbation caused by the priapulid worm *Halicryptus spinulosus* (v Seibold) Sarsia 79, 109–117

Rasmussen, B, Bose, P K, Sarkar, S, Banerjee, S, Fletcher, I R, McNaughton, N J, 2002a 1 6 Ga U–Pb zircon age for the Chorat Sandstone, lower Vindhyan, India possible implications for early evolution of animals Geology 30, 103–106

Rasmussen, B, Bengtson, S, Fletcher, I R, McNaughton, N J, 2002b Discoidal impressions and trace-like fossils more than 1200 million years old Science XX, 1112–1115

Sarkar, S, Banerjee, S, Bose, P K, 1996 Trace fossils in the Mesoproterozoic Koldaha shale, central India and their implications Neues Jahrb Geol Palaontol, Monatsh 7, 425–436

Seilacher, A, 1964 Biogenic sedimentary structures In Imbrie, J, Newell, N (Eds), Approaches to Paleoecology John Wiley and Sons, New York, pp 296–316

Seilacher, A, 1970 *Cruziana* stratigraphy of 'non-fossiliferous' Palaeozoic sandstone In Crimes, T P, Harper, J C (Eds), Trace Fossils Seel House Press, Liverpool, pp 447–476

Seilacher, A, 1978 Use of trace fossil assemblages for recognizing depositional environments In Basan, P B (Ed), Trace Fossil Concepts, SEPM Short Course, vol 5, pp 167–181

Seilacher, A, 1982 Distinctive features of sandy tempestites In Einsele, G, Seilacher, A (Eds), Cyclic and Event Stratification Springer-Verlag, Berlin, pp 333–349

Seilacher, A, 1997 Fossil art An exhibition of the Geologisches Institut Tuebingen University The Royal Tyrrell Museum of Paleontology, Drumheller

Seilacher, A, 1999 Biomat-related lifestyle in the Precambrian Palaios 14, 86–93

Seilacher, A, Pfluger, F, 1994 From biomats to benthic agriculture a biohistoric revolution In Krumbein, W E, Paterson, D M, Stal, L J (Eds), Biostabilization of Sediments Bibliotheks- und Informationssystem der Universitat Oldenburg, Oldenburg, pp 97–105

Seilacher, A, Bose, P K, Pfluger, F, 1998 Triploblastic animals more than 1 billion years ago trace fossil evidence from India Science 282, 80–83

Seilacher, A, Meschede, M, Bolton, E W, Luginsland, H, 2000 Precambrian "fossil" *Vermiforma* is a tectograph Geology 28, 235–238

Seilacher-Drexler, E, Seilacher, A, 1999 Undertraces of sea pens and moon snails and possible fossil counterparts Neues Jahrb Geol Palaontol Abh 214, 195–210

Sepkoski, J J, 1982 Flat-pebble conglomerates, storm deposits, and the Cambrian bottom fauna In Einsele, G, Seilacher, A (Eds), Cyclic and Event Stratification Springer-Verlag, Berlin, pp 371–385

Severin, K P, Culver, S J, Blanpied, C, 1982 Burrows and trails produced by *Quinqueloculina impressa* Reuss, a benthic foraminifer, in fine-grained sediments Sedimentology 29, 897–901

Valentine, J W, 1994 Late Precambrian bilaterians grades and clades Proc Natl Acad Sci U S A 91, 6751–6757

Wetzel, A, Aigner, T, 1986 Stratigraphic completeness tiered trace fossils provide a measuring stick Geology 14, 234–237

Wray, G A, Levinton, J S, Shapiro, L H, 1996 Molecular evidence for deep Precambrian divergences among metazoan phyla Science 274, 568–573

Ziebis, W, Forster, S, Huettel, M, Jorgensen, B B, 1998 Complex burrows of the mud shrimp *Callianassa truncata* and their geochemical impact in the sea bed Nature 382, 619–622

Available online at www.sciencedirect.com

ELSEVIER          Palaeogeography, Palaeoclimatology, Palaeoecology 220 (2005) 31–46

www.elsevier com/locate/palaeo

# Fossilization modes in the Chengjiang Lagerstätte (Cambrian of China): testing the roles of organic preservation and diagenetic alteration in exceptional preservation

Maoyan Zhu[a,*], Loren E. Babcock[b], Michael Steiner[c]

[a]Nanjing Institute of Geology and Palaeontology, Chinese Academy of Sciences, Nanjing 210008, China
[b]Department of Geological Sciences, The Ohio State University, Columbus, OH 43210, USA
[c]Technische Universtat Berlin, ACK 14, Ackerstrasse 71-76, Berlin 13355, Germany

Received 3 December 2002, accepted 1 March 2003

## Abstract

One important, and poorly understood, issue concerning the taphonomy of nonmineralizing organisms in Cambrian Burgess Shale-type deposits is the diagenetic pathways by which organisms have become exceptionally preserved Scanning electron microscopy (SEM) analyses of exceptionally preserved fossils from the lower Cambrian Chengjiang deposit of China demonstrate that nonmineralized tissue is preserved in a variety of ways, and further suggest that there was some taxonomic control over the precipitation of authigenic minerals during early diagenesis Organic preservation is limited to certain decay-resistant structures such as heavily sclerotized spines on the claws of anomalocaridids Labile organic tissues of nearly all nonmineralizing animals, however, are preserved by thin films of diagenetic minerals Microstructural studies, energy-dispersive X-ray (EDX) analyses, and elemental mapping of these minerals indicate that mineral films include apatite, pyrite, and Fe-rich aluminosilicates The results support the assumption that early diagenetic mineralization such as phosphatization and pyritization played a key role in the preservation of nonmineralized organisms Pyritization seems to be the most important process by which nonmineralizing Chengjiang organisms are preserved in exceptional condition Precipitation of Fe-rich aluminosilicates, which occurred following pyritization, also played a role in the preservation of some Chengjiang fossils Phosphatization in the Chengjiang Lagerstatte was evidently rare, and limited taxonomically or to animals having large masses of organic material
© 2004 Elsevier B V All rights reserved

*Keywords* Chengjiang Lagerstatte, Nonmineralizing fossils, Fossilization, Taphonomy, Cambrian, China

## 1. Introduction

Deposits of exceptional preservation (or Konservat–Lagerstatten, Seilacher et al, 1985) provide important insight into the evolutionary history of the

* Corresponding author
  *E-mail address* myzhu@nigpas cn (M Zhu)

0031-0182/$ - see front matter © 2004 Elsevier B V All rights reserved
doi 10 1016/j palaeo 2003 03 001

biosphere Significantly, such deposits help to compensate for the strong shelly fossil bias of much of the fossil record A relatively large number of deposits of exceptional preservation (Conway Morris, 1985; Babcock et al, 2001) occur in Cambrian strata, and these help to reveal major aspects of the remarkable ecological and phylogenetic radiation of metazoans during the Cambrian This radiation "event," or series of events, ranks as one of the most dramatic chapters in the history of life (Allison and Briggs, 1993, Briggs, 1991)

One of the most important unresolved issues concerning exceptional preservation in the Cambrian is the mode of preservation of organisms in deposits of exceptional preservation (which are often referred to as Burgess Shale-type deposits) Over the last two decades, attention has focused on various aspects of the taphonomy of nonmineralizing organisms, particularly those that tend to dominate in deposits of exceptional preservation Major strides have been made in our understanding of certain preburial and postburial processes, as well as in our understanding of the environmental circumstances under which exceptional preservation is likely However, it is clear that factors influencing exceptional preservation are complex Fossils of nonmineralizing organisms have been reported from various lithologies (Wollanke and Zimmerle, 1990), and their preservation spans a range of preservation modes (Allison, 1988b)

Taphonomic analyses (e g , Seilacher et al , 1985, Wollanke and Zimmerle, 1990, Brett, 1990; Babcock et al , 2001) indicate that a number of related biological, chemical, and sedimentary factors favor the preservation of nonmineralizing organisms Those factors include (1) elimination of postmortem breakdown by biologic agents such as scavengers and microbes, (2) rapid burial of remains in fine sediment or by in situ burial (obrution or sediment smothering); and (3) absence of bioturbation Postmortem breakdown by biologic agents seems to be inhibited by chemical conditions including oxygen-deficient bottom water, redox boundary close to the sediment–water interface, fresh water input to a marine basin, hypersalinity, and perhaps other conditions (Babcock et al , 2001) Postburial processes leading to exceptional preservation are still largely unresolved, although oxygen-deficient conditions

seem to be an important factor in preservation (Allison, 1988b), particularly where diagenetic mineralization is involved (e g , Briggs et al , 1991, 1996) Butterfield (1990, 1995) hypothesized that most Burgess Shale-type fossils are preserved as organic (kerogenized) films resulting from interactions between clay minerals and original (organic) biomolecules Other workers, however, have reported that early diagenetic replication of nonmineralized tissue by minerals played an important role in the exceptional preservation of Cambrian Burgess Shale-type fossils (Orr et al , 1998; Babcock and Zhang, 2001; Zhu et al , 2001)

Experimental taphonomy provides insight into the complex processes of fossilization of nonmineralizing fossils (e g , Briggs and Kear, 1993, 1994a,b, Briggs, 1995; Hof and Briggs, 1997, Babcock et al , 2001), but testing hypotheses derived from experimental methods requires information from the examination of nonmineralized fossils Exceptionally preserved fossils from the lower Cambrian Chengjiang Lagerstatte provide excellent material for such investigation To date, only a few groups of Chengjiang organisms have been investigated for their preservation styles Chen and Erdtmann (1991) and Jin et al (1991) indicated that Fe-oxides and Fe-aluminosilicates are the predominant minerals responsible for replicating delicate tissues during diagenesis (Jin et al , 1991, p 41) Leslie et al (1996) reported that phosphate was responsible for the replication of some organisms However, many specimens analyzed by previous authors are deeply weathered, and it remains unclear whether the detected minerals were formed authigenically shortly after burial or quite recently by weathering processes In order to understand better the postburial processes affecting nonmineralizing organisms from the Chengjiang deposit, we present here chemical and microstructural results from selected specimens exhibiting little evidence of weathering

## 2. Materials and methods

Fossils used in this study (Figs 1–7) were collected from two classic localities yielding fossils of the Chengjiang Biota Maotianshan, Chengjiang County, and Meishucun, Jinning County, Yunnan

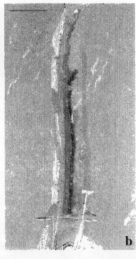

Fig. 1. Taphonomic details of worms from the Chengjiang Biota, Yunnan, China. (a) *Maotianshania* showing surface covered with thick Fe-oxide film; dark line of Fe-oxide film extending longitudinally down the middle of body indicates the structure of the gut. Specimen from Maotianshan, Chengjiang County; scale bar=1 mm; NIGPAS-ELRC-Z0009. (b) *Palaeoscolex* showing surface covered with thick mineralized film of apatite; gut structure is indicated with a dark film of pyrite. Specimen from Meishucun, Jinning County; scale bar=5 mm; NIGPAS-ELRC-Z0010.

Province, China. Most nonmineralizing fossils collected from these sites are in weathered yellow-colored shale. When fresh, the shale is gray to gray-green in color, and fossils are difficult to find in the field owing to a low contrast of the fossils against the matrix. In order to investigate the preservational state of the nonmineralizing organisms, specimens selected for scanning electron microcopy (SEM) analysis are those that visually show little or no obvious evidence of weathering.

Specimens used for SEM study were pieces removed from larger samples. They were analyzed separately in the scanning electron microscope (SEM) laboratories of the Nanjing Institute of Geology and Palaeontology, Chinese Academy of Sciences (NIG-PAS), and the Technische Universität Berlin (TU Berlin). Samples analyzed in the NIGPAS laboratory are small fragments coated with Pt for high-resolution electron microphotography under an Oxford Instruments AN-10000 SEM. Elemental analyses were carried out using an energy-dispersive X-ray (EDX) analyzer (JSM-35CF attached to the AN-10000 SEM); specimens were coated with carbon prior to EDX analysis. Elemental mapping and EDX analysis of some larger specimens (up to 100×70 mm) were carried out using a Kevex Delta V energy-dispersive elemental analyzer with a Quantex light element detector coupled to a Hitachi S-2700 SEM at TU Berlin. Samples analyzed at TU Berlin were not coated prior to analysis.

Illustrated specimens have been reposited in the museum of the Nanjing Institute of Geology and Palaeontology (Ealy Life Research Center), Chinese Academy of Sciences, Nanjing, China (NIGPAS-ELRC).

## 3. Preservation

Nearly all exceptionally preserved specimens collected from the classic Chengjiang sections at Maotianshan, Ma'anshan, and Haikou (Jinning) are deeply weathered specimens from the Maotianshan Shale Member of the Yu'anshan Formation (*Eoredlichia–Wutingaspis* Zone). Comparative specimens from localities lacking deep weathering indicate that specimens from yellowish shales of the classic sections have become depleted in organic carbon and carbonate. Total organic carbon (TOC) values for samples from the Haikou section (Chengjiang County) range up to 2.1 wt.% TC. TC values range up to 1.4 wt.% in drill core ZK 23/4 from Chengjiang. Carbon values range from 0.35 to 1.3 wt.% for $C_{carb}$, and from 0.1 to 0.3 wt.% for TOC in samples from the Dapotou section.

Surfaces of many deeply weathered fossils from classic Chengjiang sections are smooth, and light gray to yellowish or brown in color. EDX analyses indicate the presence of aluminosilicates on both the surfaces of analyzed fossils and in the surrounding matrix. However, many of the weathered fossils are covered with a thin pale brown mineral film (Fig. 1a). EDX analysis indicates that the film is composed of an Fe-rich aluminosilicate.

In collections from sections at Maotianshan, voids in some fossils, such as voids between the valves of bivalved arthropods and brachiopods, are filled with chamosite Fe-oxides with rose-like or radiating long crystals (Zhu, 1992, plate 14; Fig. 7). Some specimens show dark grains randomly covering the surfaces of Chengjiang fossils. The dark grains show a uniform gelatinous structure under

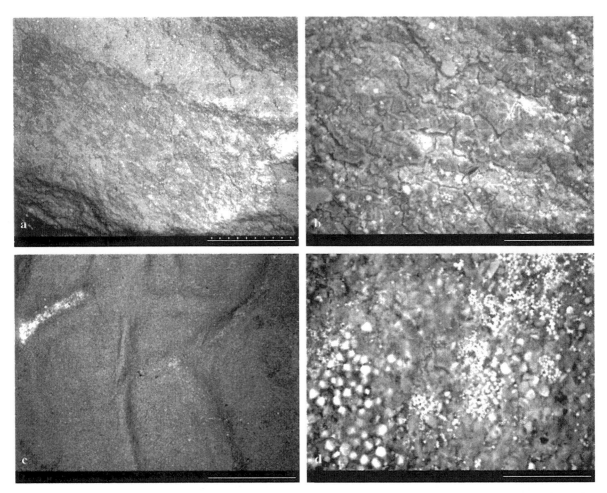

Fig. 2. Backscattered scanning electron micrographs (BSE–SEM) showing taphonomic details of arthropods from the Chengjiang Biota, Maotianshan, Chengjiang County, China. (a and b) *Anomalocaris*: uncoated fragments from black-colored spines attached to a frontal-grasping appendage showing organic film on the surface. Specimen from Maotianshan, Chengjiang County; scale bar, a=300 μm; scale bar, b=60 μm; NIGPAS-ELRC-Z0011. (c and d) *Waptia*: part of uncoated carapace showing irregular wrinkles of exoskeleton with brightly colored iron oxide (c), scale bar=1.2 mm; exoskeleton showing iron oxide microspherules (d), scale bar=30 μm; NIGPAS-ELRC-Z0012.

high SEM magnification (Zhu, 1992, plate 13; Figs. 5 and 6). EDX analyses indicate that they are composed of a Fe-rich aluminosilicate, which is interpreted to be late diagenetic chamosite (Jin et al., 1991). We regard the presence of chamosite to be of no particular relevance for determining the early diagenetic history of the nonmineralized fossils, however.

Fossils showing little or no evidence of weathering still retain thick materials on their surfaces, and these provide opportunities for investigating original features of the material on the fossil surfaces. So far, organic carbon, pyrite, apatite, and Fe-rich clay

minerals have been identified on exceptionally preserved Chengjiang fossils.

### 3.1. Organic carbon

Even though most Chengjiang fossils show pale yellow to pale brown aluminosilicate or Fe-aluminosilicate films on their surfaces, some tissues are preserved as dark films (notably gut traces of the priapulid worms *Maotianshania* and *Palaeoscolex* (Fig. 1), and the alimentary canal of *Stellostomites*). EDX analyses indicate that these dark-colored tissues are composed predominantly of Fe minerals

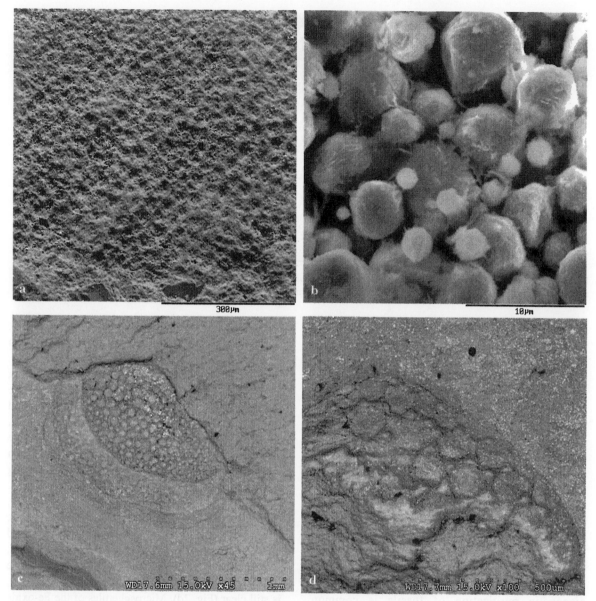

Fig. 3. BSE–SEM images of animals from the Chengjiang Biota, Maotianshan, China, showing taphonomic details. (a and b) *Maotianshania*: a worm, showing iron oxide microspherules; scale bars indicated below images; NIGPAS-ELRC-Z0013. (c and d) *Microdictyon*: a lobopod trunk and spicules showing pyrite microspherules; scale bars indicated below images; NIGPAS-ELRC-30061.

(Figs. 4 and 5). Elemental maps of specimens of *Stellostomites* and *Palaeoscolex*, however, also show that faint traces of carbon are present. In these examples, carbon is concentrated in such places as the alimentary canals (Fig. 5). The elemental map for carbon of a *Palaeoscolex* specimen (Fig. 5) indicates that carbon content is higher on the surface of fossils than in the surrounding matrix. Thus, it seems that some of the original organic materials of the organism have been preserved.

Sclerotized spines of the grasping appendages of *Anomalocaris* are ordinarily preserved as a thick black film. A sample from one of the terminal

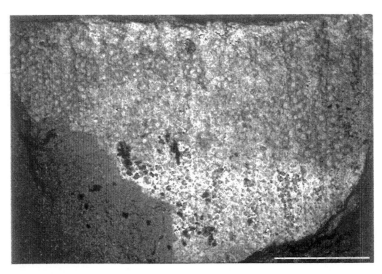

Fig. 4. BSE–SEM image of uncoated fragment of the worm *Palaeoscolex* showing mineralized film (light color) on the cuticular surface; clay matrix is dark gray; the gut trace is bright in this image. Sample is part of the specimen illustrated in Fig. 1b; scale bar=600 μm.

spines of the claw was studied under the SEM without coating. EDX and elemental analyses indicate that the black film is composed of carbon (Fig. 2a and b).

### 3.2. Pyrite

Most fossils of nonmineralizing organisms in the Chengjiang Lagerstätte show preservation by means of a brownish mineral film containing Fe-oxides. Some specimens with thicker and darker mineral films were analyzed under the SEM in order to determine the structure and composition of the Fe-rich mineralization. The microstructure of diagenetic Fe minerals coating these specimens is of two types: microspherules and framboids. Both morphotypes were probably precipitated as pyrite. Early diagenetic pyritization of fossils is common throughout the Phanerozoic (Allison, 1988a; Canfield and Raiswell, 1991). Pyrite often shows a variety of crystal shapes (Underwood and Bottrell, 1994) similar to those observed as microspherules and framboids present on Chengjiang specimens. EDX analyses indicate that the Fe-oxide-rich coatings do not presently contain sulfur; sulfur has probably been lost though diagenesis or weathering. Loss of sulfur is inferred to have occurred through weathering (see Littke et al., 1991; Moses and Herman, 1991).

In an analyzed specimen of the arthropod *Waptia*, thick, black mineralization occurs preferentially in an irregularly wrinkled area of the carapace (Fig. 2c). Wrinkling of this type is common in nonmineralizing arthropods shortly after death as specimens lie at the sediment–water interface (Babcock et al., 2001), and may be enhanced through postburial compaction. SEM–EDX analysis without prior coating of the specimen reveals no difference in composition between the surface of carapace and the surrounding matrix, which consists mostly of Si, Al, K, Mg, and O. Backscattered imaging (Fig. 2c and d), however, shows that minerals in the wrinkled area are Fe-oxide, and have a bright appearance in the backscattered image. Although the specimen is slightly weathered, the original spherical morphology of the diagenetic Fe mineral, which is assumed have originally been pyrite, is still clear (Fig. 2d). Microspherules of the Fe-oxide pseudomorph after pyrite show various shapes and sizes ranging from 0.3 to 1 μm in diameter.

SEM analysis of a specimen of the worm *Maotianshania* without coating indicates that microspherules having diameters of approximately 1–10 μm cover the entire surface of the cuticle (Fig. 3a and b). EDX analysis indicates that the microspherules are now composed of Fe-oxides.

Results obtained from a specimen of the lobopod *Microdictyon* are similar to those obtained from *Maotianshania*. In the *Microdictyon* example, pyrite

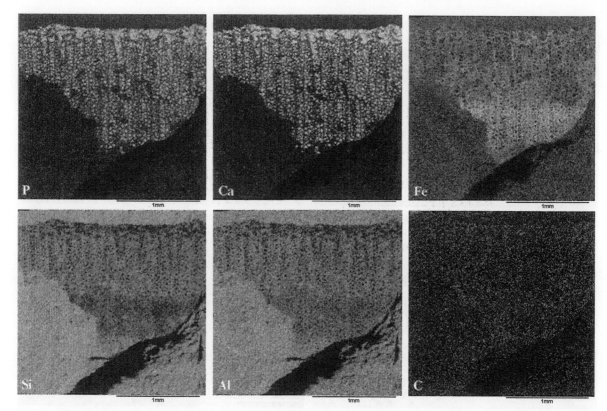

Fig. 5. Elemental maps of uncoated fragment of the worm *Palaeoscolex*. Image shows that the surface of *Palaeoscolex* is composed predominantly of P and Ca; Fe occurs preferentially in the position of the gut; Si and Al are not present on the surface of fossil; little C is present on the surface of fossil. Sample is part of the specimen illustrated in Fig. 1b; scale bars indicated in images.

microspherules cover not only the honeycomb-shaped sclerites, which have a primary phosphatic composition (Leslie et al., 1996), but also cover the non-mineralizing trunk (Fig. 3c and d).

Chancelloriid sclerites analyzed under the SEM (Fig. 7) show framboidal Fe-rich diagenetic mineralization, which, like the microspherules described above, was probably precipitated as pyrite. Framboidal pyrite on chancelloriids has been observed only as fillings of the void spaces within the hollow sclerites (Fig. 7d). Framboids are overgrown by late diagenetic clay minerals, indicating that precipitation of pyrite framboids preceded precipitation of authigenic aluminosilicate minerals.

### 3.3. Apatite

SEM and EDX analyses on a fossil of the priapulid worm, *Palaeoscolex*, show the presence of diagenetic Fe, Ca, and P, which are interpreted to be altered pyrite and apatite (Figs. 1b, 4, 5, and 6). The unweathered specimen is gray and contrasts little with the gray color of the shale matrix. The surface of the fossil has a thick coating of dark gray mineralization (Fig. 1b), which is also evident in backscattered imaging of a small fragment (Fig. 4). One fragment of the *Palaeoscolex* specimen was analyzed after being coated with Pt. SEM images show that the surface of the fossil is covered with a thin layer of microspherules ranging in diameter from 0.1 to 0.3 μm (Fig. 6). EDX analysis reveals the fossil to be composed of Si, Al, O, P, Ca, Mg, Fe, etc.; unfortunately, it is difficult to distinguish P and Pt peaks on EDX readout because the sample was coated by Pt. Another fragment of the specimen was analyzed without any coating. However, elemental maps (Fig. 5) show that high concentrations of P and Ca are associated with the small nodes and transverse ridges on the cuticular

Fig. 6. Scanning electron micrograph (SEM) of coated specimen of the worm *Palaeoscolex* showing detailed surface structure and diagenetic mineralization; part of specimen illustrated in Fig. 1b. Annulations and granular surface (a); phosphatic (apatite) microspherules covering the surface (b–d); scale bar=1 mm; scale bar, b=100 μm; scale bar, c=10 μm; scale bar, d=1 μm.

surface, and that concentrations of Fe are limited to the alimentary canal (Figs. 4 and 5). Phosphate does not co-occur with iron oxide in the gut trace. Elemental maps do not show the presence of Si, Al, and K within the fossil. These elements seem to be related instead to clay minerals in the matrix.

Results from the investigated cuticle of *Palaeoscolex* show diagenetic phosphatization similar to that recorded for paleoscolecid worms from elsewhere. Diagenetic phosphate on the Chengjiang specimens was precipitated preferentially on the nodes and other areas of epicuticular ornamentation, and this pattern of occurrence closely resembles that noted for Cambrian and Ordovician palaeoscolecid cuticles reported from various lithologies (Kraft and Mergl, 1989; Hinz et al., 1990; Müller and Hinz-Schallreuter, 1993; Zhang and Pratt, 1996; Conway Morris, 1997). It is likely that the original biochemical structure of the cuticular matrix of palaeoscolecid worms helped

to increase the chances that phosphatization would occur early in their diagenetic history.

### 3.4. Fe-rich aluminosilicates

Numerous chancelloriid specimens from a single layer at Maotianshan show well-preserved sclerites having diagenetic crystals internally, and provide a key to understanding the relative sequence of diagenetic changes that these fossils have undergone. SEM imaging shows the hollow interiors of chancelloriid sclerites to have been lined initially with cubic crystals ranging in diameter from 5 to 10 μm (Fig. 7a and b). The cubic crystals, in turn, have inside them smaller framboids (1–2 μm in diameter; Fig. 7d). EDX analyses indicate that the framboids are iron oxide (presumably often pyrite). The pyrite cubes are covered with crystalline clay minerals (Fe-rich aluminosilicates; Fig. 7c). The structure of diagenetic

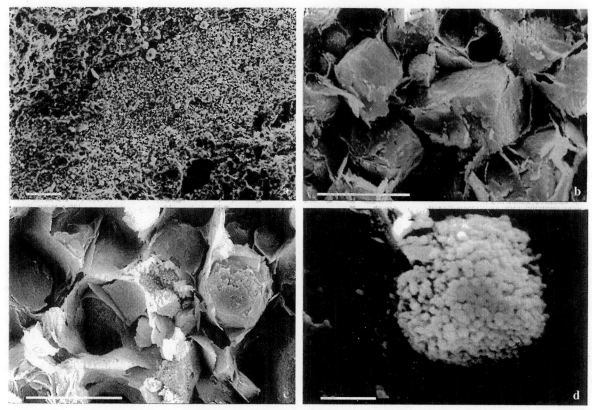

Fig 7 SEM images of a gold-coated chancellorid sclerite from the Chengjiang Biota, Maotianshan, Chengjiang County, China, showing granular diagenetic mineralization (pyrite with Fe-rich aluminosilicate overgrowths) in the internal cavity of sclerite (a), cubic pyrite grains and pyrite framboids inside the sclerite (b), clay mineral overgrowths on the framboids inside the sclerite (c), and pyrite framboids underlying the clay mineral overgrowths (d), scale bar, a=100 μm, scale bar, b=10 μm, scale bar, c=8 57 μm, scale bar, d=1 μm, NIGPAS-ELRC-Z0014

minerals inside the chancellorid sclerites suggests that precipitation of diagenetic pyrite was followed by precipitation of Fe-rich clay minerals.

## 4. Interpretation

SEM and EDX analyses indicate that diagenetic mineralization played a key role in the exceptional preservation of nonmineralizing animals from the Chengjiang deposit (cf Allison, 1988b) Diagenesis may have begun while remains were present at the sediment–water interface and prior to full burial (cf Wilby et al., 1996a,b; Borkow and Babcock, 2003), however, most diagenesis probably occurred following burial Pyritization, and possibly phosphatization, likely occurred early in the diagenetic history of the fossils Precipitation of Fe-rich aluminosilicates fol-

lowed the precipitation of pyrite Experimental analyses have demonstrated that decay of organic matter rapidly changes the chemical microenvironment surrounding a decaying carcass, thus leading to mineralization (e g, Sagemann et al, 1999) However, diagenesis of nonmineralized skeletal tissues and internal soft tissues is controlled by a complex of factors, and the timing of fossilization of nonmineralizing organisms is therefore likely to have varied according to such factors as taxonomic group, composition of body parts, sediment type, Eh–pH conditions, and availability of microbes responsible for mediating diagenetic activity

Experimental work demonstrates that phosphatization is an early diagenetic process, and results from Chengjiang do not conflict with that interpretation Phosphatization of soft and other nonmineralized tissues has been shown to be underway as early as 2 weeks following death (Allison, 1988c, Briggs et al, 1993, Briggs and Kear, 1993, 1994, Hof and Briggs, 1997), although in some examples, apatite may preserve morphological details that are easily lost within hours of the onset of decay (Martill et al, 1992, Wilby and Whyte, 1995)

The process of phosphatization is reasonably well known in terms of its generalities As currently understood, release of $CO_2$ and $H_2S$ during initial phases of microbial decay of a carcass in a dysoxic or anoxic setting (such as within the sediment column) will result in a chemical microenvironment of low pH immediately surrounding the carcass This condition leads to the rapid precipitation of francolite as microspherules At sites of active phosphogenesis today, microbial mats at the sediment–water interface may fulfill the role of altering the pH so as to favor precipitation of the apatite mineral francolite. It has been suggested that microbial mats may have been a prerequisite for phosphogenesis in ancient sediments, particularly in examples where soft tissues are inferred to have been preserved by the outlines of phosphatized microbial mats surrounding vertebrate fossils (Briggs et al, 1997, Wilby et al, 1996a,b) Microbes metabolize soft tissues and release phosphorus, and in turn phosphorus is fixed by microbes, resulting in phosphatization of microbes This process of autolithification was demonstrated experimentally (Lucas and Prevot, 1984, Hirschler et al, 1990; Prevot and Lucas, 1990), and was supported by evidence of muscle fibers preserved through phosphatized microspherules in Cretaceous fishes (Martill, 1988) and Jurassic bivalves (Wilby and Whyte, 1995) We interpret similar apatite microspherules on the surface of *Palaeoscolex* as microbially mediated phosphatization rather than phosphatized bacteria cells because the microspherules are too small in size (0 1–0 3 μm in diameter) to be considered autolithified bacteria Autolithified bacteria normally have diameters ranging from one to several microns (Hirschler et al, 1990; Briggs and Kear, 1993)

Precipitation of apatite is limited by the concentration of phosphorus in sediment pore waters Generally, phosphorus is rare in sediments, but pore water concentrations will increase as the result of microbial metabolism of organic material (Lucas and Prevot, 1984) Microbial metabolism of large organic tissue masses will release more phosphorus and favor the precipitation of apatite, whereas metabolism of small tissue masses would not release sufficient acidic biproducts to result in decreasing pH For this reason, phosphatization of fossils is clearly susceptible to some taxonomic control (Briggs and Wilby, 1996, Wilby and Whyte, 1995) Even within the carcass of a single organism, differences in diagenetic mineralization can be expected as different types of tissue undergo decay (Briggs and Wilby, 1996) The presence of phosphatization in only a few groups of Chengjiang organisms indicates a differential taxonomic control on phosphatization or diagenesis Phosphatization of *Palaeoscolex* may be due partly to the thickness of the surface cuticle, as it would provide considerable labile tissue for decay

Conditions leading to phosphate precipitation are partly limited by time Microenvironments having low pH values initially tend to persist for only a few hours (Martill, 1988) because ongoing decay processes will end chemical conditions conducive to phosphate precipitation As indicated by Benmore et al (1983), phosphorus released from the decay of organic material will first be adsorbed by iron hydroxides (FeOOH), then released again to pore waters when iron hydroxides become reduced at the redox boundary; this leads to precipitation of calcium phosphate at the redox interface The redox interface at the carcass shifts rapidly as decomposition of organic matter proceeds Further sulfate reduction leads to a high ambient concentration of sulfide and bicarbonate ions, which can result in the precipitation of pyrite and calcium carbonate Thus, precipitation of calcium phosphate is inhibited by a kinetic factor That these processes operated in the exceptional preservation of Chengjiang *Palaeoscolex* is evinced by the co-occurrence of diagenetic iron minerals and apatite on the cuticular surface and also by the presence of iron minerals in the gut of *Palaeoscolex* It is likely that precipitation of pyrite followed the phosphatization of *Palaeoscolex*

Phosphatized fossils are known from a number of geological occurrences Among the fine examples of exceptional preservation by means of phosphatization are algae and embryos from the Neoproterozoic Doushantuo Formation in South China (Xiao and Knoll, 1999, Chen et al, 2000), three-dimensionally preserved arthropods from a middle Cambrian phosphorite in Australia (Walossek et al, 1993), three-dimensionally preserved arthropods from upper Cambrian concretions (Orsten) in the Alum Shale of Sweden (Muller, 1990), muscles of squids from the Jurassic Oxford Clay (Allison, 1988c), and fish with preserved muscle masses from concretions in the Cretaceous Santana Formation of Brazil (Martill, 1988)  In addition to occurrences such as these, there are also numerous instances of phosphatization of soft tissues or nonmineralized skeletal matter from sediments poor in iron content, such as thinly laminated limestones (Plattenkalk deposits; Wilby et al , 1995). Nonmineralizing organisms preserved in siliciclastic mudrocks tend to have a proportionately greater association with diagenetic iron mineralization (especially pyrite or siderite) than do similar organisms preserved in carbonate sediments

Pyritized fossils, including steinkerns and replacements of skeletal material, are relatively common through the Phanerozoic stratigraphic record (e g , Allison, 1988b; Fisher and Hudson, 1982)  Pyritization of recalcitrant organic remains (e g , plants, Kenrick and Edwards, 1988, graptolites, Underwood and Bottrell, 1994), however, is relatively rare (Briggs et al , 1991, 1996; Wilby et al , 1996a,b). Briggs et al (1991) noted that if soft tissues are preserved in pyrite and other minerals, usually only their outlines survive because the crystals are coarse and form too late in the taphonomic process to replicate the finest details Exceptions to this observation do occur, as an example, pyritization of soft tissues, including structural details (muscle fibers) in squid fossils, has been reported from the Jurassic (Wilby et al , 1996a,b)  In the Chengjiang deposit, pyrite replication of skeletal or soft tissues seems to have occurred much more frequently, or at least in many more taxonomic groups, than did phosphatization The relatively high frequency of thinly pyritized organisms may be due to high quantities of dissolved iron in clay sediments at the time of burial (see Briggs et al., 1996) Because the clay sediments of Chengjiang largely represent distal tempestites or sediments transported from shallow coastal areas such as river mouths, iron hydroxides adsorbed by these terrestrial clays are normally high (Zhu et al , 2001)

Precipitation of pyrite occurs immediately following precipitation of phosphate in the sulfate reduction zone below the redox boundary in modern sediments (Berner, 1984) Experiments using organic material of worms and plants indicate that decay-induced pyritization is an extremely rapid process (within 80 days) under closed conditions (Allison, 1988a; Grimes et al , 2001)  In general, closed conditions in fine deposits such as clays limit chemical exchange between sediment pore waters and bottom seawater, leading to two effects  (1) release of $H_2S$, $HS^-$, and $S^{2-}$ from sulfate reduction of organic material and incorporation with $Fe^{2+}$ from reduction of iron hydroxides to form pyrite, and (2) reduction of sulfate because of an insufficient supply of sulfate to sediment pore waters, combined with an increase of bicarbonate in sediment pore waters from decay of organics, leading to formation of siderite or ferroan carbonate In the Chengjiang deposit, microspherules of pyrite covering the surfaces of various nonmineralizing organisms indicate that pyritization occurred rather quickly Fine details of surface structure are preserved (Fig 3a), which is consistent with an early timing of pyrite diagenesis Discovery of framboidal pyrite in the internal voids of chancellorid sclerites is further evidence that pyritization occurred quite early in the taphonomic process in the Chengjiang deposit (compare Canfield and Raiswell, 1991) Pyrite microspherules may be related to bacterial autolithification (Davis and Briggs, 1995, Borkow and Babcock, 2003), which is characterized by replication of bacteria that coated decaying organic remains (Ferris et al , 1987, 1988) This preservational mode appears to have occurred in other deposits of exceptional preservation, including soft tissues from the Eocene Messel Shale (Wuttke, 1983, Franzen, 1990) A mineralized biofilm, of the sort that may play a major role in exceptional preservation, has been observed in modern environments (Labrenz et al , 2000)

Iron carbonate can play a role similar to that of apatite in soft tissue or nonmineralized skeletal preservation Iron carbonate (siderite and ferroan calcite), which is of early diagenetic origin, has been reported in the fossilization of dinosaur skin in Lower Cretaceous lacustrine limestones (Briggs et al , 1997), and in the fossilization of other nonmineralized skeletal or other tissues in lake sediments (Wuttke, 1983; Allison, 1988d) Some iron oxides preserved on the surfaces of Chengjiang fossils may have an iron carbonate origin, but early diagenetic iron carbonates are generally limited to freshwater settings or settings where freshwater influx occurs in marginal marine areas (e g , Carboniferous Mazon Creek-type deposits, Baird et al , 1985) Early diagenetic iron carbonate in freshwater sediments develops because of a limited supply of sulfate, resulting in insufficient sulfide for pyritization Although the supply of sulfate from bottom seawater is generally limited in clay deposits, the inferred rapid influx of clay in the Chengjiang deposit may have initially filled pore waters with seawater prior to compaction Microbial sulfate reduction in organic-rich microenvironments surrounding decaying carcasses could have provided sufficient sulfide for pyritization of fossils. Study of fossils from the Jurassic Kimmeridge Clay has revealed the presence of diagenetic pyrite, siderite, and ferroan calcite in organic-rich marine mudrocks In the Kimmeridge Clay, framboidal pyrite formed first, siderite formed next, and ferroan calcite formed last (Macquaker et al., 1997) From preliminary studies on Chengjiang material reported here, it is probably too early to speculate whether all iron oxide staining on Chengjiang fossils is due to pyrite However, the diagenetic sequence reported by Macquaker et al (1997) from the Kimmeridge Clay supports the interpretation that much of the iron oxide on Chengjiang fossils is due to weathering of pyrite The presence of siderite or ferroan calcite is not entirely unexpected, especially given the inferred coastal marine depositional setting of much of the Chengjiang deposit (Zhu et al , 2001) Sporadic fluvial input and tidal pumping (Babcock et al , 2001) may have occasionally carried enough freshwater out on the shelf to cause the precipitation of iron carbonates at times

Iron-rich aluminosilicates do not seem to be a primary means of replicating organic tissues in the Chengjiang shales, but they do seem to play a relatively important secondary role The diagenetic timing of clay minerals in the Chengjiang deposit is demonstrated by the overgrowth of Fe-rich aluminosilicates over framboidal pyrite in interior voids of chancellorid sclerites These clay minerals thus represent a diagenetic stage that occurred following the onset of pyrite precipitation In open-marine, phosphatized stromatolites from elsewhere (Sanchez-Navas et al , 1998), authigenetic Fe-rich clay minerals formed during early diagenesis Together, these lines of evidence suggest that films of amorphous Fe-rich aluminosilicates developed on the surfaces of non-

mineralizing organisms in the Chengjiang Lagerstatte relatively early, probably soon after pyrite was precipitated The precipitation of the amorphous Fe–Si–Al mineral phase subsequent to the initial development of framboidal pyrite is evinced by a transitional morphology of minerals in voids of the chancellorid sclerites (Fig 7b and c) Clay formation resulting from organic degradation of organisms has been reported in lingulid shells (Williams and Cusack, 1997), although it is uncertain whether this particular diagenetic pathway operated on Chengjiang fossils. Reported replication of the Burgess Shale fossils by clay minerals (Orr et al , 1998) further supports the view that authigenetic clay minerals played an important role in exceptional preservation during the Cambrian

In summary, nonmineralizing organisms from Chengjiang analyzed to date show that exceptional preservation has occurred by at least four methods (1) as organic carbon; (2) through precipitation of phosphate, (3) through precipitation of pyrite (and possibly iron carbonates), and (4) through precipitation of Fe-rich aluminosilicates Phosphate, in the form of apatite, precipitates on some organisms early in the diagenetic process, and it is followed in time by precipitation of pyrite Precipitation of Fe-rich aluminosilicates occurs last in time sequence Original compositional differences among taxa and even within individual carcasses apparently led to different diagenetic pathways Decay-resistant, nonmineralized skeletal tissues, such as heavily sclerotized, chitinous spines attached to the grasping appendages of anomalocaridids, may be preserved as organic films in shale This type of preservation seems to be relatively rare in the Chengjiang deposit despite experimental taphonomic work showing that chitin has a comparatively high preservation potential relative to other tissues in nonmineralized arthropods (Bass et al , 1995) Precipitation of calcium phosphate (apatite) occurred in such examples as the cuticle of *Palaeoscolex*, where there was likely to have been a phylogenetically based link to phosphatic preservation Phosphatization also seems to have occurred preferentially in tissues of relatively large organic mass Precipitation of pyrite tended to occur on body parts of relatively small organic mass, and beginning after the onset of phosphate precipitation Phosphate and pyrite, however, do not co-occur in all fossils A more common association is that of pyrite and Fe-rich aluminosilicates In studied examples, surfaces of the fossils are covered by a film of minerals composed of pyrite, overlain by Fe-rich aluminosilicates

## 5. Discussion

Recent studies have led to divergent views about taphonomic processes in Cambrian deposits of exceptional preservation On the basis of some specimens from the middle Cambrian Burgess Shale, Butterfield (1990, 1995) proposed a model for the organic carbon preservation of nonmineralizing organisms, and applied the model in interpreting the secular distribution of the Burgess Shale-type fossils through geological history. He emphasized the importance of interactions between organic molecules or enzymes and iron-rich clay minerals (nontronite, iron-rich smectite), and concluded that all Burgess Shale-type fossils are preserved as organic films Other work, however, shows that authigenetic mineralization was fundamental to preserving fossils in exceptional state (Orr et al, 1998) Elemental maps of some Burgess Shale specimens indicate that bodily remains were replicated by authigenetic clay minerals (Orr et al, 1998) A similar style of preservation was reported from the Ordovician Soom Shale of South Africa (Gabbott, 1998) Conway Morris (1990) reported that organic carbon preservation in the Burgess Shale of Canada is limited to relatively few fragments of carbonized skeletons and some tenuous clay–organic "pigmentation" structures Butterfield's (1990) model was also challenged by Towe (1996), who noted that (1) except for a relatively limited number of specimens, fossils from the Burgess Shale are not composed of organic carbon films, (2) the mineralogy of the Burgess Shale is predominantly illite, with subordinate amounts of chlorite and kaolinite, making it similar in composition to most Paleozoic shales, and implying that the type of clay present is not the major factor governing the secular distribution of Burgess Shale-type preservation, and (3) original TOC in the Burgess clays may have been high initially, but microbial metabolism reduced TOC values so that TOC values in the lithified equivalent are now low If the primary TOC of the Burgess Shale were high, according to Butterfield's (1990) model, it should

have been preserved through organic carbon–clay interactions

Although organic carbon–clay interactions of the type discussed above may not be the primary reason for exceptional preservation in Cambrian deposits of exceptional preservation, clay–organic interactions nevertheless play another important role in the preservation of nonmineralizing organisms Wollanke and Zimmerle (1990) emphasized the thixotropic sol–gel behavior of clay–organic interactions fine-grained sediments are initially fluid (sol), but develop a "sticky" plasticity when left undisturbed Entombment of bodily remains in such materials could allow for early injection of fluidized sediments into body cavities, followed by later gel stabilization that produces faithful impressions of the organisms The replicated organisms become sealed prior to compression and cementation During initial phases of replication, outlines of organisms become sufficiently "firm" that they remain even if the organic matter is later removed autolytically, removed by bacterial action, replaced diagenetically, or even preserved as "fused decay products" consisting of cuticle and a clay fabric It is this process that probably led to Towe's (1996) observation that the fidelity of structural preservation of cuticular organic matter in Burgess Shale fossils is secondary to aluminosilicate replacement and to the exquisite gross morphology outlined by fine-grained sediments and replaced by their diagenetic counterparts

Taphonomic analyses of Chengjiang fossils show that both diagenetic minerals (apatite, pyrite, and Fe-rich aluminosilicates) and organic films are important for preserving nonmineralized skeletons or internal soft tissues Preservation by means of organic films seems limited to only those tissues that had a high resistance to decay A clay matrix for burial of remains is inferred to have played an important role in the preservation of nonmineralizing organisms, but its role may have been mostly related to its thixotropic sol–gel behavior Developing impressions of non-mineralized parts in thixotropic clays is an early step toward replication of bodily remains by diagenetic minerals Preservational differences related to ontogeny in medusiform fossils from the middle Cambrian Kaili Lagerstatte of Guizhou Province, China, provide an indication of the importance of diagenetic mineralization following replication in a thixotropic clay matrix (Zhu et al, 1999) Juveniles of medusiform animals are rare in the Kaili deposit One possible explanation is that their rarity can be attributed to the absence of labile tissues, which may result from having masses of organic matter that are too small to produce decay products sufficient for precipitation of authigenetic minerals Nonmineralized tissues could not be easily preserved as fossils without precipitation of authigenetic minerals, even though impressions of the tissues may have been developed in the mud matrix

Chengjiang fossils tend to show little three-dimensional preservation Fossils normally have minor relief against bedding, and distinct, full relief such as that recorded from phosphatized fossils in concretions of the Alum Shale of Sweden (Muller and Walossek, 1987) is unusual in the Chengjiang deposit Where fine three-dimensional preservation does occur, it is usually related to internal cavities of animals with mud infillings such as the alimentary canals of *Stellostomites*, worms, and arthropods Mud fillings of alimentary canals can show rounded cross sections without significant compaction (e g , *Maotianshania* illustrated by Babcock et al , 2001, Fig 5), indicating that consolidation of the mud within body cavities occurred earlier than lithification of the matrix Organic breakdown processes acting during the lives of host animals may have established chemical microenvironments within the body cavities that differed from those outside the bodies and led to early diagenetic activity Mud-filled alimentary canals appear similar in color and grain size to the matrix They differ, however, from gut traces lacking mud fillings, which are typically black or dark gray, presumably due to the presence of organic matter Therefore, there appears to be no obvious relationship between composition of the mud and the preservation of organic matter in the Chengjiang deposit

## Acknowledgements

We thank Chen Junyuan, Bernd-D Erdtmann, Jean Vannier, and Zhang Junming for productive discussions Mao Yongqian (NIGPAS) and Jorg Nissen (TU Berlin) provided technical support Laboratory work by Zhu at TU Berlin was supported by an exchange program between the Max-Planck-Gesellschaft (MPG) of Germany and the Chinese Academy of Sciences Research was supported by grants from the NSFC (grant 40232020), the Ministry of Science and Technology (grant G200077702), and the Chinese Academy of Sciences Hundred Talents program to M Zhu, and by National Science Foundation grants 9405990, 0073089, 0106883 OPP-0229757, and OPP-0346829 to L E Babcock M Steiner was supported by Deutsche Forschungsgesellschaft (DFG)

## References

Allison, P A , 1988a The role of anoxia in the decay and mineralization of proteinaceous macro-fossils Paleobiology 14, 139–154

Allison, P A , 1988b Konservat–Lagerstatten cause and classification Paleobiology 14, 331–344

Allison, P A , 1988c Phosphatized soft-bodied squids from the Jurassic Oxford Clay Lethaia 21, 403–410

Allison, P A 1988d Taphonomy of the Eocene London Clay Palaeontology 31, 1079–1100

Allison, P A , Briggs, D E G , 1993 Exceptional fossil record distribution of soft-tissue preservation through the Phanerozoic Geology 21, 527–530

Babcock, L E , Zhang, W T, 2001 Stratigraphy, paleontology and depositional setting of the Chengjiang Lagerstatte In Peng, S C , Babcock, L E , Zhu, M Y (Eds ), Cambrian System of South China University of Science and Technology of China Press, pp 66–86

Babcock, L E , Zhang, W T, Leslie, S A , 2001 The Chengjiang Biota record of the early Cambrian diversification of life and clues to exceptional preservation of fossils GSA Today 11 (2), 4–9

Baird, G C , Shabica, C W, Anderson, J L , Richardson Jr , E S , 1985 Biota of a Pennsylvanian muddy coast habitats within the Mazonian delta complex, northern Illinois J Paleontol 59, 253–281

Bass, M , Briggs, D E G , van Heemst, J D H , Kear, A J , de Leeuw, J W , 1995 Selective preservation of chitin during the decay of shrimp Geochim Cosmochim Acta 59, 945–951

Benmore, R A , Coleman, M L , Mcarther, J M , 1983 Origin of sedimentary francolite from its sulphur and carbon isotope composition Nature 302, 516–518

Berner, R A , 1984 Sedimentary pyrite, an update Geochim Cosmochim Acta 48, 605–615

Borkow, P S , Babcock, L E , 2003 Turning pyrite concretions outside-in, role of biofilms in pyritization of fossils Sed Rec 1 (3), 4–7

Brett, C E , 1990 Obrution deposits In Briggs, D E G , Crowther, P R (Eds ), Palaeobiology A Synthesis Blackwell Scientific Publications, pp 239–243

Briggs, D E G , 1991 Extraordinary fossils Am Sci 79, 130–141

Briggs, D E G , 1995 Experimental taphonomy Palaios 10, 539–550

Briggs, D E G, Kear, A J, 1993 Decay and preservation of polychaetes taphonomic thresholds in soft-bodied organisms Paleobiology 19, 107–135

Briggs, D E G, Kear, A J, 1994a Decay of *Branchiostoma* implications for soft-tissue preservation in conodonts and other primitive chordates Lethaia 26, 275–287

Briggs, D E G, Kear, A J, 1994b Decay and mineralization of shrimps Palaios 9, 431–456

Briggs, D E G, Wilby, P R, 1996 The role of the calcium carbonate–calcium phosphate switch in the mineralization of soft-bodied fossils J Geol Soc 153, 665–668

Briggs, D E G, Bottrell, S H, Raiswell, R, 1991 Pyritization of soft-bodied fossils Beecher's Trilobite Bed, Upper Ordovician, New York State Geology 19, 1221–1224

Briggs, D E G, Kear, A J, Martill, D M, Wilby, P R, 1993 Phosphatization of soft-tissue in experiments and fossils J Geol Soc 150, 1035–1038

Briggs, D E G, Raiswell, R, Bottrell, S H, Hatfield, D, Bartels, C, 1996 Controls on the pyritization of exceptionally preserved fossils an analysis of the Lower Devonian Hunsruck Slate of Germany Am J Sci 296, 633–663

Briggs, D E G, Wilby, P R, Perez-Moreno, B P, Sanz, J L, Fregenal-Martinet, M, 1997 The mineralization of dinosaur soft tissue in the Lower Cretaceous of Las Hoyas, Spain J Geol Soc 154, 587–588

Butterfield, N J, 1990 Organic preservation of non-mineralizing organisms and the taphonomy of the Burgess Shale Paleobiology 16, 172–286

Butterfield, N J, 1995 Secular distribution of Burgess-Shale-type preservation Lethaia 28, 1–13

Canfield, D E, Raiswell, R, 1991 Pyrite formation and fossil preservation In Allison, P A, Briggs, D E G (Eds), Taphonomy Releasing Data Locked in Fossil Record Plenum, pp 337–387

Chen, J -Y, Erdtmann, B -D, 1991 Lower Cambrian Lagerstatte from Chengjiang, Yunnan, China insight for reconstructing early metazoan life In Simonneta, A M, Conway Morris, S (Eds), The Early Evolution of Metazoan and the Significance of Problematic Taxa Cambridge University Press, pp 57–76

Chen, J -Y, Oliveri, P, Li, C -W, Zhou, G -Q, Gao, F, Hagadorn, J W, Peterson, K J, Davidson, E, 2000 Precambrian animal diversity putative phosphatized embryos from the Doushantuo Formation of China PNAS 97 (9), 4457–4462

Conway Morris, S, 1985 Cambrian Lagerstatten their distribution and significance Philos Trans R Soc Lond, B 311, 48–65

Conway Morris, S, 1990 Burgess shale In Briggs, D E G, Crowther, P R (Eds), Palaeobiology A Synthesis Blackwell Scientific Publications, pp 270–274

Conway Morris, S, 1997 The cuticular structure of the 495-Myr-old type species of the fossil worm *Palaeoscolex, P piscatorum* (?Priapulida) Zool J Linn Soc 119, 69–82

Davis, P G, Briggs, D E G, 1995 Fossilization of feathers Geology 23, 783–786

Ferris, F G, Fyfe, W S, Beveridge, T J, 1987 Bacteria as nucleation sites for authigenic minerals in a metal-contaminated lake sediment Chem Geol 63, 225–232

Ferris, F G, Fyfe, W S, Beveridge, T J, 1988 Metallic ion binding by *Bacillus subtilis* implications for the fossilization of microorganisms Geology 16, 149–152

Fisher, I St J, Hudson, J D, 1982 Pyrite geochemistry and fossil preservation in shales Philos Trans R Soc Lond, B 311, 167–170

Franzen, J L, 1990 Grube messel In Briggs, D E G, Crowther, P R (Eds), Palaeobiology A Synthesis Blackwell Scientific Publications, pp 289–293

Gabbott, S E, 1998 Taphonomy of the Ordovician Soom Shale Lagerstatte an example of soft tissue preservation in clay minerals Palaeontology 41, 631–667

Grimes, S T, Brock, F, Rickard, D, Edwards, D, Briggs, D E G, Parkes, R J 2001 Understanding fossilization experimental pyrilization of plants Geology 29, 123–126

Hinz, I, Kraft, P, Mergl, M, Muller, K J, 1990 The problematic *Hadimopanella, Kaimenella, Milaculum* and *Utahphospha* identified as spicules of Palaeoscolecida Lethaia 23, 217–221

Hirschler, A, Lucas, J, Hubert, J C, 1990 Bacterial involvement in apatite genesis FEMS Microbiol Ecol 73, 211–220

Hof, C H J, Briggs, D E G, 1997 Decay and mineralization of mantis shrimps (Stomatopoda Crustacea)—a key to their fossil record Palaios 12, 420–438

Jin, Y -G, Wang, H -Y, Wang, W, 1991 Palaeoecological aspects of brachiopods from Chiungchussu Formation of Early Cambrian age, eastern Yunnan, China In Jin, Y G, Wang, J G, Xu, S H (Eds), Palaeoecology of China, vol 1 Nanjing University Press, pp 25–47

Kenrick, P, Edwards, D, 1988 The anatomy of Lower Devonian *Glossingia breconensis* Heard based on pyritised axes, with some comments on the permineralisation process Bot J Linn Soc 97, 95–123

Kraft, P, Mergl, M, 1989 Worm-like fossils (Palaeoscolecida, ?Chaetognatha) from the Lower Ordovician of Bohemia Sb Geol Ved, Paleontol 30, 9–36

Labrenz, M, Druschel, G K, Thomsen-Ebert, T, Gilbert, B, Welch, S A, Kemner, K M, Logan, G A, Summons, R E, Stasio, G D, Bond, P L, Lai, B, Kelly, S D, Banfield, J F, 2000 Formation of sphalerite (ZnS) deposits in natural biofilms of sulfate-reduced bacteria Science 290, 1744–1747

Leslie, S A, Babcock, L E, Zhang, W T, 1996 Community composition and taphonomic overprint of the Chengjiang Biota (Early Cambrian, China) Sixth North American Paleontological Convention Abstracts of Papers, Special Publication-Paleontological Society, vol 8, pp 237

Littke, R, Klussmann, U, Krooss, B, Leythaeuser, D, 1991 Quantification of loss of calcite, pyrite, and organic matter due to weathering of Toarcian black Shales and effects on kerogen and bitumen characteristics Geochim Cosmochim Acta 55, 3369–3378

Lucas, J, Prevot, L, 1984 Apatite synthesis by bacterial activity from phosphatic organic matter and several calcium carbonates in natural freshwater and seawater Chem Geol 42, 101–118

Macquaker, J H S, Curtis, C D, Coleman, M L, 1997 The role of iron in mudstone diagenesis comparison of Kimmeridge Clay Formation mudstones from onshore and offshore (UKCS) localities J Sediment Res 67, 871–878

Martill, D M , 1988 Preservation of fish in the Cretaceous Santana Formation of Brazil Palaeontology 31, 1–18

Martill, D M , Wilby, P R , Williams, N , 1992 Elemental mapping a technique for investigating delicate phophatized fossil soft tissues Palaeontology 35, 869–874

Moses, C O , Herman, J S , 1991 Pyrite oxidation at circumcentral pH Geochim Cosmochim Acta 55, 47–482

Muller, K J , 1990 Upper Cambrian 'Orsten' In Briggs, D E G , Crowther, P R (Eds ), Palaeobiology A Synthesis Blackwell Scientific Publications, pp 274–277

Muller, K J , Hinz-Schallreuter, I , 1993 Palaeoscolecid worms fiom the Middle Cambrian of Australia Palaeontology 36, 549–592

Muller, K J , Walossek, D , 1987 Morphology, ontogeny, and life habit of *Agnostus pisiformis* from the Upper Cambrian of Sweden Fossils Strata 19, 1–124

Orr, J P , Briggs, D E G , Kearns, S L , 1998 Cambrian Burgess Shale animals replicated in clay minerals Science 281, 1173–1175

Prevot, L . Lucas, J , 1990 Phosphate In Briggs, D E G , Crowther, P R (Eds ), Palaeobiology A Synthesis Blackwell Scientific Publications, pp 256–257

Sagemann, J . Bale, S J , Briggs, D E G , Parkes, R J , 1999 Controls on the formation of authigenic minerals in association with decaying organic matter an experimental approach Geochim Cosmochim Acta 63, 1083–1095

Sanchez-Navas, A , Martin-Algarra, A , Nieto, F , 1998 Bacterially-mediated authigenesis of clays in phosphate stromatolites Sedimentology 45, 519–533

Seilacher, A , Reif, W E , Westphal, F , 1985 Sedimentological, ecological, and temporal patterns of fossil Lagerstatten Philos Trans R Soc Lond , B 311, 5–23

Towe, K M , 1996 Fossil preservation in the Burgess Shale Lethaia 29, 107–108

Underwood, C J , Bottrell, S H , 1994 Diagenetic controls on multiphase pyritization of graptolites Geol Mag 131, 315–327

Walossek, D , Hinz-Schallreuter, I , Shergold, J H , Muller, K , 1993 Three-dimensional preservation of arthropod integument from the Middle Cambrian of Australia Lethaia 26, 7–15

Wilby, P R , Whyte, M A , 1995 Phophatized soft-tissues in bivalves from the Portland Roach of Dorset (Upper Jurassic) Geol Mag 132, 117–120

Wilby, P R , Briggs, D E G , Viohl, G , 1995 Controls on the phosphatization of soft tissues in plattenkalks 2nd International Symposium on Lithographic Limestones, Cuenca, Spain, Extended Abstracts Universidad Autonoma de Madrid, pp 165–166

Wilby, P R , Briggs, D E G , Bernier, P , Gaillard, C , 1996 Role of microbial mats in the fossilization of soft tissues Geology 24, 787–790

Wilby, P R , Briggs, D E G , Riou, B , 1996 Mineralization of soft-bodied invertebrates in a Jurassic metalliferous deposit Geology 24, 847–850

Williams, A , Cusack, M , 1997 Lingulid shell mediation in clay formation Lethaia 29, 349–360

Wollanke, G , Zimmerle, W , 1990 Petrographic and geochemical aspect of fossil embedding in extraordinarily well preserved fossils deposits Mitt Geol -Palontol Inst Univ Hamb 69, 77–97

Wuttke, M , 1983 'Weichteil-Erhaltung' durch litifiyierte Mikroorganismem bei mittel-eozanen Vetebraten aus den Olschiefen der 'Grube Messel' bei Darmstadt Senckenb Lethaea 64, 509–527

Xiao, S , Knoll, A , 1999 Fossil preservation in the Neoproterozoic Doushantuo phosphorite Lagerstatte, South China Lethaia 32, 219–240

Zhang, X -G , Pratt, B R , 1996 Early Cambrian palaeoscolecid cuticles from Shaanxi, China J Paleontol 70, 275–279

Zhu, M -Y , 1992 Taphonomy of Chengjiang Fossils, Yunnan PhD dissertation, Nanjing Institute of Geology and Palaeontology (unpublished), pp 1–159

Zhu, M -Y , Erdtmann, B -D , Zhao, Y -L , 1999 Taphonomy and ecology of the early–middle Cambrian Kaili Lagerstatte in Guizhou, China Acta Palaeontol Sin 38, 28–57 (Suppl )

Zhu, M -Y , Zhang, J -M , Li, G -X , 2001 Sedimentary environments of the early Cambrian Chengjiang Biota sedimentology of the Yu'anshan Formation in Chengjiang County, eastern Yunnan Acta Palaeontol Sin 40, 80–105 (Suppl )

Available online at www.sciencedirect.com

Palaeogeography, Palaeoclimatology, Palaeoecology 220 (2005) 47–67

ELSEVIER

www.elsevier.com/locate/palaeo

# Paleoecology of benthic metazoans in the Early Cambrian Maotianshan Shale biota and the Middle Cambrian Burgess Shale biota: evidence for the Cambrian substrate revolution

Stephen Q. Dornbos[a,*], David J. Bottjer[a], Jun-Yuan Chen[b]

[a]Department of Earth Sciences, University of Southern California, Los Angeles, CA 90089-0740, USA
[b]Nanjing Institute of Geology and Palaeontology, Chinese Academy of Sciences, Nanjing 210008, PR China

Received 6 December 2001, accepted 1 November 2003

## Abstract

As the depth and intensity of bioturbation increased through the Proterozoic–Phanerozoic transition, the substrates on which marine benthos lived changed from being relatively firm with a sharp sediment–water interface to having a high water content and blurry sediment–water interface. Microbial mats, once dominant on normal marine Proterozoic seafloors, were relegated to stressed settings lacking intense metazoan activity. This change in substrates has been termed the agronomic revolution [Seilacher, A., Pfluger, F., 1994. From biomats to benthic agriculture: a biohistoric revolution. In: Krumbein, W.E. (Ed.), Biostabilization of Sediments. Bibliotheks und Informationsystem del Carl von Ossietzky Universitat, Oldenburg, 97–105], and the impact of the subsequent development of the mixed layer on benthic metazoans has been termed the Cambrian substrate revolution [Bottjer, D.J., Hagadorn, J.W., Dornbos, S.Q., 2000. The Cambrian substrate revolution. GSA Today 10 (9), 1–7]. Because the Early Cambrian was a transitional time in this substrate revolution, it is hypothesized that benthic metazoans adapted to typical Proterozoic-style soft substrates co-existed with benthic metazoans adapted to more typical Phanerozoic-style soft substrates. Paleoecological examination of the benthic metazoans of the Early Cambrian Maotianshan Shale (Chengjiang) biota and Middle Cambrian Burgess Shale biota, combined with examination of core samples from the rocks in which the Maotianshan Shale biota is preserved, was performed in order to test this hypothesis. Results of the core analysis reveal that the rocks in which the Maotianshan Shale biota was preserved, and therefore the substrate on which it lived, contains almost no evidence for bioturbation of any kind, while the immediately underlying rocks of similar lithology are dominated by intermediate levels of horizontal bioturbation and contain some evidence for the beginning of mixed layer development. These results suggest that benthic metazoans of the Maotianshan Shale biota lived on a typical Proterozoic-style soft substrate, and that environmental conditions associated with the exceptional preservation of the Maotianshan Shale biota may have also suppressed bioturbation levels. Results of the paleoecological analysis indicate that the Maotianshan Shale and Burgess Shale biotas contain mobile and sessile benthic metazoans adapted to typical Proterozoic-style soft substrates and others adapted to

* Corresponding author. Present address: Department of Geosciences, University of Wisconsin-Milwaukee, P.O. Box 413, Milwaukee, WI 53201, USA. Fax: +1 414 229 5452.
  E-mail address: sdornbos@uwm.edu (S.Q. Dornbos).

typical Phanerozoic-style soft substrates In addition, the Maotianshan Shale biota is dominated by benthic suspension feeding genera adapted to typical Proterozoic-style soft substrates, indicating that this biota lived during the early stages of the Cambrian substrate revolution The Burgess Shale biota, however, contains more benthic suspension feeding genera adapted to typical Phanerozoic-style soft substrates than the Maotianshan Shale biota, which suggests that the soft substrate on which they lived had more advanced levels of mixed layer development This research suggests that the adaptive radiation of benthic metazoans during the "Cambrian explosion" was driven in part by the Cambrian substrate revolution, as benthic metazoans were forced to adapt to the development of the mixed layer in subtidal siliciclastic soft substrate environments during the Cambrian

*Keywords* Maotianshan Shale, Chengjiang, Burgess Shale, Cambrian, Paleoecology, Benthic metazoans, Bioturbation

## 1. Introduction

One of the signatures of the Proterozoic–Phanerozoic transition is the increase in bioturbation levels as metazoans began to inhabit the infaunal realm during the "Cambrian explosion" Changes in the extent and type of bioturbation in siliciclastics and carbonates through the Proterozoic–Phanerozoic transition have been well-documented (e g Droser, 1987; Droser and Bottjer, 1988; McIlroy and Logan, 1999), and these changes led to a significant transition in dominant soft substrate types in subtidal siliciclastic environments (e g Hagadorn and Bottjer, 1999, Seilacher, 1999, Seilacher and Pfluger, 1994) The ecological impact of this substrate transformation on benthic metazoans, however, is only beginning to be examined (e g Bottjer et al, 2000, Dornbos and Bottjer, 2000; Parsley and Prokop, 2001, Parsley and Zhao, 2002; Babcock, 2003, Lefebvre and Fatka, 2003)

The first undisputed trace fossil evidence for metazoan activity is found at the end of the last Neoproterozoic glaciation, about 594 Mya (Narbonne et al, 1994, Brasier and McIlroy, 1998) It was not until the Cambrian, however, that trace fossil diversity increased dramatically (Crimes, 1992) This increase in trace fossil diversity is coupled with an increase in the intensity and depth of bioturbation, as observed in both carbonates and siliciclastics (e g Droser, 1987, Droser and Bottjer, 1988, McIlroy and Logan, 1999)

The Cambrian increase in bioturbation, particularly vertical bioturbation, caused a transformation of subtidal siliciclastic environments from typical Proterozoic-style soft substrates to typical Phanerozoic-style soft substrates, each dominated by different processes of sedimentary fabric formation (Fig 1)

Typical Proterozoic-style soft substrates are characterized by a low level of strictly horizontal bioturbation, a low water content, a relatively sharp sediment–water interface, and common seafloor microbial mat development Typical Phanerozoic-style soft substrates, on the other hand, are characterized by significant horizontal and vertical bioturbation, a high water content, a diffuse sediment–water interface, and the absence of well-developed seafloor microbial mats (e g Droser, 1987, Droser and Bottjer, 1988; Droser et al, 1999, 2002, Hagadorn and Bottjer, 1999; Seilacher, 1999, Seilacher and Pfluger, 1994) A crucial characteristic of these typical Phanerozoic-style soft substrates is the presence of a well-developed mixed layer, the soupy upper few centimeters of seafloor sediment that are homogenized by bioturbation (e g Ekdale et al, 1984) It is important to note that the mixed layer developed in a spatially and temporally variable way during the Cambrian, such that both typical Proterozoic-style and typical Phanerozoic-style soft substrates existed throughout much of the Cambrian (e g. Dornbos et al, 2004). Bioturbation levels in some Middle Cambrian siliciclastic deposits, for instance, such as the relatively deep water Kaili Formation of China (Parsley and Zhao, 2002), are still extremely low (ii1–2)

Seilacher and Pfluger (1994) termed this substrate transition the "agronomic revolution", and defined it as a transition from Proterozoic "matgrounds", with well-developed seafloor microbial mats, to Phanerozoic "mixgrounds", with high levels of horizontal and vertical bioturbation and lacking well-developed seafloor microbial mats They also postulated that Neoproterozoic benthic metazoan lifestyles centered on seafloor microbial mats (Seilacher, 1999, Seilacher and Pflüger, 1994) These lifestyles included mat

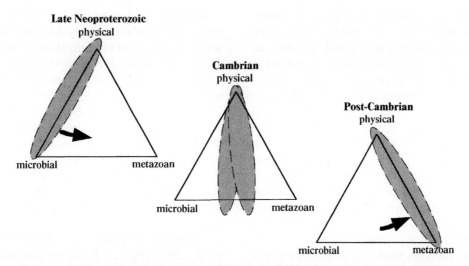

Fig 1 Change in dominant processes controlling the sedimentary fabric of normal marine neritic soft substrates during the Proterozoic–Phanerozoic transition These processes are primary physical and microbial processes, as well as secondary metazoan bioturbation The movement of the ovals within the triangle indicates changes in the relative dominance of these processes During this transition, the dominant processes move from a combination of physical and microbial processes to a combination of physical and metazoan processes Benthic metazoan adaptations to these transitional soft substrates is the focus of this study (modified from Hagadorn and Bottjer, 1999, Bottjer et al , 2000)

encrusters, which lived attached to the mats, mat scratchers, which grazed on the mats, mat stickers, which lived with their lower ends inserted in the mats, and undermat miners, which burrowed horizontally underneath the mats and fed on decomposing mat remnants (Seilacher, 1999, Seilacher and Pfluger, 1994)

While the paleoecology of Neoproterozoic benthic metazoans is still not well understood, the paleoecology of Cambrian benthic metazoans is easier to interpret and extremely informative because they were living during the Proterozoic–Phanerozoic substrate transition It is these Cambrian fossils that currently allow for a thorough examination of the ecological and evolutionary effects of this substrate transition on nonburrowing benthic metazoans, effects termed the "Cambrian substrate revolution" by Bottjer et al (2000)

Previous work on the Cambrian substrate revolution has focused on the evolutionary response of Cambrian nonburrowing benthic echinoderms and the ecological response of early grazing molluscs to these substrate changes (Bottjer et al , 2000, Dornbos and Bottjer, 2000, Parsley and Prokop, 2001, Parsley and Zhao, 2002, Lefebvre and Fatka, 2003, Dornbos et al , 2004), as well as the contribution of burrowing

trilobites to this substrate transition (Babcock, 2003). Early sessile suspension feeding echinoderms display a variety of evolutionary responses to the Cambrian substrate revolution The enigmatic Early Cambrian helicoplacoid echinoderms lived as sediment stickers but lacked soft substrate adaptations typical of Phanerozoic suspension feeders, such as attachment structures or a root-like holdfast, suggesting that they were unable to adapt to the substrate changes taking place in the Cambrian The Cambrian substrate revolution may have therefore led to their extinction (Dornbos and Bottjer, 2000) The other two groups of early sessile suspension feeding echinoderms, the edrioasteroids and eocrinoids, some of which were sediment stickers or attachers in the Early–Middle Cambrian, either lived attached to hard substrates or had evolved attaching stems by the Late Cambrian, consistent with an evolutionary response to the Cambrian substrate revolution (Bottjer et al , 2000, Dornbos and Bottjer, 2000, Parsley and Prokop, 2001; Parsley and Zhao, 2002, Lefebvre and Fatka, 2003)

Both monoplacophorans and polyplacophorans are groups of benthic grazing molluscs with long histories, and show an environmental distribution through the Phanerozoic that is consistent with a response to the

Cambrian substrate revolution (Bottjer et al , 2000) During the terminal Proterozoic and Cambrian, these grazing molluscs and their soft-bodied ancestors, as represented by radular grazing traces, were present in nearshore hard substrate and neritic soft substrate environments (e g Dornbos et al , 2004) However, they are now present only in the deep sea and nearshore hard substrate environments (Bottjer et al , 2000), settings in which surficial microbial mats still exist for ingestion This environmental distribution of early grazing molluscs through the Phanerozoic is consistent with a response to the disappearance of well-developed seafloor microbial mats in neritic soft substrate environments during the Cambrian substrate revolution (Bottjer et al , 2000)

While this previous work reveals ecological and evolutionary patterns that are consistent with the Cambrian substrate revolution hypothesis, the goal of this research is to further test for these ecological effects on benthic metazoans by examining in detail the adaptive morphologies and paleoecology of benthic suspension feeders and mobile benthic metazoans in the soft-bodied Maotianshan Shale (Early Cambrian) and Burgess Shale (Middle Cambrian) biotas, as well as the rocks in which the Maotianshan Shale biota is preserved Furthermore, because the Cambrian is a time of transition between typical Proterozoic-style soft substrates and typical Phanerozoic-style soft substrates one might predict that· (1) the Maotianshan Shale and Burgess Shale biotas would contain benthic metazoans adapted to both substrate types, (2) the Burgess Shale biota might have a greater proportion of benthic suspension feeding genera adapted to typical Phanerozoic-style soft substrates than the Maotianshan Shale biota, and (3) there would be some evidence for the beginning of mixed layer development in the rocks in which the Maotianshan Shale biota is preserved or those underlying them

## 2. Geologic setting

The soft-bodied Maotianshan Shale (Chengjiang) biota is preserved in the Lower Cambrian Yuanshan Formation in the Chengjiang area of Yunnan Province, China (Chen and Zhou, 1997) (Fig 2) The Yuanshan Formation is the first Lower Cambrian stratigraphic unit in the region that bears trilobites It represents a

150 m thick, shallowing upward sequence of shallow marine siliciclastics probably deposited in a prodelta setting, perhaps just offshore of an estuary (Chen and Zhou, 1997, Babcock et al , 2001; Bergstorm, 2001) The Yuanshan Formation consists of three lithological members, the lowermost Black Shale Member (about 20 m thick), the Maotianshan Shale Member (about 60 m thick, with extraordinary preservation of soft-bodied fossils), and the uppermost Siltstone Member (about 60 m thick) (Fig 2) Immediately underlying the Yuanshan Formation is the pre-trilobite Shiyantou Formation, a 50 m thick shallowing upward sequence of siliciclastic deposits, with 10 m of black shale at its base and mudstone and siltstone with sandstone interbeds in its uppermost 40 m (Chen and Zhou, 1997; Babcock et al , 2001) (Fig 2) Based on their relative stratigraphic positions in this shallowing upward sequence, the Shiyantuo Formation and Maotianshan Shale Member probably represent somewhat different depositional environments The Shiyantuo Formation was probably deposited in a slightly deeper environment than the Maotianshan Shale Member This depth difference was relatively minor, however, as both are shallow marine siliciclastic deposits

Although the details of the preservation of the soft-bodied fossils in the Maotianshan Shale biota are still controversial, with possibilities ranging from obrution events to tidal pulses being examined (e g Chen and Zhou, 1997, Babcock et al , 2001), all evidence indicates that they were buried with minimal transport (Chen and Zhou, 1997) Many infaunal forms are found preserved in life position perpendicular to bedding (Chen and Zhou, 1997) This minimal transport suggests that the rocks in which the soft-bodied fossil biota is preserved represent the substrate on which they were living, making a detailed characterization of that substrate desirable

The soft-bodied Burgess Shale biota, meanwhile, is preserved in the Middle Cambrian Stephen Formation of British Columbia, Canada The Stephen Formation is about 150 m thick at the Burgess Shale locality and consists of dark shales usually interpreted as being deposited in a relatively deep basinal setting (e g Briggs et al , 1994) The Burgess Shale biota is typically thought to have been preserved adjacent to a large carbonate escarpment, represented by the Cathedral Formation, over which obrution events would sweep and bury an assemblage of fossils in deep water

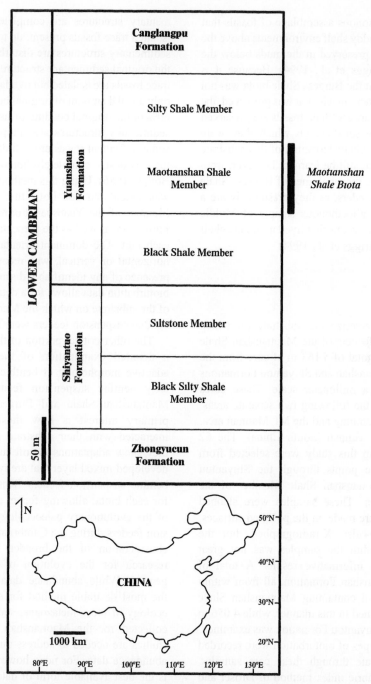

Fig 2 General Lower Cambrian stratigraphy of Yunnan Province, China showing the stratigraphic location of the Maotianshan Shale biota and map of China marking the location (with a star) of the Maotianshan Shale biota (Luo et al , 1984, Chen and Zhou, 1997, Babcock et al , 2001)

mud (e g Briggs et al , 1994) It is thought that the carbonate platform represented by the Cathedral Formation had been covered in siliciclastic muds by the

time the of Burgess Shale, meaning that Burgess Shale animals most likely lived on muddy shelf seafloors (e g Briggs et al , 1994) This preservation process resulted in an allocthonous assemblage of fossils that had been living in muddy shelf environments above the escarpment, but was preserved in the muds below the escarpment (e g  Briggs et al , 1994)  Because it is generally thought that the Burgess Shale biota was not living in the environment in which it was preserved, the rocks in which the Burgess Shale fossils are preserved do not represent the substrate on which they were living, making direct characterization of this substrate impossible  Much can still be learned, however, from studying the adaptive morphologies of Burgess Shale benthic suspension feeders, as they presumably are a fairly representative allocthonous sample of Middle Cambrian benthic metazoans that lived in muddy shelf environments (e g  Briggs et al , 1994)

## 3. Methods

In order to characterize the substrate on which benthic suspension feeders of the Maotianshan Shale biota were living, a total of 5 163 m of core from the Lower Cambrian Yuanshan and Shiyantou Formations were examined on a millimeter scale  These cores were obtained from the following two separate areas. the Shanjia Village, Jinning and the Mt  Maotian area, Chengjiang, central Yunnan (South China)  The 62 samples examined in this study were selected from random stratigraphic points through the Shiyantou Formation and Maotianshan Shale Member of the Yuanshan Formation  These samples were slabbed and observations were made on the polished surfaces  The samples were also X-radiographed, but the density contrast within the samples was not great enough to produce informative results  A total of 1 107 m of the Yuanshan Formation, all from within the soft-bodied fossil containing Maotianshan Shale Member, was examined in this manner, while 4 056 m of the underlying Shiyantou Formation was examined

The levels and types of bioturbation were recorded on a millimeter scale through these core samples, utilizing the ichnofabric index method of Droser and Bottjer (1986)  This semiquantitative method, designed for determination of bioturbation levels when looking at the vertical face of an outcrop or core, groups the extent of bioturbation into six categories, called ichnofabric indices (ii), based on visual observation. ii1—no bioturbation observed, original sedi-

mentary structures are completely preserved, ii2—isolated trace fossils present, up to 10% of the original sedimentary structures are disturbed, ii3—10–40% of the original sedimentary structures are disturbed, most trace fossils are isolated but overlap occasionally, ii4—only a small amount of original bedding is visible, 40–60% of the original bedding is disturbed, ii5—original sedimentary structures are completely disturbed, but some individual traces are still discernible, ii6—sediment is nearly completely homogenized (Droser and Bottjer, 1986)  Using this methodology, the average ii was calculated for both the Maotianshan Shale Member of the Yuanshan Formation and Shiyantou Formation, as well as the percentages of each ii within each unit  The dominant orientation of bioturbation, horizontal or vertical, was recorded, as well as the presence of any identifiable discrete trace fossils  This bioturbation data allows for a detailed characterization of the substrate on which the Maotianshan Shale biota benthic suspension feeders were living

The other critical portion of this research involved a detailed examination of the paleoecology and adaptive morphology of benthic metazoans, particularly benthic suspension feeding genera, in the Maotianshan Shale and Burgess Shale biotas  Of primary interest is how these metazoan genera interacted with their substrate and whether or not they show adaptations to soft substrates with a well-developed mixed layer that are typical of Phanerozoic benthic suspension feeders  This data was compiled for each biota, allowing for an accurate examination of the evolutionary paleoecology of benthic suspension feeders during the Cambrian radiation, as well as consideration of the broader implications of this research for the evolution of early metazoans in general  While abundance data would certainly be the most desirable method for examining the paleoecology of these metazoans, such data has yet to be collected for the Maotianshan Shale biota  Future studies are needed to address this unfortunate lack of abundance data  For now, however, generic diversity is the best available form of data

## 4. Shiyantou Formation and Maotianshan Shale member core analysis

Although a total of 5 463 m of core were examined from the Shiyantou and Yuanshan Members, the cross-bedded, very fine to fine sandstone interbeds in these core samples were virtually unbioturbated (average ii=1 0), because they represent rapidly deposited event beds In order to allow for a direct comparison of the siltstone and mudstone facies of the Shiyantou Formation and the Maotianshan Shale Member of the Yuanshan Formation, therefore, these sandstone interbeds were not considered when comparing the two units, or when calculating the average ii's or ii percentages for each member These sandstones account for 2 193 m of the stratigraphy of the Shiyantou Member samples, leaving 1 863 m of siltstones and mudstones in this member for comparison with the 1 107 m of siltstones and mudstones of the overlying Maotianshan Shale Member No sandstone interbeds were found in the Maotianshan Shale Member core samples, although they are known to occur within this unit

The Shiyantou Formation core samples are characterized by generally moderate levels (average ii=2 9) of horizontal bioturbation In addition to mottled biofabrics (Fig 3A), this bioturbation typically consists of cross sections of horizontal burrows preserved as small (<3–5 mm in width) ovals (Fig 3B), oval cross sections of even smaller (<1 mm in width) horizontal burrows (Fig 3C), and discrete traces identifiable as the shallow infaunal horizontal feeding trace *Teichichnus* (Fig 3D), which is typically less than 1 cm wide in these samples There is no evidence for any larger discrete burrows, and very limited evidence for any vertical bioturbation (three small vertical burrows less than 1 7 cm in depth and 3 mm in width were observed)

The ii percentages for the Shiyantou Formation reveal not only the relative dominance of moderate levels of bioturbation (ii3), but also the broad spectrum of bioturbation levels found in these rocks These ii percentages, and the amount of stratigraphy they represent, are as follows ii1—16% (29 7 cm), ii2—13% (25 cm), ii3—49% (90 4 cm); ii4—12% (23 5 cm), and ii5—10% (17 7 cm) (Fig 4) Note that while 49% of the core sample stratigraphy is moderately bioturbated (ii3), significant amounts of the stratigraphy are also unbioturbated (16% ii1) and completely bioturbated (10% ii5) (Fig 4) These percentages indicate that the substrate that these rocks represent was subjected to highly variable levels of horizontal bioturbation, with moderate levels occur-

ring most frequently Even with this amount of spatial and temporal variability in horizontal bioturbation levels, however, this substrate generally lacked any sign of a mixed layer except for the 10% of its stratigraphy in which it was thoroughly bioturbated (ii5) Also of interest is that 29% of the strata in these samples showed very little (ii2) or no (ii1) evidence at all for any bioturbation (Fig 4)

This bioturbation data indicates that the substrates represented by the Shiyantou Formation siltstones and mudstones, while typically modified by moderate levels of horizontal bioturbation (ii3), generally did not have a well-developed mixed layer Consistent with the prediction of the Cambrian substrate revolution hypothesis, however, the beginnings of mixed layer development are preserved within these rocks as vertically restricted intervals of thorough bioturbation (ii5) that account for just 10% of the stratigraphy in these samples (Fig 3D) This substrate was apparently experiencing the early stages of a transition from a typical Proterozoic-style soft substrate to a typical Phanerozoic-style substrate Accordingly, one might expect to find benthic suspension feeders adapted to both substrate types co-existing in the Maotianshan Shale biota

In marked contrast to the underlying Shiyantou Formation, the Maotianshan Shale Member, in which the soft-bodied fossils are preserved, is almost completely unbioturbated The average ii for these 1 107 m of stratigraphy is only 1 0 Bioturbation in the core samples examined is limited to three occurrences of individual horizontal burrows visible as gray oval cross sections, leading to ii percentages of 99% ii1 and 1% ii2 (Fig 4) No evidence of moderate to thorough bioturbation levels (ii3 to ii5) was found in these rocks These extremely low levels of bioturbation suggest that whatever environmental factors contributed to the exceptional preservation of the soft-bodied fossil biota also suppressed bioturbation levels in the Maotianshan Shale Member siltstones and mudstones

With these extremely low levels of bioturbation, the substrate represented by these rocks would have been relatively firm and seafloor microbial mats may have been present However, careful examination of the sedimentology of the Maotianshan Shale Member reveals little evidence for such seafloor microbial mats Samples from the Maotianshan Shale Member

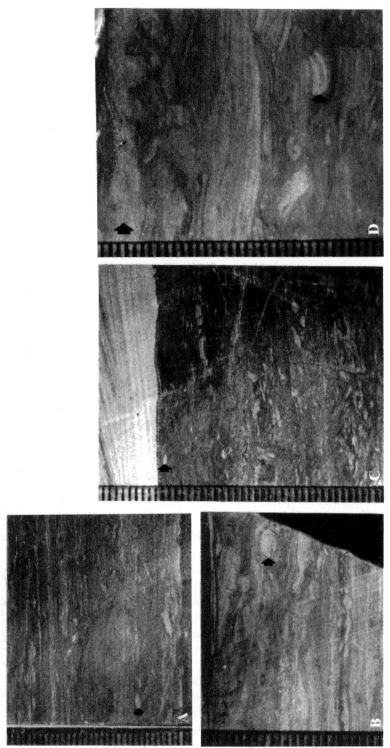

Fig. 3. Photographs of typical Shiyantou Formation bioturbation in core samples. Scale on the left side of photographs is in millimeters. (A) Mottled beds with some individual horizontal burrows seen as gray ovals (arrow). (B) Oval cross section of a larger horizontal burrow (arrow) amidst smaller burrows and mottled beds. (C) Oval cross sections of sub-millimeter horizontal burrows (arrow) and mottled beds capped by a cross-bedded very fine sandstone with no evidence of bioturbation. (D) Discrete *Teichichnus* trace fossils (small arrow) below possible evidence of mixed layer development in a 2–3 mm thick zone of sediment homogenized by bioturbation (large arrow).

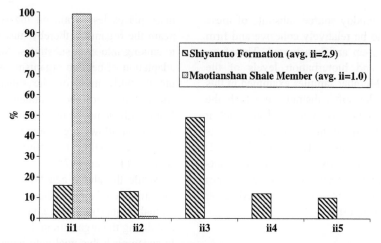

Fig. 4. Graph of ii percentages of the Shiyantou Formation and the Maotianshan Shale Member of the Yuanshan Formation mudstones and siltstones. See text for definitions of ichnofabric indices (ii) and discussion.

generally consist of thin-bedded to laminated mudstones and siltstones that are gray to black in color (Fig. 5A). These thin beds are typically much less than 1 cm thick, but are up to 2 cm thick in places, while the laminae are sub-millimeter in thickness. Many of these thin beds appear to have been rapidly deposited as they are typically graded and sometimes exhibit small-scale cross-bedding. Many of these thin event beds also contain extremely small (typically less

than 1 mm in length) black to dark gray mudstone intraclasts, which are elongate but irregular in form, thereby exhibiting flexible behavior (Fig. 5B). This combination of cohesiveness and flexibility may suggest that these intraclasts were bound by seafloor microbial mats (e.g. Pflüger and Gresse, 1996; Schieber, 1999), but these characteristics may also just be the result of reworking of cohesive seafloor muds. Even lacking solid evidence for seafloor

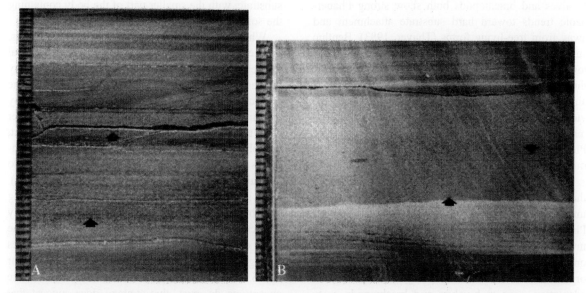

Fig. 5. Photographs of typical Maotianshan Shale Member core samples. Scale on the left side of photographs is in millimeters. (A) Thin beds and laminae with no bioturbation. Note the black mudstone intraclasts common in several of the beds (arrows). (B) Graded bed containing black mudstone intraclasts (arrows) amidst thins beds and laminae.

microbial mats, the muddy source substrate of these intraclasts still had to be relatively cohesive and firm in order to be ripped-up and transported Therefore, the sedimentology and bioturbation levels of the Maotianshan Shale Member samples examined in this study indicate that the soft substrate on which the Maotianshan Shale biota was living, and in which it was preserved, was relatively firm This conclusion is consistent with the findings of Droser et al (2002) that Cambrian siliciclastic substrates, particularly silts and muds, were more cohesive than at later times because of the embryonic state of mixed layer development in the Cambrian

## 5. Adaptations of benthic suspension feeders to typical Phanerozoic-style soft substrates

Phanerozoic immobile benthic suspension feeders typically show several adaptations that allow for their survival on unconsolidated substrates, and it is important to examine these adaptations carefully in order to assess the paleoecology of benthic suspension feeders from the Maotianshan Shale and Burgess Shale biotas Perhaps the most successful of these adaptations is the ability to attach to hard substrates, such as the skeletons of other animals Benthic bivalves and brachiopods both show strong Phanerozoic trends toward hard substrate attachment and away from free-living forms (Thayer, 1983) Benthic suspension feeding echinoderms show a similar trend toward attachment and away from free-living forms through the Cambrian (Bottjer et al , 2000, Dornbos and Bottjer, 2000) Hard substrate attachment provides a stable refuge from soft, soupy substrates, making it a powerful metazoan adaptation to typical Phanerozoic-style soft substrates with well-developed mixed layers

Despite their decrease in generic abundance through the Phanerozoic (Thayer, 1983), free-living immobile forms also show adaptations to typical Phanerozoic soft substrates All of these adaptations involve either decreasing the amount of strain placed on the substrate by the organism or stabilizing the organism in the soft substrate (Thayer, 1975) The first adaptation, seen in bivalves and brachiopods, is to decrease overall body mass by growing thin, non-costate shells (Thayer, 1975) This decrease in body

mass places less strain on the soft substrate underneath the organism, thereby making it more resistant to sinking into such substrates Another soft substrate adaptation of benthic organisms is to broadly distribute its body mass on the substrate, an adaptation called the "Snowshoe" strategy by Thayer (1975) Commonly achieved by having a wide, thin body, this adaptation allows organisms to virtually float on top of a soft substrate by diffusing their strain on the seafloor (Thayer, 1975)

While the above adaptations focus on decreasing the amount of strain placed on the seafloor by the animal in question, the remaining adaptations rely on stabilizing the organism in the soft substrate One way to accomplish this goal is to have a root-like holdfast that extends into the soft substrate, thereby stabilizing the animal (Sprinkle and Guensburg, 1995) Ordovician crinoids utilized this adaptation as they inhabited soft substrate settings during the Ordovician radiation (Sprinkle and Guensburg, 1995) Another way in which an animal can stabilize itself on a typical Phanerozoic soft substrate is to live as a deep sediment sticker, with a large skeletal element that extends deep into the seafloor and stabilizes the animal (Thayer, 1975, Seilacher, 1999) This adaptation was termed the "Iceberg" strategy by Thayer (1975), because most of the animal extends into the substrate, with the smaller part of the body containing the soft parts perched above the soft seafloor

When combined with one of the above strategies, another effective technique for survival on typical Phanerozoic-style soft substrates is to dramatically increase body size, making bioturbation levels in the soft substrate irrelevant to survival (Thayer, 1983) By growing to a large size very quickly, at least faster than the biological turnover rate of the sediment, immobile benthic suspension feeders can keep themselves above the sediment–water interface, helping to neutralize the effects of bioturbation (Thayer, 1983) This strategy is utilized by the modern attaching endobyssate bivalve Pinna, which is normally restricted to stabilized or hard substrates, but in some cases can survive on well-bioturbated soft substrates through unusually rapid growth, up to 15 cm in their first 6 months (Thayer, 1975, 1983) Although notable exceptions do exist, this strategy does suggest that free-living immobile benthic suspension feeders adapted to typical Phanerozoic-style soft substrates are generally, but certainly not always, larger than those well-adapted for survival on typical Proterozoic-style soft substrates It is important to remember, however, that large benthic metazoans can clearly be well-adapted to typical Proterozoic-style soft substrates if they lack stabilizing or strain reducing adaptations, which would be particularly important for a large benthic metazoan to have in well-bioturbated settings

## 6. Adaptations of benthic suspension feeders to typical Proterozoic-style soft substrates

In contrast to the benthos discussed above, immobile benthic suspension feeders well-adapted to typical Proterozoic-style soft substrates lack adaptations that decrease the amount of strain on the seafloor or stabilize the animal in the soft substrate One strategy of benthic suspension feeders well-adapted to typical Proterozoic-style substrates is to live as shallow sediment stickers (Seilacher, 1999, Dornbos and Bottjer, 2000) Unlike the deep sediment stickers discussed above, these animals have no large, stabilizing skeletal elements that extend deep into the seafloor Instead, only a small portion of their body extends into the seafloor (Dornbos and Bottjer, 2000) Benthic suspension feeders with this lifestyle were well-adapted for survival on typical Proterozoic-style soft substrates because they lack the stabilizing adaptations seen in those benthos that were well-adapted to typical Phanerozoic-style soft substrates with well-developed mixed layers

Immobile benthic suspension feeders that lived freely on the seafloor as sediment resters, but lack adaptations that would reduce their strain on the seafloor, were well-adapted to living on typical Proterozoic-style soft substrates Benthic invertebrates with this lifestyle probably could not have survived on soft substrates with well-developed mixed layers, especially considering that even sediment resters with adaptations to well-bioturbated soft substrates, such as a broad body mass distribution, decrease in generic abundance through the Phanerozoic as bioturbation levels steadily increase (Thayer, 1983)

Another lifestyle indicative of adaptation to typical Proterozoic-style soft substrates is soft substrate attachment Any immobile benthic invertebrate that

simply lived attached to an unconsolidated soft substrate, such as the Neoproterozoic probable cnidarian Corumbella (Leslie et al , 2001), was clearly well-adapted to a typical Proterozoic-style soft substrate with a low water content and seafloor microbial mats Any such benthic suspension feeder could probably not survive on a soft substrate with a well-developed mixed layer

## 7. Paleoecology of Maotianshan Shale biota benthic suspension feeders

Utilizing the criteria discussed above (Table 1), the adaptive morphologies of Maotianshan Shale biota benthic suspension feeders were carefully examined in order to determine whether they were adapted to survive on either typical Proterozoic-style soft substrates, which would have been relatively firm and often bound by seafloor microbial mats, or typical Phanerozoic-style soft substrates with a well-developed mixed layer This examination indicates that the Maotianshan Shale biota contains sediment resters, shallow sediment stickers, and hard substrate attachers (Table 2) One example of a sediment rester is the sack-shaped demosponge Crumillospongia, which exhibits no adaptations to typical Phanerozoic-style soft substrates (Fig 6A) While one specimen has been noted to be over 10 cm in height, Crumillo-

Table 1
Criteria used to interpret the life modes of benthic suspension feeders examined in this study

| Proterozoic-style | Phanerozoic-style |
|---|---|
| (1) Lower end pointed for shallow insertion into sediment | (1) Evidence of attachment to hard substrates |
| (2) Lower end blunt or flat for sediment resting | (2) Presence of extensive root-like holdfast |
| (3) Evidence for attachment to seafloor sediment | (3) Broad body mass distribution (wide, thin, and flat) |
| | (4) Presence of long skeletal extension for insertion into substrate |

In the first column, number 1 corresponds to shallow sediment stickers (sss), number 2 to sediment resters (sr), and number 3 to sediment attachers (sa) In the second column, number 1 corresponds to hard substrate attachers (hs), number 2 to root-like holdfasts (rlh), number 3 to snowshoe strategists (sno), and number 4 to iceberg strategists (ice)

Table 2

Maotianshan Shale fauna benthic suspension feeding genera with brief morphological descriptions and interpreted life modes

| Genus | Description | Life mode |
|---|---|---|
| *Allantospongia* | globular sponge | sediment rester |
| *Allonia* | conical animal with pointed lower end | shallow sediment sticker |
| *Archotuba* | conical animal attached to hard substrates | hard substrate attacher |
| *Cambrorhytium* | conical animal attached to hard substrates | hard substrate attacher |
| *Choia* | globular sponge | sediment rester |
| *Choiaella* | globular sponge | sediment rester |
| *Crumillospongia* | sac-like sponge | sediment rester |
| *Dinomischus* | animal with blunt spine on lower end | shallow sediment sticker |
| *Heliomedusa* | flat brachiopod | sediment rester |
| *Iotuba* | long, thin animal with rounded lower end | shallow sediment sticker |
| *Leptomitella* | conical sponge with pointed lower end | shallow sediment sticker |
| *Leptomitus* | conical sponge with pointed lower end | shallow sediment sticker |
| *Paraleptomitella* | conical sponge with pointed lower end | shallow sediment sticker |
| *Quadrolaminiella* | conical sponge with pointed lower end | shallow sediment sticker |
| *Saetospongia* | globular sponge | sediment rester |
| *Takakkawia* | conical animal with pointed lower end | shallow sediment sticker |
| *Tricitispongia* | globular sponge | sediment rester |
| *Xianguangia* | cnidarian with blunt lower end | sediment rester |

Only interpretable genera are included in this table

*spongia* is typically only around 2 cm in height (Rigby, 1986), making this genus even more susceptible to the effects of high bioturbation levels and the resultant mixed layer *Crumillospongia*, then, seems to be a good example of a Maotianshan Shale biota benthic suspension feeder well-adapted for survival on a typical Proterozoic-style soft substrate

Another such Maotianshan Shale benthic suspension feeder is *Takakkawia*, a conical demosponge with a narrow pointed lower end well-adapted for insertion into the seafloor (Fig 6B) *Takakkawia*, therefore, most likely lived as a shallow sediment sticker, but lacks any adaptations to either stabilize it in the soft substrate or decrease its strain on the seafloor, as exhibited by benthic suspension feeders adapted to typical Phanerozoic-style soft substrates with a well-developed mixed layer Its adaptive morphology, as

well as its small size (typically less than 3 cm in height (Rigby, 1986), both indicate that *Takakkawia* was well-adapted to typical Proterozoic-style soft substrates

As predicted earlier, benthic suspension feeders well-adapted for survival on typical Phanerozoic-style soft substrates also exist in the Maotianshan Shale biota One example is the funnel-shaped possible cnidarian *Cambrorhytium* that lived attached to brachiopod shells or other hard substrates (Chen and Zhou, 1997) (Fig 6C) Because it lived attached to available hard substrates such as the skeletons of other animals, *Cambrorhytium* was clearly well-adapted for survival in environments with typical Phanerozoic-style soft substrates with a well-developed mixed layer

When the adaptive morphologies of all 19 immobile benthic suspension feeding genera in the Maotianshan Shale biota are analyzed in a similar manner, it becomes obvious that an overwhelming majority (88% or 15 genera) are well-adapted to typical Proterozoic-style soft substrates (Fig 7, Table 3) The remainder (12% or 2 genera) are well-adapted to typical Phanerozoic-style soft substrates (Fig 7, Table 3) The adaptive morphologies of two genera are uninterpretable because of limitations in the fossil evidence The dominance of benthic suspension feeding genera adapted to typical Proterozoic-style soft substrates in the Maotianshan Shale biota suggests that the Proterozoic–Phanerozoic substrate transition was only in its early stages during this time These results are consistent with the core sample analysis of the Maotianshan Shale Member, which indicates that the Maotianshan Shale biota was living on a relatively firm soft substrate with no mixed layer

It is interesting to note that 11 of the 15 Maotianshan Shale benthic suspension feeding genera well-adapted to typical Proterozoic-style soft substrates are sediment resting and shallow sediment sticking sponges (Fig 8) The free-living nature of these sponges differs greatly from typical modern shallow water sponges, which are dominated by hard substrate encrusters None of the Maotianshan Shale biota sponges are hard substrate encrusters, resulting in a general Phanerozoic trend toward hard substrate attachment that is consistent with the Cambrian substrate revolution In addition, these differences between modern shallow marine sponges and Early

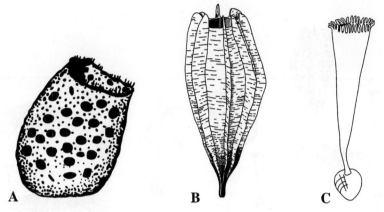

Fig. 6. Examples of Maotianshan Shale biota benthic suspension feeders. (A) Sediment resting demosponge *Crumillospongia*, typically around 2 cm in height (modified from Rigby, 1986). (B) Shallow sediment sticking demosponge *Takakkawia*, typically around 3–4 cm in height (modified from Rigby, 1986). (C) Hard substrate attacher *Cambrorhytium*, a possible cnidarian typically 2–3 cm in height (modified from Chen and Zhou, 1997).

Cambrian ones shows that the results of this paleoecological study are not merely the product of a taxonomic effect brought on by the dominance of sponges in the Maotianshan Shale biota.

The prediction of the Cambrian substrate revolution hypothesis that benthic suspension feeders adapted to Proterozoic-style soft substrates and those adapted to Phanerozoic-style soft substrates would both be found in the Maotianshan Shale biota is upheld by this paleoecological analysis. Although benthic suspension feeders adapted to typical Phanerozoic-style soft substrates comprise only 12% of

Maotianshan Shale benthic suspension feeders, they are still present and indicate that the Proterozoic–Phanerozoic soft substrate transition was underway during this time. The core sample analysis of the underlying Shiyantou Formation strongly supports this interpretation because it indicates that the mixed

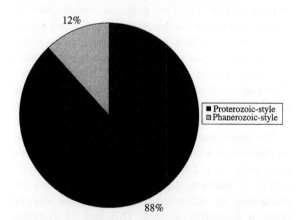

Fig. 7. Pie chart showing the percentages of benthic suspension feeding genera adapted to typical Proterozoic-style and typical Phanerozoic-style soft substrates in the Maotianshan Shale biota. Only genera for which paleoecological interpretations were possible are included.

Table 3

Adaptive morphologies of Maotianshan Shale fauna benthic suspension feeding genera

| Proterozoic-style | Phanerozoic-style | Unknown |
|---|---|---|
| *Allantospongia* (sr) | *Archotuba* (hs) | *Halichondrites* |
| *Allonia* (sss) | *Cambrorhytium* (hs) | *Hazelia* |
| *Choia* (sr) | | |
| *Choiaella* (sr) | | |
| *Crumillospongia* (sr) | | |
| *Dinomischus* (sss) | | |
| *Heliomedusa* (sr) | | |
| *Iotuba* (sss) | | |
| *Leptomitella* (sss) | | |
| *Leptomitus* (sss) | | |
| *Paraleptomitella* (sss) | | |
| *Quadrolaminiella* (sss) | | |
| *Saetospongia* (sr) | | |
| *Takakkawia* (sss) | | |
| *Tricitispongia* (sr) | | |
| *Xianguangia* (sr) | | |

Columns indicate the type of soft substrate that the benthic suspension feeding genera are adapted to, either typical Proterozoic-style soft substrates (left column), typical Phanerozoic-style soft substrates (middle column), or unknown (right column). "sr"= sediment rester. "sss"=shallow sediment sticker. "hs"=hard substrate attacher. See text for discussion.

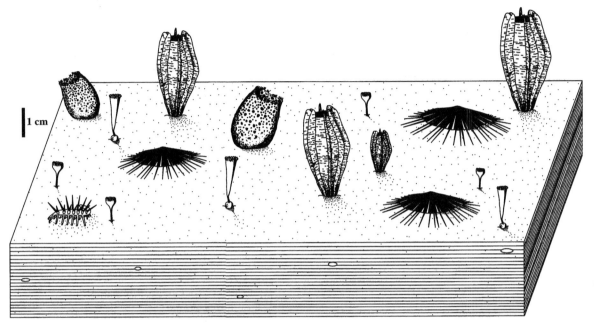

Fig. 8. Box diagram reconstructing Maotianshan Shale biota benthic suspension feeding genera, most of which are discussed in the text, in life position on top of a virtually unbioturbated substrate with just a few horizontal burrows seen as oval cross sections a few millimeters in diameter. An example of the sediment resting demosponge *Crumillospongia* is in the upper left corner. An example of the shallow sediment sticking demosponge *Takakkawia* is in the upper right corner. An example of the sediment resting demosponge *Choia* is next to this *Takakkawia*. An example of the hard substrate attaching possible cnidarian *Cambrorhytium* is in the lower right corner. Examples of the enigmatic shallow sediment sticker *Dinomischus* are in the lower left corner next to the lobopodian *Hallucigenia*.

layer was beginning to develop in these environments before the preservation of the Maotianshan Shale biota.

## 8. Paleoecology of Burgess Shale biota benthic suspension feeders

While there are distinct differences between the preservational environments of the Maotianshan Shale and Burgess Shale biotas, it is still informative to examine the paleoecology of Burgess Shale benthic suspension feeders considering that the Burgess Shale biota provides one of the best windows into animal life in the Cambrian. In any case, the environmental differences between the Maotianshan Shale and Burgess Shale biota may actually provide for an interesting comparison between the two biotas because they share many taxa, leading to speculation by some that the Burgess Shale biota may represent a deep water refuge for Early Cambrian holdovers (Conway Morris, 1998).

The same criteria used to analyze the adaptive morphologies of Maotianshan Shale biota benthic suspension feeders were used to analyze the adaptive morphologies of Burgess Shale benthic suspension feeders (Table 1). The goal, as above, was to determine whether benthic suspension feeding genera in the Burgess Shale biota were adapted to either typical Proterozoic-style soft substrates, or typical Phanerozoic-style soft substrates. Much like the Maotianshan Shale biota, the Burgess Shale biota contains shallow sediment stickers, sediment resters, and hard substrate attachers (Table 4).

The small round to elliptical demosponge *Choia* is an example of a sediment rester in the Burgess Shale biota (Fig. 9A). *Choia*, typically only a few millimeters in diameter (Rigby, 1986), also shows no stabilizing or strain-reducing adaptations despite being a free-living sponge. This lack of common adaptations to typical Phanerozoic soft substrates together with its small size indicates that *Choia* could not have survived on a heavily bioturbated soft substrate with a well-developed mixed layer. *Choia*

Table 4

Burgess Shale fauna benthic suspension feeding genera with brief morphological descriptions and interpreted life modes

| Genus | Description | Life mode |
| --- | --- | --- |
| Cambrorhytium | conical animal attached to hard substrates | hard substrate attacher |
| Capsospongia | sponge with pointed lower end | shallow sediment sticker |
| Chancelloria | conical animal with pointed lower end | shallow sediment sticker |
| Choia | globular sponge | sediment rester |
| Crumillospongia | sac-like sponge | sediment rester |
| Diagoniella | sponge with pointed lower end | shallow sediment sticker |
| Dinomischus | animal with blunt spine on lower end | shallow sediment sticker |
| Diraphora | brachiopod attached to hard substrates | hard substrate attacher |
| Echmatocrinus | animal attached to hard substrates | hard substrate attacher |
| Eiffelia | sponge attached to hard substrates | hard substrate attacher |
| Gogia | echinoderm attached to hard substrates | hard substrate attacher |
| Leptomitella | conical sponge with pointed lower end | shallow sediment sticker |
| Leptomitus | conical sponge with pointed lower end | shallow sediment sticker |
| Mackenzia | brachiopod attached to hard substrates | hard substrate attacher |
| Micromitra | brachiopod attached to hard substrates | hard substrate attacher |
| Nisusia | brachiopod attached to hard substrates | hard substrate attacher |
| Takakkawia | conical animal with pointed lower end | shallow sediment sticker |
| Thaumaptilon | cnidarian with rounded lower end | shallow sediment sticker |
| Vauxia | branching sponge with pointed lower end | shallow sediment sticker |
| Walcottidiscus | echinoderm with flat lower surface | sediment rester |
| Wapkia | sponge with pointed lower end | shallow sediment sticker |

Only interpretable genera are included in this table

was therefore well-adapted to typical Proterozoic-style soft substrates

An example of a shallow sediment sticker in the Burgess Shale biota is the small conical to sack-shaped hexactinellid sponge *Diagoniella* (Fig 9B) Reaching a maximum height of just 15 mm and showing no adaptations to stabilize itself in the soft substrate on which it lived (Rigby, 1986, Briggs et al ,

1994), *Diagoniella* was well-adapted for survival on a typical Proterozoic-style soft substrate with a sharp sediment–water interface and seafloor microbial mats It almost certainly could not have survived on a soft substrate with a well-developed mixed layer

The rhynchonelliform brachiopod *Nisusia* is an example of a Burgess Shale benthic suspension feeder that was well-adapted for survival on typical Phanerozoic-style soft substrates (Fig 9C) *Nisusia* lived as a hard substrate attacher, using its pedicle to attach to sponges, particularly the demosponge *Pirania* (Rigby, 1986) Because it lived attached to sponges, *Nisusia* neutralized any effect that high levels of bioturbation may have had on its ecology, making it well-adapted to typical Phanerozoic-style soft substrates with a well-developed mixed layer

This paleoecological analysis was performed on all Burgess Shale benthic suspension feeding genera in which an adaptive morphological interpretation was possible Of these 32 genera, the adaptive morphologies of 11 genera are uninterpretable mostly because many of them are known from only a very few fragmented specimens (Rigby, 1986) Of the 21 genera for which an interpretation is possible, however, 62% (13 genera) are well-adapted to typical Proterozoic-style soft substrates, while 38% (8 genera) are well-adapted to typical Phanerozoic-style soft substrates (Fig 10; Table 5) Although a direct reconstruction is not possible, these results at least suggest that the soft substrate on which the Burgess Shale biota lived was a transitional mosaic of relatively firm substrates and substrates with some mixed layer development

Some interesting observations can be made when comparing the benthic suspension feeders of the Maotianshan Shale and Burgess Shale biotas Although the Burgess Shale biota is generally thought to have been preserved in a much deeper marine setting than the Maotianshan Shale biota, it contains an allocthonous assemblage of metazoans from shallower muddy shelf environments that has a much higher number of benthic suspension feeders adapted to typical Phanerozoic-style soft substrates, all hard substrate attachers This difference reveals an interesting trend in that an Early Cambrian shallow marine setting probably had considerably less mixed layer development than the Middle Cambrian muddy shelf environments in which the Burgess Shale biota

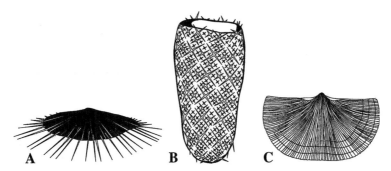

Fig. 9. Examples of Burgess Shale biota benthic suspension feeders. (A) Sediment resting demosponge *Choia*, typically 8–12 mm in central-disc width (modified from Rigby, 1986). (B) Shallow sediment sticking hexactinellid sponge *Diagoniella*, typically up to 15 mm in height (modified from Briggs et al., 1994). (C) Hard substrate attaching orthid brachiopod *Nisusia*, typically around 22 mm in width (modified from Briggs et al., 1994).

benthic metazoans lived. This observation may provide more evidence for the onshore–offshore nature of the Cambrian increase in bioturbation levels (e.g. Hagadorn and Bottjer, 1999), and is consistent with increased mixed layer development in subtidal siliciclastic soft substrate environments through the Cambrian.

Similar to the Maotianshan Shale biota, shallow sediment sticking and sediment resting sponges dominate the Burgess Shale benthic suspension feeders well-adapted to typical Proterozoic-style soft substrates. In addition, 7 of the 13 genera in this category are present in both biotas. The most crucial differences between the two biotas are actually found in the benthic suspension feeders well-adapted to typical Phanerozoic-style soft substrates. A relatively diverse group of hard substrate attachers including

echinoderms (*Gogia*), brachiopods (*Diraphora, Micromitra*, and *Nisusia*), possible cnidarians (*Cambrorhytium* and *Mackenzia*), and one sponge genus (*Eiffelia*) comprise the Burgess Shale benthic suspension feeders that are well-adapted to typical Phanerozoic-style soft substrates (Table 5). What this diverse group has in common is that they all lived attached to the hard skeletons of other animals, including sponges, brachiopods, priapulid worms, and hyolithids. In the Maotianshan Shale biota, meanwhile, only two genera, the possible cnidarian *Cambrorhytium* and the enigmatic *Archotuba*, fit into this

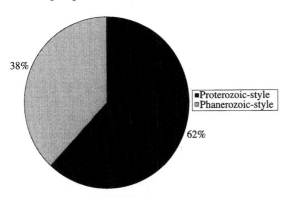

Fig. 10. Pie chart showing the percentages of benthic suspension feeding genera adapted to typical Proterozoic-style and typical Phanerozoic-style soft substrates in the Burgess Shale biota. Only genera for which paleoecological interpretations were possible are included.

Table 5

Adaptive morphologies of Burgess Shale fauna benthic suspension feeding genera

| Proterozoic-style | Phanerozoic-style | Unknown |
| --- | --- | --- |
| *Capsospongia* (sss) | *Cambrorhytium* (hs) | *Canistrumella* |
| *Chancelloria* (sss) | *Diraphora* (hs) | *Falospongia* |
| *Choia* (sr) | *Echmatocrinus* (hs) | *Fieldospongia* |
| *Crumillospongia* (sr) | *Eiffelia* (hs) | *Halichondrites* |
| *Diagoniella* (sss) | *Gogia* (hs) | *Hamptonia* |
| *Dinomischus* (sss) | *Mackenzia* (hs) | *Hazelia* |
| *Leptomitella* (sss) | *Micromitra* (hs) | *Moleculospina* |
| *Leptomitus* (sss) | *Nisusia* (hs) | *Pirania* |
| *Takakkawia* (sss) | | *Protospongia* |
| *Thaumaptilon* (sss) | | *Sentinelia* |
| *Vauxia* (sss) | | *Stephenospongia* |
| *Walcottidiscus* (sr) | | |
| *Wapkia* (sss) | | |

Columns indicate the type of soft substrate that the benthic suspension-feeding genera are adapted to, either typical Proterozoic-style soft substrates (left column), typical Phanerozoic-style soft substrates (middle column), or unknown (right column). "sr"= sediment rester. "sss"=shallow sediment sticker. "hs"=hard substrate attacher. See text for discussion.

category, both of which lived attached to brachiopod shells (Table 3) This data suggests that, in general, many Early and Middle Cambrian benthic suspension feeders began seeking refuge from increasing bioturbation levels in siliciclastic soft substrate settings by attaching to the skeletons of other animals, dead or alive It also appears, through comparison of the two biotas, that this mode of life became more common amongst benthic suspension feeders through the Early and Middle Cambrian

While the specific examples of benthic suspension feeders well-adapted to typical Proterozoic soft substrates discussed above are all relatively small in size, it is important to note that some Maotianshan Shale and Burgess Shale biota benthic suspension feeders adapted to such substrates are actually quite large For instance, the large conical demosponges *Leptomitus* and *Leptomitella*, which can reach over 30 cm in length (Rigby, 1986), are found in both biotas and are interpreted as shallow sediment stickers adapted to typical Proterozoic soft substrates This interpretation is based on their conical morphology, with *Leptomitella* having a sharply pointed lower end, and their lack of any stabilizing or strain reducing adaptations despite their large size, which would seemingly cause them to strongly require such adaptations Indeed, in the post-Cambrian Paleozoic hard substrate attaching and root-stabilized crinoids are the dominant high level tierers in neritic environments (Sprinkle and Guensburg, 1995; Ausich and Bottjer, 2001), as opposed to the free-living or soft substrate attaching forms of the Neoproterozoic and Cambrian

## 9. Paleoecology of mobile benthic metazoans in the Maotianshan Shale and Burgess Shale biotas

While the substrate interactions of mobile benthic metazoans are more difficult to interpret than those of immobile benthic suspension feeders, some interpretations are still possible Because of this difficulty in interpretation, however, a quantitative assessment of the paleoecology of the Maotianshan Shale and Burgess Shale biota mobile benthic metazoans is not possible at this time Instead, there are criteria that will help to identify key individual examples of mobile benthic metazoans adapted to both substrate types in these biotas For instance, mobile benthic metazoans

well-adapted to typical Proterozoic-style soft substrates would include animals that are dependent on seafloor microbial mats as a trophic resource, such as "mat scratchers" (Seilacher, 1999), which scratched microbial mats for food without destroying them Also included in this category would be benthic metazoans that are dependent on a relatively firm, stable substrate with a sharp sediment–water interface for survival

Using these criteria, mobile benthic metazoans that appear to have been adapted to typical Proterozoic-style soft substrates are present in the Maotianshan Shale and Burgess Shale biotas The unusual lobopodians, resembling onychophorans and found in both biotas, walked on the seafloor with numerous short (<1 cm) jointless legs, perhaps living as scavengers or grazing on sponges or algae (e g Briggs et al , 1994; Chen and Zhou, 1997) The strange lobopodian species *Hallucigenia*, found in both biotas, has two rows of dorsal spines and is only 0 5 to 3 cm long (Fig 11A), while *Microdictyon*, found in only the Maotianshan Shale biota, has flat oval plates on its sides and is of similar size (e g Briggs et al , 1994, Chen and Zhou, 1997) It seems unlikely that these small animals could have survived on a typical Phanerozoic-style soft substrate with a diffuse sediment–water interface and high water content resulting from a well-developed mixed layer, especially considering the muddy seafloors on which they lived They instead appear to have been well-adapted for living on a relatively firm Proterozoic-style soft substrate with a sharp sediment–water interface and low water content

While the lobopodians were probably dependent on the firmness and stability of typical Proterozoic-style soft substrates, there are examples of Burgess Shale mobile benthic metazoans that were probably dependent on the seafloor microbial mats found on this type of soft substrate as a trophic resource One example, assuming that it lived on a muddy substrate, which is the common interpretation (e g Conway Morris, 1985, Briggs et al , 1994), is the small, 3 5 to 55 mm long, enigmatic metazoan *Wiwaxia* (Fig 11B) With its dorsal side covered with flat sclerites and spines and its ventral side unarmored, *Wiwaxia* is commonly interpreted to have crawled along the sediment using muscular contractions while feeding on the surface of the substrate with its two rows of

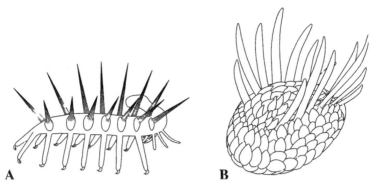

Fig 11 Examples of Maotianshan Shale and Burgess Shale biota mobile benthic metazoans (A) The Maotianshan Shale and Burgess Shale lobopodian *Hallucigenia*, typically 0 5 to 3 cm in length (modified from Briggs et al , 1994) (B) The enigmatic Burgess Shale biota mobile benthic metazoan *Wiwaxia*, typically 3 5 to 55 mm in length (modified from Briggs et al , 1994)

anterior teeth (Conway Morris, 1985), a grazing mode of life typically restricted to hard substrates today Especially considering the muddy substrate on which *Wiwaxia* lived, this mode of life would have been dependent on a relatively firm typical Proterozoic-style soft substrate with seafloor microbial mats, which may have served as an important trophic resource for *Wiwaxia*

Although taxonomically controversial, another possible example of a Burgess Shale mobile benthic metazoan well-adapted to typical Proterozoic-style soft substrates may be the possible mollusc *Scenella* Originally interpreted as a helcionelloid mollusc with a single cap-shaped dorsal shell (Rasetti, 1954), but later reinterpreted as a pelagic chondrophorine cnidarian (Yochelson and Gil Cid, 1984, Babcock and Robison, 1988), *Scenella* is one of the most abundant fossils found in the Burgess Shale An important reason for this reinterpretation of Burgess Shale *Scenella* as pelagic cnidarians is their preservation in a deep, muddy environment (Yochelson and Gil Cid, 1984) Their presence in such an environment was considered inconsistent with the seafloor grazing life mode of helcionelloid molluscs, a life mode typically restricted to nearshore hard substrates today (Yochelson and Gil Cid, 1984) While the taxonomy of Burgess Shale *Scenella* needs further work, it is important to note that our increasing knowledge of the nature of Cambrian subtidal siliciclastic substrates explains their apparently inconsistent environmental distribution because relatively firm substrates with seafloor microbial mats, typical Proterozoic-style soft substrates, would not be unexpected in such environments during the Middle

Cambrian Accordingly, preservation of seafloor grazing molluscs would not be unexpected in such deep, muddy environments during this time.

Mobile benthic metazoans well-adapted for survival on typical Phanerozoic-style soft substrates are also probably found in the Maotianshan Shale and Burgess Shale biotas Although it is difficult to determine the precise life mode of many of the Maotianshan Shale and Burgess Shale biota trilobites, it is likely that some of them lived as benthic deposit feeders, using their legs to dig into the upper few centimeters of the substrate and ingest organic detritus This trilobite feeding behavior is preserved in trace fossils such as *Cruziana* and *Rusophycus*, common in siliciclastics throughout the Cambrian Not only were deposit feeding trilobites well-adapted to survival on typical Phanerozoic-style soft substrates, but they also contributed greatly to the Cambrian substrate revolution itself by bioturbating the upper few centimeters of the sediment (Babcock, 2003) It seems clear, therefore, from this paleoecological analysis that mobile benthic metazoans adapted to both typical Proterozoic-style soft substrates and typical Phanerozoic-style soft substrates coexisted in the Maotianshan Shale and Burgess Shale biotas, as would be expected during this time of dramatic siliciclastic soft substrate transition

## 10. Discussion

The results of this research support the Cambrian substrate revolution hypothesis by upholding the three predictions of the hypothesis outlined earlier· (1) the Maotianshan Shale and Burgess Shale biotas do contain benthic metazoans adapted to both substrate types, (2) the Burgess Shale biota does have a greater number of benthic suspension feeding genera adapted to typical Phanerozoic-style soft substrates than the Maotianshan Shale biota, and (3) there is some evidence for the beginning of mixed layer development in the Shiyantou Formation mudstones and siltstones, which underlie the rocks in which the soft-bodied fossil biota is preserved

The results of this research also have broader implications for the early evolution and ecology of metazoans that warrant some discussion First of all, this research, as well as previous work on the Cambrian substrate revolution, suggests that the seemingly strange morphology of many early metazoans, such as helicoplacoid echinoderms and *Wiwaxia*, may be partially attributed to the fact that they were well-adapted to non-actualistic environmental settings Because modern subtidal siliciclastic shelf environments are not dominated by well-developed seafloor microbial mats but, instead, are characterized by intense bioturbation and a well-developed mixed layer, similar metazoans with comparable lifestyles do not exist in such environments today So rather than these unusual forms being simply early evolutionary "experiments", as is commonly stated, they were actually just well-adapted to environments that no longer exist in subtidal siliciclastic settings today

Another implication of this research is that the metazoan invasion of the infaunal realm appears to have been an important agent of natural selection on benthic metazoans during the "Cambrian explosion" This infaunalization and the increased bioturbation it produced began to drive the evolution of immobile benthic metazoans toward attachment abilities and away from free-living lifestyles, setting the stage for this continued trend through the Phanerozoic as well as the development of the Paleozoic Fauna, in which a diverse array of hard substrate attachers, particularly crinoids, played an important role (e g Sprinkle and Guensburg, 1995). Because benthic metazoans were forced to adapt to this soft substrate transition, the Cambrian substrate revolution helped to fuel the adaptive radiation of metazoans during the "Cambrian explosion".

## 11. Conclusions

The core analysis portion of this research reveals that the Lower Cambrian Shiyantou Formation mudstones and siltstones are characterized by moderate levels (average $n=2 9$) of horizontal bioturbation preserved as oval cross sections less than 5 mm in width, and also commonly contain the horizontal feeding trace fossil *Teichichnus* Evidence for the beginnings of mixed layer development is also found in these rocks, as predicted by the Cambrian substrate revolution hypothesis The Shiyantou Formation mudstones and siltstones are dominated by n3, but a wide range of bioturbation levels are also present in this unit This Shiyantou Formation bioturbation data indicates that the substrate represented by these rocks was a transitional mixture of typical Proterozoic-style soft substrates and typical Phanerozoic-style soft substrates

In contrast, the mudstones and siltstones of the Maotianshan Shale Member, which contains the soft-bodied biota, contain almost no evidence for bioturbation of any kind (average $n=1 0$) This lack of bioturbation may indicate that whatever environmental conditions led to the exceptional preservation of the Maotianshan Shale biota also suppressed bioturbation levels In addition, small (sub-millimeter), elongate, and irregular black mudstone intraclasts in thin (up to 2 cm thick) graded beds in the Maotianshan Shale Member provide evidence for the presence of firm seafloors because of their cohesiveness This data from the mudstones and siltstones of the Yuanshan Member indicates that the fine-grained substrate on which the Maotianshan Shale biota lived was relatively firm

Adaptive morphological and paleoecological analysis indicates that Maotianshan Shale biota benthic suspension feeders are dominated by genera well-adapted to survival on typical Proterozoic-style soft substrates (88%), while only two genera (12%) are well-adapted to typical Phanerozoic-style soft substrates These numbers indicate that the Cambrian substrate revolution was only in its early stages when the Early Cambrian Maotianshan Shale biota was preserved

Burgess Shale biota benthic suspension feeders, meanwhile, contain a majority of genera well-adapted to typical Proterozoic-style soft substrates (62%), but

S Q Dornbos et al / Palaeogeography, Palaeoclimatology, Palaeoecology 220 (2005) 47–67

also contain a significant number (38%) of genera well-adapted to typical Phanerozoic-style soft substrates. Although no direct data exists, these numbers suggest that the soft substrate on which the Burgess Shale biota lived was a transitional mosaic of typical Proterozoic-style soft substrates and typical Phanerozoic-style soft substrates

Although preserved in different environments, a comparison of Maotianshan Shale and Burgess Shale benthic suspension feeders suggests that mixed layer development was more advanced in the substrate on which the Burgess Shale biota lived Increased mixed layer development in Middle Cambrian shelf muds, on which the Burgess Shale biota lived, when compared to Early Cambrian shallow marine muds provides more evidence for the general, although spatially and temporally variable, increase in mixed layer development in subtidal siliciclastic soft substrate environments through the Cambrian In addition, as predicted by the Cambrian substrate revolution hypothesis, the Burgess Shale biota contains a greater number of benthic suspension feeding genera adapted to typical Phanerozoic-style soft substrates than the Maotianshan Shale biota Furthermore, mobile benthic metazoans adapted to both typical Proterozoic-style soft substrates and Phanerozoic-style soft substrates co-existed in the Maotianshan Shale and Burgess Shale biotas.

Broader implications of this research include the probability that many early benthic metazoans may appear so unusual simply because they were well-adapted to non-actualistic environments, not because they were just early evolutionary "experiments" In addition, the Cambrian substrate revolution forced Cambrian benthic metazoans to adapt to mixed layer development in siliciclastic soft substrate environments, thereby helping to fuel the "Cambrian explosion"

## Acknowledgements

The authors would like to thank Frank Corsetti, Bob Douglas, Al Fischer, Donn Gorsline, James Hagadorn, and Ben Waggoner for helpful discussions during this research Thanks to James Hagadorn for valuable comments on an early draft of this manuscript Two anonymous reviewers provided helpful suggestions and insights A grant to DJB from the National Geographic Society and grants to SQD from Sigma Xi, the University of Southern California Department of Earth Sciences, and the University of Southern California Wrigley Institute of Environmental Studies funded this research. The grants to JYC from National Science Foundation of China (grant nos 40132010 and 990902) and from National Ministry of Science and Technology of China (grant no 20007700) are gratefully acknowledged

## References

Ausich, W I, Bottjer, D J, 2001 Sessile invertebrates In Briggs, D E G, Crowther, P R (Eds ), Palaeobiology, vol II Blackwell, Oxford, pp 384–386

Babcock, L E, 2003 Trilobites in Paleozoic predator–prey systems, and their role in reorganization of early Paleozoic ecosystems In Kelley, P H, Kowalewski, M, Hansen, T A (Eds ), Predator–Prey Interactions in the Fossil Record Plenum Publishers, New York, pp 55–92

Babcock, L E, Robison, R A, 1988 Taxonomy and Paleobiology of some Middle Cambrian Scenella (Cnidaria) and Hyolithids (Mollusca) from Western North America University of Kansas Paleontological Contributions University of Kansas Paleontological Institute, Lawrence

Babcock, L E, Zhang, W, Leslie, S A, 2001 The Chengjiang biota record of the Early Cambrian diversification of life and clues to exceptional preservation of fossils GSA Today 11 (2), 4–9

Bergstrom, J, 2001 Chengjiang In Briggs, D E G, Crowther, P R (Eds ), Palaeobiology, vol II Blackwell, Oxford, pp 337–340

Bottjer, D J, Hagadorn, J W, Dornbos, S Q, 2000 The Cambrian substrate revolution GSA Today 10 (9), 1–7

Biasier, M D, McIlroy, D, 1998 Neonereites uniserialis from c 600Ma year old rocks in western Scotland and the emergence of animals J Geol Soc (Lond ) 155, 5–13

Briggs, D E G, Erwin, D H, Collier, F J, 1994 The Fossils of the Burgess Shale Smithsonian Institution Press, Washington

Chen, J, Zhou, G, 1997 Biology of the Chengjiang fauna In Chen, J, Cheng, Y, Van Iten, H (Eds ), The Cambrian Explosion and the Fossil Record National Museum of Natural Science, Taichung, Taiwan, pp 11–106

Conway Morris, S, 1985 The Middle Cambrian metazoan Wiwaxia corrugata (Matthew) from the Burgess Shale and Ogygopsis Shale, British Columbia, Canada Philos Trans R Soc Lond, B 307, 507–582

Conway Morris, S, 1998 The Crucible of Creation Oxford University Press, Oxford

Crimes, T P, 1992 The record of trace fossils across the Proterozoic–Cambrian boundary In Lipps, J, Signor, P W (Eds ), Origin and Early Evolution of the Metazoa Plenum Press, New York, pp 177–202

Dornbos, S Q, Bottjer, D J, 2000 The evolutionary paleoecology of the earliest echinoderms helicoplacoids and the Cambrian substrate revolution Geology 28 839–842

Dornbos, S Q, Bottjer, D J, Chen, J Y, 2004 Evidence for seafloor microbial mats and associated metazoan lifestyles in Lower Cambrian phosphorites of Southwest China Lethaia 37, 127–137

Droser, M L, 1987 Trends in depth and extent of bioturbation in Great Basin Precambrian–Ordovician strata, California, Nevada, and Utah PhD thesis, University of Southern California

Droser, M L, Bottjer, D J, 1986 A semiquantitative classification of ichnofabric J Sediment Petrol 56, 558–559

Droser, M L, Bottjer, D J, 1988 Trends in depth and extent of bioturbation in Cambrian carbonate marine environments, western United States Geology 16, 233–236

Droser, M L, Gehling, J G, Jensen, S, 1999 When the worm turned concordance of Early Cambrian ichnofabric and trace-fossil record in siliciclastic rocks of South Australia Geology 27, 325–328

Droser, M L, Jensen, S, Gehling, J G, Myrow, P M, Narbonne, G M, 2002 Lowermost Cambrian ichnofabrics from the Chapel Island Formation, Newfoundland implications for Cambrian substrates Palaios 17, 3–15

Ekdale, A A, Muller, L N, Novak, M T, 1984 Quantitative ichnology of modern pelagic deposits in the abyssal Atlantic Palaeogeogr Palaeoclimatol Palaeoecol 45, 189–223

Hagadorn, J W, Bottjer, D J, 1999 Restriction of a Late Neoproterozoic biotope suspect-microbial structures and trace fossils at the Vendian–Cambrian transition Palaios 14, 73–85

Lefebvre, B, Fatka, O, 2003 Palaeogeographical and palaeoecological aspects of the Cambro-Ordovician radiation of echinoderms in Gondwanan Africa and peri-Gondwanan Europe Palaeogeogr Palaeoclimatol Palaeoecol 195, 73–97

Leslie, S A, Babcock, L E, Grunow, A M, Sadowski, G R, 2001 Paleobiology and paleobiogeography of Corumbella, a late Neoproterozoic Ediacaran-grade organism North American Paleontological Convention Program and Abstracts Paleobios 21, 83–84

Luo, H L, Jiang, Z W, Wu, X C, Song, X L, Ouyang, L, 1984 Sinian–Cambrian Boundary Stratotype Section at Meishucun, Jinning, Yunnan, China Yunnan Renming Publishing House, Kunming

McIlroy, D, Logan, G A, 1999 The impact of bioturbation on infaunal ecology and evolution during the Proterozoic–Cambrian transition Palaios 14, 58–72

Narbonne, G M, Kaufman, A J, Knoll, A H, 1994 Integrated chemostratigraphy and biostratigraphy of the Windermere Supergroup, northwestern Canada implications for Neoproterozoic correlations and the early evolution of animals Geol Soc Amer Bull 106, 1281–1282

Parsley, R L, Prokop, R J, 2001 Functional morphology and paleoecology of Middle Cambrian echinoderms from marginal Gondwana basins of Bohemia Geological Society of America Annual Meeting Abstracts with Programs, vol 33, p 247

Parsley, R L, Zhao, Y L, 2002 Eocrinoids of the Middle Cambrian Kaili biota, Taijiang County, Guizhou, China Geological Society of America Annual Meeting Abstracts with Programs, vol 34, p 81

Pfluger, F, Gresse, P G, 1996 Microbial sand chips—a non-actualistic sedimentary structure Sediment Geol 102, 263–274

Rasetti, F, 1954 Internal shell structures in the Middle Cambrian gastropod Scenella and the problematic genus Stenothecoides J Paleontol 28, 59–66

Rigby, J K, 1986 Sponges of the Burgess Shale (Middle Cambrian), British Columbia Palaeontogi Can 2, 1–105

Schieber, J, 1999 Microbial mats in terrigenous clastics the challenge of identification in the rock record Palaios 14, 3–12

Seilacher, A, 1999 Biomat-related lifestyles in the Precambrian Palaios 14, 86–93

Seilacher, A, Pfluger, F, 1994 From biomats to benthic agriculture a biohistoric revolution In Krumbein, W E (Ed ), Biostabilization of Sediments Bibliotheks and Informationsystem del Carl von Ossietzky Universitat, Oldenburg, pp 97–105

Sprinkle, J, Guensburg, T E, 1995 Origin of echinoderms in the Paleozoic evolutionary Fauna the role of substrates Palaios 10, 437–453

Thayer, C W, 1975 Morphologic adaptations of benthic invertebrates to soft substrata J Mar Res 33, 177–189

Thayer, C W, 1983 Sediment-mediated biological disturbance and the evolution of marine benthos In Tevesz, M J S, McCall, P L (Eds ), Biotic Interactions in Recent and Fossil Communities Plenum Press, New York, pp 479–625

Yochelson, E L, Gil Cid, M D, 1984 Reevaluation of the systematic position of Scenella Lethaia 17, 331–340

Available online at www.sciencedirect.com

Palaeogeography, Palaeoclimatology, Palaeoecology 220 (2005) 69–88

www.elsevier.com/locate/palaeo

# Palaeoecology of the Early Cambrian Sinsk biota from the Siberian Platform

Andrey Yu. Ivantsov[a], Andrey Yu. Zhuravlev[b,*], Anton V. Leguta[a],
Valentin A. Krassilov[a], Lyudmila M. Melnikova[a], Galina T. Ushatinskaya[a]

[a]*Palaeontological Institute, Russian Academy of Sciences, ul. Profsoyuznaya 123, Moscow 117997, Russia*
[b]*Área y Museo de Paleontología, faculdad de Ciences, Universidad de Zaragoza, C/ Pedro Cerbuna, 12, E-50009, Zaragoza, Spain*

Received 1 February 2002, accepted 15 January 2004

## Abstract

The Sinsk biota (Early Cambrian, Botoman Stage, Siberian Platform) inhabited an open-marine basin within the photic zone, but in oxygen-depleted bottom waters. Its rapid burial in a fine-grained sediment under anoxic conditions led to the formation of one of the earliest Cambrian Lagerstatte. All the organisms of the biota were adapted to a life under dysaerobic conditions. It seems possible that the adaptations of many Cambrian organisms, which composed the trophic nucleus of the Sinsk Algal Lens palaeocommunity to low oxygen tensions allowed them to diversify in the earliest Palaeozoic, especially during the Cambrian. Nowadays these groups comprise only a negligible part of communities and usually survive in settings with low levels of competition. Nonetheless, the organization of the Algal Lens palaeocommunity was not simple, it consisted of diverse trophic guilds. The tiering among sessile filter-feeders was well developed with the upper tier at the 50 cm level. In terms of individuals, the community was dominated by sessile filter-feeders, vagrant detritophages, and diverse carnivores/scavengers. The same groups, but in slightly different order, comprised the bulk of the biovolume. vagrant epifaunal and nektobenthic carnivores/scavengers, sessile filter-feeders, and vagrant detritophages. The Algal Lens and Phyllopod Bed (Burgess Shale) Lagerstatten share a number of common features including a representativeness of certain groups, a relative percentage of fauna in terms of individuals and biovolumes, feeding habits, and substrate relationships.
© 2005 Published by Elsevier B.V.

*Keywords:* Palaeoecology, Early Cambrian, Individual abundance, Biovolume, Trophic nucleus

## 1. Introduction

Siberian Platform sites containing organisms of exceptional preservation are confined to the outer shelf to slope-basin facies of the Yudoma-Olenek basin (Fig. 1). The basin faced the open ocean and to a

* Corresponding author
  *E-mail address:* ayzhur@mail.ru (A.Y. Zhuravlev)

0031-0182/$ - see front matter © 2005 Published by Elsevier B.V.
doi:10.1016/j.palaeo.2004.01.022

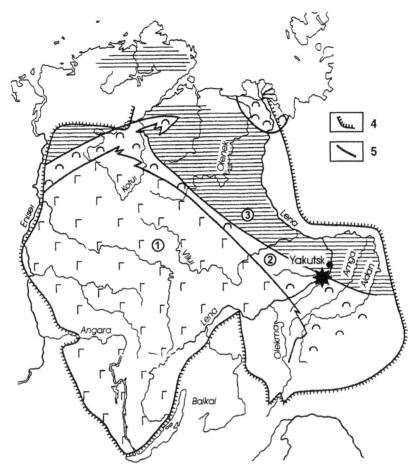

Fig 1 Palaeogeographic map of the Siberian Platform in the Early Cambrian Asterisk marks the Algal Lens Sinsk Lagerstatte (1) Turukhans-Irkutsk-Olekma saliniferous basin (inner shelf), (2) Anabar-Sinsk reefal belt, (3) Yudoma-Olenek basin (outer shelf), (4) boundary of the Siberian Platform, (5) major facies boundaries (modified after Savitskiy and Astashkin, 1979)

certain extent was comparable to the outer detrital belt of Laurentia famous of its Burgess-Shale-type localities Excavations of the Early Cambrian Sinsk Formation in the middle courses of the Lena River brought out a large number of relatively diverse and well-preserved fossils (Ivantsov et al , 1996, Ivantsov, 1998, 1999, Melnikova, 1998, 2000) The principal localities occur on the right bank of the Lena River in the vicinity of the mouths of the Achchagyy-Tuoydakh and Ulakhan-Tuoydakh rivers

The main fossiliferous horizon is restricted to a calcareous shale bed at the lower Sinsk Formation of 0 5 m in thickness (Fig 2) The shale is dark brown, thinly laminated with calcitic and clayey laminae Only organisms with rigid enough covers (envelopes

and cuticles) such as seaweeds, palaeoscolecidans, some cnidarians, arthropods and spicular sponges are preserved The Sinsk biota does not represent a typical Burgess-Shale-type preservation which often keeps tiny details of soft integuments Nonetheless, the Sinsk localities should be attributed to Lagerstatten because they yield forms which are not usually fossilized and which contribute to a disproportionate amount and quality of palaeobiological information

The age of the Sinsk biota is *Bergeroniellus gurarii* Zone of the Early Cambrian Botoman Stage (Astashkin et al , 1990) This biota is among the oldest soft-bodied Early Cambrian together with the late Atdabanian Sirius Passet fossils of Greenland and early Botoman Chengjiang fauna of China (Conway

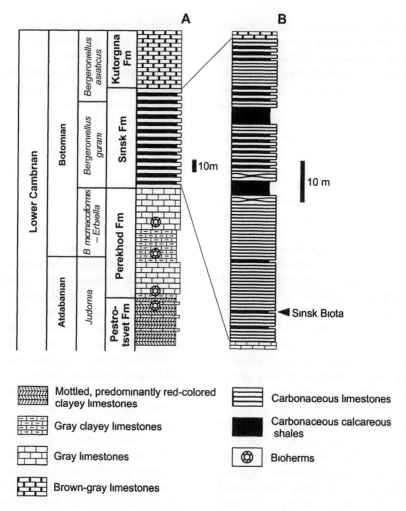

Fig 2 Sections of the Sinsk Formation on the right bank of the Lena River middle courses (A) Generalised section of the western Yudoma-Olenek basin, (B) section of the Sinsk Formation located 2 5 km downstream the Achchagyy-Tuoydakh River mouth, the level of the Sinsk Lagerstatten is indicated

Morris et al , 1987, Chen et al , 1989) The latter is approximately coeval with the Sinsk biota

Despite of an apparent heterogeneity in fossil composition among different Sinsk sites (Algal Lens, Tuoydakh, and others) it is obvious that they represent oryctocenoses resulting from a catastrophic burial of organisms composing natural communities Thus, Sinsk localities and especially the Algal Lens Lagerstatte provide a rare opportunity to reconstruct an ancient community in detail

The Algal Lens oryctocoenosis itself supplies the principal data set of this study because it is the richest of the Sinsk Lagerstatten It occurs on the right bank of the Lena River, 300 m downstream the Ulakhan-Tuoydakh River mouth The total excavated area encircles over 5 m$^2$ The locality yields cyanobacteria, algae, sponges, cnidarians, brachiopods, arthropods, and other fossil groups

Many typical Cambrian groups of animals preserved in the Algal Lens Lagerstatte became rare, including cephalorhynch worms, tardipolypodians, bradoriids, coeloscleritophorans, and eldonioideans The cephalorhynchs comprise a phylum of primitive worms which include modern priapulids, loriciferans, kinorhynchs, and nematomorphs as well as a number of fossil groups of class rank, such as the Ancalogonida and Louisellida (Malakhov and Adrianov, 1995) Of these, only louisellids (one species) are present in the Sinsk Lagerstatten. Palaeoscolecidans also bear a number of characteristics typical of cephalorhynchs, including cuticular structure, features of the proboscis, alimentary tract, and caudal end (Conway Morris and Robison, 1986, Barskov and Zhuravlev, 1988, Marss, 1988, Muller and Hinz-Schallreuter, 1993, How and Bergstrom, 1995, Zhang and Pratt, 1996; Conway Morris, 1997) The entire set of features is consistent with features represented with the phylum Cephalorhyncha, to which palaeoscolecidans are assigned as a class (Conway Morris and Robison, 1986) Palaeoscolecidans are diverse in the Sinsk Lagerstatten, with five species from only the Algal Lens oryctocoenosis

Tardipolypodians are ancient lobopodian animals often compared with onychophorans or tardigrades, but probably represent a separate extinct phylum (Chen and Zhou, 1997). Bradoriids, which were once considered to be ostracode-like crustaceans, are now classified as several independent groups of primitive bivalve arthropods due to structure of their appendages (Hou et al, 1996) Coelosclentophorans are Cambrian in age, sclerite-bearing animals which can be a stem-group of brachiopods, mollusks, and annelids (Bengtson and Missarzhevsky, 1981, Conway Morris and Peel, 1995) The Sinsk Lagerstatten contains two different coelosclentophorans· a sessile radially symmetrical calcareous spiny chancellorid, and vagrant bilaterally symmetrical organic-sclerital wiwaxiid Eldonioideans can represent either a stem-group of radial-symmetrical planktic deuterostomians or extremely deviated lophophorates (Conway Morris, 1993, Dzik et al, 1997, Friend et al, 2002)

The material under discussion is housed in the Palaeontological Institute of the Russian Academy of Sciences (Moscow), collection PIN 4349

## 2. Environments

### 2 1 Burial facies

The combination of palaeomagnetic, sedimentological, and geochemical data revealed that the marine basin inhabited by Sinsk communities occurred at low latitudes and was characterized by a humid warm climate (Savitskiy et al, 1972; Savitskiy and Astashkin, 1979, Nikolaeva, 1981, Smethurst et al, 1998)

The early-middle Botoman Sinsk Formation consists of mostly dark bituminous platy to flaggy, thinly laminated, variously argillaceous limestones Both slump bedding and breccia are observed in places, as well as unidirectional current marks and aligned trilobite carapaces Dark bituminous limestones are composed of clotted to finely peloidal varieties often altered to a neomorphic microcrystalline limestone Gradational sorting of primary peloids and disseminated organic matter is well expressed Microclotted limestone of the Algal Lens locality contains relatively less organic matter and argillaceous admixture than average Sinsk Formation rocks The limestones are mostly calcisiltites and calcilutites consisting of redeposited aleuritic size peloids, the texture of which depends entirely on the degree of recrystallization and the amount of carbonaceous and argillaceous admixture

The fine-grained sediment was likely accumulated under calm conditions below storm wave base The complete absence of bioturbation as well as high organic and pyritic content further reveal dysaerobic conditions developed within the sediment and near-bottom water column (Byers, 1977; Coniglio and Dix, 1992, Anderson et al, 1994) The gradational structure reflects the deposition from suspension Each layer accumulated rapidly during a single, short depositional event. The bedding structures, especially the bed foot texture indicate a deposition resulting from a loading of unidirectional fading flows Judging by a low thicknesses and sharp erosional contacts at bed foots, with a gradational sorting and thin (aleuritic) composition of fabrics, low density suspension flows are inferred as the source of deposits (Eberli, 1991, Piper and Stow, 1991) The fine (submillimetric) horizontal lamination is also typical of such conditions

X-ray fluorescence microscopy of Sinsk limestones has revealed an enrichment in minor and trace elements, which being corrected by elements to Al $(10^4)$ ratio, reach the following values· V=176, As= 32, Cr=19, Cu=67, Ni=32 Such geochemical manifestations imply oxygen-depleted conditions during the accumulation of the Sinsk Formation and may be taken into account for the explanation of the preservation of the Sinsk Lagerstatten (Zhuravlev and Wood, 1996) Anoxia might limit bioturbation and scavenger activity which resulted in the excellent preservation of rigid carcasses, but certainly was not a strong enough factor to prevent a rapid enzymatic degradation of organic structures (Allison, 1988, Butterfield, 1990) The preservation of microscopical details of cephalorhynch and tardipolypodian muscle fibres and of lingulate mantle cells in Sinsk phosphate replicas also infer tissue decay under oxygen-depleted conditions (Hof and Briggs, 1997)

Bearing in mind the distal position of the Sinsk Formation, the aforementioned strata are interpreted as microturbidites which were deposited from low density turbid flows generated by storms within relatively shallow water conditions, transported down-slope, and accumulated below storm wave base under dysaerobic conditions. The peloids were probably the products of destruction of calcimicrobial reefs, and larger fragments of such reefs are common in the underlying Perekhod Formation representing more shallow water facies.

In summary, the Sinsk Lagerstatten are among the typical Lagerstatten, where the preservation of organic remains is due to the following factors (1) the presence of tough, resistant cuticles, (2) a dysoxic milieu of accumulation, and (3) rapid burial in (4) a fine grained sediment (Fig 3)

## 2 2 Pre-burial facies

Despite the dissimilar position of Cambrian low latitude Lagerstatten within basin in which they were formed (Siberian and Laurentian—at an open shelf, Chengjiang—at an open shelf influenced in places by estuarine or fluvial sources, Paseky—in a brackish lagoon, and Emu Bay—probably, within a closed basin), they preserved comparable biotas (Piper, 1972, Chlupáč, 1995; Lindstrom, 1995, Elrick and Hinnov, 1996, Nedin, 1997; Babcock and Zhang, 2001)

The abundance and diversity of seaweeds and calcified cyanobacteria are indicative of the photic zone Algae and cyanobacteria are particularly diverse in the Algal Lens and Chengjiang oryctocoenoses where they comprise over one-third of all fossils (Babcock and Zhang, 2001) These groups are numerous in the Phyllopod Bed of Burgess Shale as well (Conway Morris, 1986) In all three Lagerstatten, algae and cyanobacteria (mostly, Marpolia) are preserved as countless coverings Only the Sirius Passet and Emu Bay sites completely lack algae The Sirius Passet locality perhaps yields the deepest of similar biotas, because it does not express any features of transportation of its fossils (Vidal and Peel, 1993) On the contrary, the Emu Bay biota appeared to live under utterly unfavourable conditions of a shallow stagnant basin that resulted in its extremely low

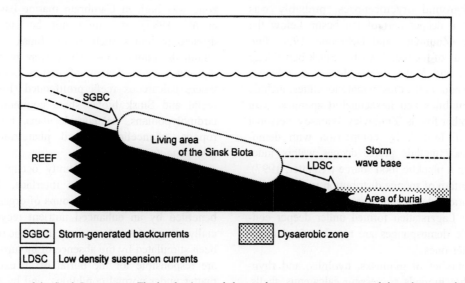

Fig 3 Taphonomy of the Sinsk Lagerstatten The benthic biota inhabiting photic zone were transported downslope into deeper dysaerobic environment

diversity (11 species), and in a complete absence of Cambrian animals typical of normal marine conditions (sponges, brachiopods, mollusks, hyoliths, and echinoderms; Nedin, 1995) The composition of the brackish Paseky Shale biota is inferior, where 3 arthropod species are present, in addition to some acritarchs and filamentous cyanobacteria *Marpolia* (Chlupáč, 1995, Fatka and Konzalová, 1995; Steiner and Fatka, 1996)

The presence of hexactinellids, brachiopods including the rhynchonellate one, as well as mollusks and hyoliths in the Sinsk Lagerstatten shows, that the communities preserved as Algal Lens and Tuoydakh oryctocenoses inhabited a normal marine basin

The modern seaweeds resembling the Sinsk algae are common under low intertidal to shallow subtidal conditions They grow on rocky shores within bays free of severe storms The restriction of the Sinsk living communities to the photic zone is followed from the presence of kunmingellids (bradoriids) with pronounced eyes (Shu et al, 1999) Additionally, all trilobites but miomeric *Delgadella*, as well as trilobitomorph *Phytophilaspis* possessed developed vision The abundance of green algae allows us to suggest that the biota rarely lived below a hundred metres, but rather inhabited depths of 30–40 m of shallow subtidal zone

In contrast to the Algal Lens palaeocommunity, the genuine biota of the Sinsk Formation, which comprises background oryctocenoses, probably was restricted to a deeper part of the basin, below the photic zone (Zhuravlev and Debrenne, 1996 Fig 1 3) Seawards of the reefal Anabar-Sinsk belt (along the Yudoma and Maya rivers), fossil assemblages depauperate and, in the most remote localities, include miomeric trilobites and hexactinellid sponges, with only a few other fossils (Zhuravlev, Ivantsov, personal observations) Indeed, by comparison with demosponges, hexactinellids prefer deeper waters independently on a planktic food source (Krautter, 1997). A similar pattern of sponge distribution is observed in Sinsk localities definite hexactinellids (*Lenica*) are restricted to Lagerstatten formed under deeper conditions while demosponges are typical of relatively shallow water ones

A low number of mollusks, hyoliths, and rhynchonellate brachiopods possessing calcareous shells, as well as infaunal filter-feeders and detritophages

probably imply dysaerobic conditions (Byers, 1977; Morris, 1979, Savdra and Bottjer, 1986) Spicular sponges, which are able to survive anoxia, are very common There is a high diversity and great abundance of cephalorhynch worms, including priapulids, and lingulate brachiopods are evident for low oxygen tensions because these animals rely upon hemerythrins for oxygen transportation (Runnegar and Curry, 1992) Brachiopods are very resistant to anoxic conditions and turbidity (Thayer, 1981; Tunncliffe and Wilson, 1988; Ushatinskaya, 2001) Priapulids burrow in the mud even within anaerobic milieu (van der Land, 1970) Besides, wide pleurae of many Sinsk trilobites are the feature of well-developed gills, which are necessary in environments lacking enough oxygen (Fortey and Wilmot, 1991) Plentiful Sinsk bradoriids bore thin, densely porous carapaces pierced by a condensed net of branching vessels They may have possessed paddle-like exopodites, which ventilated space under the carapace, especially in the vicinity of the main gas-exchange areas, crowded with vessels (Vannier et al, 1997; Shu et al., 1999) The entire set of bradoriid features indicates their adaptation to dysaerobic conditions Indeed, in the Upper Cambrian of Scania (Sweden) bradoriids (*Cyclotron* sp ) are restricted exclusively to strata accumulated at low oxygen levels (Clarkson et al, 1998)

If the upper boundary of the oxygen minimum zone was high in Cambrian marine basins (Wright et al., 1987), it would not be surprising that dysaerobic biotas such as the Sinsk one occurred within the photic zone These were biotops spreading out several dozens of meters below the belt where calcareous reefs proliferated Both adjacent reefal and Sinsk biotas share common species of tardipolypodians, palaeoscolecidans, bivalved arthropods, chancelloriids, and pharetronid sponges (*Dodecaactinella*)

The Algal Lens community occupied a biotop abutting aerobic/dysaerobic interface. This was a space favourable to some groups of organisms which benefited by an enhanced nutrient recycling, especially phosphates, nitrates, and iron The recycling has been stimulated by the absence of bioturbators, which are responsible for the burial of an excess organic matter under normal conditions, and by the presence of reducing conditions in the area of the biota burial (see above) Over 50% of primary production in such a milieu is formed by a microbial regeneration of nutrients (Huntley et al, 1991) Increased nutrient regeneration and recycling could maintain the proliferation of abundant and often monotypic producents in Sinsk palaeocommunities

The dominance of sessile filter-feeders (brachiopods, sponges, and chancellorid *Archiasterella*) in the Algal Lens palaeocommunity (Table 1, Figs 4, 5A and 6) infers permanent currents along Sinsk biotops, probably induced by storms and a slope gradient These currents could scour sea bottom sediments in such a way that only animals adapted for deeper burrowing survived, specifically cephalorhynchs The presence of inverted proboscis indicates an ability to employ a hydrostatic system for locomotion and thus, for burrowing in advanced cephalorhynch worms The same follows from an observation of developed

peripheral circular muscles of body-wall musculature, which are typical of burrowing worm-like invertebrates (Budd, 1998) The circular muscle fibers are organized in bundles of about 15–25 μm across and well preserved in Sinsk cephalorhynchs due to an early phosphatization The sclerite cover pattern also reveals palaeoscolecidan burrowing abilities One of species exhibits sclerites with hook-like nodes Another palaeoscolecidan species bore sclerites oriented by their sharp edges across the trunk. In both cases, sclerites were well-shaped for an anchoring in the sediment

The sea floor probably consisted of a mixture of calcareous sand and mud, loose enough for burrowers but lithified rapidly to provide a firm ground for seaweeds, pharetronid sponges, and large brachiopods Storm-generated currents provided food and some oxygen for Sinsk animals, while processes at the

Table 1
The composition of the Algal Lens oryctocoenosis

| Name | Affinity | N (specimens) | Biovolume | Feeding habit | Life habit |
|---|---|---|---|---|---|
| *Diagoniella* | sponge | 1 | 15 5 cm² | filter-feeder | sessile epibenthos |
| *Lenica* | sponge | 2? | 84 0 cm² | filter-feeder | sessile epibenthos |
| *Wapkia* | sponge | 4 | 10 0 cm² | filter-feeder | sessile epibenthos |
| Demosponge n gen | sponge | 3? | 18 0 cm² | filter-feeder | sessile epibenthos |
| *Cambrorhytium* | cnidarian | 4? | 0 3 cm² | filter-feeder? | sessile epibenthos |
| *Corallioscolex* | cephalorhynch | 2 | 1 8 cm² | carnivore/scavenger | burrowing infauna |
| Palaeoscolecid n gen 1 | cephalorhynch | 1 | 2 0 cm² | carnivore/scavenger | burrowing infauna |
| Louseillid n gen | cephalorhynch | 1 | 0 9 cm² | carnivore/scavenger | burrowing infauna |
| Palaeoscolecid n gen 2 | cephalorhynch | 3 | 2 2 cm² | carnivore/scavenger | burrowing infauna |
| Palaeoscolecid n gen 3 | cephalorhynch | 2 | 3 6 cm² | carnivore/scavenger | burrowing infauna |
| Palaeoscolecida | cephalorhynch | 1 | 4 2 cm² | carnivore/scavenger | burrowing infauna |
| *Microdictyon* | tardypolipodia | 1 | sclerite | carnivore/scavenger | vagrant epibenthos |
| *Xenusia* indet | tardypolipodian | 1 | 1 8 cm² | carnivore/scavenger | vagrant epibenthos |
| *Tardipolypoda* | tardypolipodian | 1 | 1 8 cm² | carnivore/scavenger | vagrant epibenthos |
| *Delgadella* | trilobite | ~10 | 1 3 cm² | filter-feeder? | vagrant epibenthos? |
| *Jakutus* | trilobite | 6 | 648 cm² | carnivore/scavenger? | vagrant epibenthos |
| *Edelsteinaspis* | trilobite | 5 | 26 9 cm² | ? | vagrant epibenthos |
| *Bergeroniaspis* | trilobite | 3 | 5 3 cm² | carnivore/scavenger? | vagrant epibenthos |
| *Bergeroniellus* | trilobite | ~10 | 28 6 cm² | carnivore/scavenger? | vagrant epibenthos |
| *Binodaspis* | trilobite | 1 | 0 3 cm² | ? | vagrant epibenthos |
| *Sinskolutella* | bradorid | 64 | 26 1 cm² | detritophag? | vagrant epibenthos |
| *Yakutingella* | bradorid | 3 | 1 1 cm² | detritophag? | vagrant epibenthos |
| *Phytophilaspis* | arthropod | 11 | 863 5 cm² | carnivore/scavenger? | vagrant epibenthos |
| *Eoobolus* | lingulate | 12 | 0 9 cm² | suspension-feeder | sessile epibenthos |
| *Linnarssonia* | lingulate | 120 | 1 8 cm² | suspension-feeder | sessile epibenthos |
| *Botsfordia* | lingulate | 37 | 1 4 cm² | suspension-feeder | sessile epibenthos |
| *Nisusia*? | rhynchonellate | 1 | 2 9 cm² | suspension-feeder | sessile epibenthos |
| *Wiwaxia* | wiwaxiid | 1 | 3 4 cm² | grazer? | vagrant epibenthos |
| *Archiasterella* | chancellorid | 3? | 36 cm² | suspension-feeder? | sessile epibenthos |
| *Eldonia* | eldonioidean | 39 | 477 8 cm² | carnivore? | nektobenthos? |

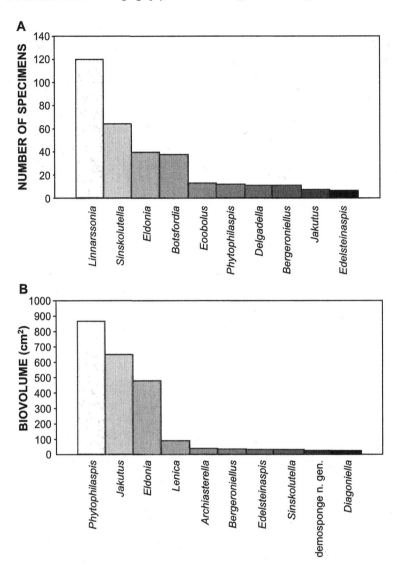

Fig 4  Relative percentages of different species in the Algal Lens oryctocoenosis  (A) Number of individuals, (B) biovolume

aerobic/dysaerobic interface supplied nutrients for abundant planktic and benthic producents (Fig 3)

## 3. Composition of the Algal Lens oryctocoenosis

The Sinsk Lagerstatten contain remains of all principal groups of organisms typical of Cambrian Lagerstatten which are seaweeds, spicular hexactinellids and demosponges, cnidarians, cephalorhynch worms, tardipolypodians, trilobites, bradoriids, other arthropods, coeloscleritophorans, and eldonioideans

The localities lack anomalocaridids, annelids, and echinoderms  Likewise, anomalocaridids rarely occur in the early Middle Cambrian Kaili Formation (South China), and echinoderms are absent from Sirius Passet, Chengjiang, and Emu Bay of South Australia which are the oldest Cambrian Lagerstatten  The absence of echinoderms can be related to their low diversification by that time  As to annelids, their relatively thin cuticle does preserve in extremely suitable conditions only (Briggs and Kear, 1993), which are not typical of the Sinsk Lagerstatten as well as of Sirius Passet, Kaili, and Emu Bay ones  Similar

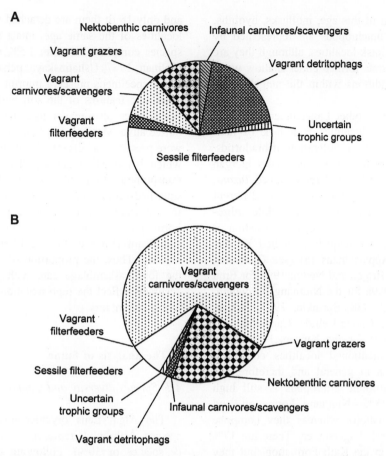

Fig. 5. Trophic nucleus of the Algal Lens palaeocommunity. (A) Individuals, (B) biovolume (in cm$^2$).

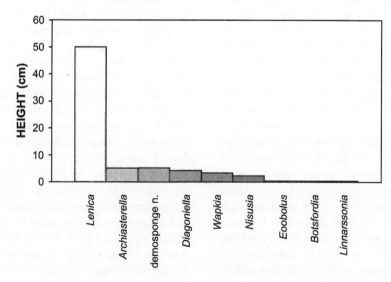

Fig 6  Tiering among epifaunal sessile filter-feeders of the Algal Lens palaeocommunity

to many Lagerstatten of this age, mollusks, hyoliths, and rhynchonellate brachiopods are represented by one–two species in Sinsk localities, although they are common in coeval strata formed under shallow water well-oxygenated conditions within the neighbouring Anabar-Sinsk belt

By comparison with other Lagerstatten, a noticeable feature of Sinsk localities is a lack of non-trilobite arthropods, and a proliferation of bradoriids There are only 4 such fossils there concilitergan trilobitomorph *Phytophilaspis*, large bivalved *Tuzoia*, and fragments of two other probably different arthropods A similar number of non-trilobite arthropods occur in Emu Bay (4), Niutitang Formation of South China (4), and inadequately studied Kinzers Formation of the Appalachians (5) (see Glaessner, 1979, Nedin, 1995, Briggs and Nedin, 1997 for Emu Bay, Zhao et al , 1999a for the Niutitang Formation, Resser and Howell, 1938; Sprinkle, 1973, Briggs, 1978, Rigby, 1987, Garcia-Bellido Capdevila and Conway Morris, 1999 for the Kinzers Formation) However, all aforementioned localities contain a relatively poor fauna in general and therefore, the proportion of non-trilobite arthropods is still high (40%—Emu Bay, 18%—Niutitang Formation, and 23%—Kinzers Formation), whereas they comprise only 11% in the Sinsk Lagerstatten There are 17% of such arthropods in the Kaili Formation, but they are not of good preservation Noteworthy, that both the Sinsk and Kaili Lagerstatten occur within dominantly carbonate facies while other Cambrian Lagerstatten are restricted to siliciclastics The argillaceous Paseky Shale (Czech Republic) formed under brackish conditions, and does not contain any trilobites at all (Chlupáč, 1995)

The total number of fossil animals in the Algal Lens oryctocoenosis is 353 counting out isolated spicules and sclerites (Table 1) Algae, cyanobacteria, and acritarchs are so plentiful that their remains are beyond the computation In places, algae completely cover an area over several square metres Intact skeletons of sponges, coeloscleritophorans, some trilobites, and *Phytophilaspis*, coiled cephalorhynch bodies, articulated brachiopod shells and bradoriid carapaces led us to suggest that these organisms were buried alive For instance, of 120 individuals of brachiopod *Linnarssonia rowelli*, 104 individuals (86 7%) represent complete shells with intact valves,

and only 16 of them are detached valves In ordinary localities of the same age, intact shells of the same species comprise less than 1 5% in a sampling unit (Pelman, 1977, Ushatinskaya, personal observations) Many specimens preserve imprints of the mantle and other fine features of the soft body structure These peculiarities also reveal that the brachiopods were buried alive Merely a few brachiopods and bradoriids were preserved as disarticulated shells and carapaces, which belonged to animals that either died before transportation to the place of their burial, or had been squashed during digestion by carnivores/scavengers In situ trilobite exuviae are common in the Sinsk Formation and their number in oryctocoenoses is significant However, the Algal Lens Lagerstatte lacks exuviae Thus, the proportion of trilobite remains in the fossil assemblage can, with a good degree of accuracy, reflect the representation of this group as a living community

## 4. The analysis of fauna

### 4 1 Species diversity and composition

The Algal Lens oryctocoenosis consists of 30 species Arthropods represent the most diverse group (9 species or 30%) Following arthropods are the Cephalorhynch worms (6 species or 20%), sponges, and brachiopods (4 species or 13 3% each) It should be noted that in the Algal Lens oryctocoenosis, as well as in other Cambrian Lagerstatten, genera are almost exclusively monospecific (Table 1, Fig 4)

By the species composition, the Algal Lens oryctocoenosis resembles the famous early Middle Cambrian Phyllopod Bed from the Burgess Shale (British Columbia) where arthropods (38%), sponges (16%), and cephalorynchs (6%) prevail similarly [Here and hereafter the Phyllopod Bed data are mostly from Conway Morris (1986) ] Problematic animals are also significant in the Phyllopod Bed (21%) In this paper, the 'problematic' label is assigned to tardipolypodians, coeloscleritophorans, cnidarians, and eldonioideans Altogether they comprise a comparable figure of 23 3% Taking into consideration the diversity of spicules which represent at least 7 more species of sponges, the structure of both oryctocoenoses appears to be systematically comparable

A similar share of the principal systematic groups is typical of other Early Cambrian Lagerstatten, though their sampling units seem to represent more time-averaged data on several different communities In Chenjiang (Haikou, Ercaicun locality, Yu'anshan Formation), arthropods constitute 43 5%, sponges—14 5%, and cephalorhynchs—10% Tardipolypodians—8 7%, chordates—4 3%, and brachiopods—4 3%, are also important [calculation of Wang et al.'s (2001) data] According to a recent computation by Babcock and Zhang (2001), arthropods account for more than three-quarter of 140 species and two-thirds of all specimens in the Chengjiang biota In the Kinzers Formation, the same three main groups give the number of 46 4% (arthropods), 3 6% (sponges), and 7 1% (cephalorhynchs), respectively (another 7 1% are of echinoderms) In Emu Bay, arthropods comprise 54 5%, anomalocaridids—27 3%, cephalorynchs—18 9% but the sampling unit itself is too small to provide a statistically significant figure The Middle Cambrian Kaili, Wheeler, and Marjum formations as well as the Spence Shale (the last three are in Utah), arthropods share from 42% to 57%, sponges—from 5% to 26%, and cephalorhynchs—from 2% to 6% (see Mao et al, 1992, Huang et al, 1994; Zhao et al, 1996, 1999b for the Kaili Formation, Robison, 1990, 1991, Robison and Wiley, 1995, Rigby et al, 1997, Sumrall and Sprinkle, 1999 for the Wheeler and Marjum formations and Spence Shale) The species composition of brachiopods (10–12%) and echinoderms (6–8%) becomes also weighty in the Middle Cambrian Lagerstatten

The principal difference of the Algal Lens and Phyllopod Bed oryctocoenoses lies in the species diversity of trilobites They are divers in the Algal Lens Lagerstatte but not variable in the Phyllopod Bed, neither in Sirius Passet, Chenjiang, nor in Emu Bay assemblages

### 4 2 Individual abundance

Lingulate brachiopods are the most abundant group within the Algal Lens oryctocoenosis (169 specimens or 47 9%) Subsequent groups are arthropods (113 specimens or 32%) [namely, bradoriids (67 specimens or 19%), trilobites (35 specimens or 9 9%), and *Phytophilaspis* (11 specimens or 3 1%)], eldonioideans (39 specimens 11%), cephalorynchs, and

sponges (10 specimens or 2 8% each) Bearing in mind the proportion of alive and dead by the time of burial individuals, one can draw even a larger brachiopod share in the natural community (Table 1, Figs 4B, 5A and 6)

In the Phyllopod Bed oryctocoenosis, arthropods lead in the number of individuals (58%). They are followed by hemichordates (17%), mollusks including hyoliths (12%), and cephalorhynchs (5%) (Briggs and Whittington, 1985, Conway Morris, 1986) Brachiopods comprise 3% only of individuals, and eldonioideans counting together with echinoderms—about 2% In Chengjiang (Haikou and Anning sites of 186 and 319 individuals in total, respectively), again arthropods form the majority of individuals (48–50%) bradoriids (27–23%), trilobites (2–21%), and others (19–7%) Other significant groups are cephalorhynchs (22–2%), brachiopods (4–24%), anomalocaridids and eldonioideans (about 1% each) (Zhang et al, 2001) Some similarity of the Algal Lens, Phyllopod Bed, and Chengjiang oryctocoenoses in the species abundance is observed in the significance of arthropods, but neither bradoriids, nor trilobites comprise a high share of the Phyllopod Bed arthropods In terms of arthropod species abundance, some Chengjiang oryctocoenoses are much closer to that of almost coeval Algal Lens· bradoriids encompass over 70% of individuals representing 7 4% only in the generic diversity (How and Bergstrom, 1991, Babcock and Zhang, 2001) In Algal Lens, these tiny bivalve arthropods encompass almost 20% of individuals in total while providing 6 7% only of the generic diversity (2 species of 2 genera) (Figs 4A,B and 6) However, trilobites are scarce in Chengjiang—1 5% of preserved arthropods only (Babcock and Zhang, 2001)

### 4 3 Biovolume

When a Recent community is analyzed, a relative biomass of each group is examined in order to measure energy flows along a trophic web It is difficult to estimate a biomass of long vanished bodies in a palaeocommunity, but some approximation is possible with use of data on biovolume (e g , Conway Morris, 1986) Conway Morris computed a biovolume of each genus as the average length of an individual multiplied by the number of individual specimens with use of formulae for standard shapes Here, the biovolume is calculated for each species Two measurements have been used for every individual. the diameter and height—for sessile forms, or the width and length—for vagrant ones The sum of these figures is expressed in square centimetres (Table 1 and Figs 4B, 5B and 6) The number of mature and juvenile individuals is counted separately because they differ significantly in size

Arthropods dominate by biovolume in the Algal Lens oryctocoenosis—70 5% Other voluminous groups are eldonioideans (21%), sponges (5 6%), coeloscleritophorans (1 7%), and cephalorhynchs (0 6%)

The Phyllopod Bed fossils make up a comparable pattern 49% of arthropods, 17% of sponges, about 15% of eldonioideans, and 10% of cephalorhynchs In both Phyllopod Bed and Algal Lens arthropod biovolumes, trilobites comprise a minor part

## 5. Palaeoecological characteristics of the Algal Lens community

### 5 1 Ethological-trophic groups

Several producent groups are widespread in the Algal Lens oryctocoenosis Among acritarchs, two monotypic assemblages are recognized These assemblages consist of either diminutive (less than 20 μm in diameter) spiny forms resembling *Micrhystridium* (Zhuravlev and Wood, 1996) or relatively large (from 50 to 300 μm in diameter) spherical smooth envelops of *Leiosphaeridia*-group acritarchs The same two major groups of phytoplankters (small acanthomorphs and large sphaeromorphs) are recognized in the Phyllopod Bed and Kaili Formation oryctocoenoses (Conway Morris, 1986; Yin and Yang, 1999) (Table 1, Fig 5A,B)

Benthic producents include abundant seaweeds (4 genera and 4 species) and filamentous cyanobacteria (*Marpolia*) Relatively large fronds of green siphoneous algae and their rhizoids dominate in number Another representative of the same group possesses a smaller bushy thallus with slender, dichotomous-fastigated branches and filiform terminal branches Only a relatively rare species which has an articulate frond is conditionally affiliated to red algae Addi-

tionally, calcified cyanobacteria (*Obruchevella*) are common

Several trophic levels can be distinguished among consumers The first level or primary consumers are represented by filter-feeders, detritophages, and possible grazers and browsers Secondary and even higher level consumers (carnivores/scavengers) can be inferred

Miscellaneous guilds of filter-feeders are represented by fine filter-feeders or filter-feeders sensu stricto, namely abundant hexactinellids, demosponges, and pharetronids, as well as by coarse filter-feeders or suspension-feeders, which are lingulate brachiopods and possibly chancelloriids However, the latter lacks the distinct features of filter-feeders (Bengtson and Hou, 2001) The examples of chancelloriid scleritomes with a bent basal part indicate that these animals have been anchored in a soft mud (Mehl, 1996) Immobile chancelloriid life style is confirmed also by an in situ immuration of their scleritomes with early marine cements (Zhuravlev, 2001)

Miniature lingulate brachiopods were attached to algal thalli as an attempt to gain levels of stronger water currents and to escape mud clogging Such settlements (*Eoobolus* brachiopod on *Margaretia* green siphoneous algae) are recognized in other Sinsk Lagerstatten, but are absent from the Algal Lens.

Cnidarians (*Cambrorhytium*) could be either tiny predators or filter-feeders like modern corals of a similar size (Sorokin, 1990). *Cambrorhytium* commonly occurs in dense settlements along ribbon-like structures possibly of algal origin The epiphytism on seaweeds is indeed common among modern small cnidarians, mostly hydrozoans, with a short life span which escaped hostile soft mobile substrata (Morri et al , 1991)

The zooplankton, which was the food source for some filter-feeders, consisted of abundant trilobite and lingulate brachiopod larvae at least as well as problematic embryos

The presence of several filter-feeding guilds was expressed in distinct tiers, where the uppermost tier (above 25 cm) was occupied by hexactinellid *Lenica*, intermediate tiers (from 3 to 5 cm) were deployed by two demosponge species (*Wapkia* and new one), by hexactinellid *Diagoniella*, and by chancelloriid *Archiasterella*, while the lowermost tier (below 1 cm) was lined up by lingulate brachiopods and by *Cambrorhytium* (Fig 6) The latter two groups were able to move to an upper tier by attaching themselves to algal thalli The majority of Sinsk brachiopods were epifaunal anchored forms by their life habits (Ush-atinskaya, 2001)

Unquestionable browsers and grazers are very rare in Cambrian strata Mobile epifaunal coelosclerito-phoran *Wiwaxia* possessing a radula probably grazed bacterial scum on algal thalli (Conway Morris, 1985). The few gastropods (*Nomgoliella*? sp ) were browsers (Kruse et al , 1995) Their remains are not distin-guished in Algal Lens itself, but such shells are very fragile and are etched from the Sinsk Formation with difficulty

Bradoriids (at least kunmingellids) and miomeric and some polymeric trilobites are assigned here to detritophages Dumpy bradoriids moved on postan-tennular appendages and stirred up the sediment surface in order to capture edible particles by the frontal postantennular spiny appendages (Hou et al , 1996, Shu et al , 1999) The natant hypostome of miomeric eodiscid trilobites implies a somewhat benthic life style and feeding on detritus including algal debris (Fortey and Owens, 1999)

Palaeoscolecidans and louisellid were probably predators like modern large cephalorhynchs and Cambrian ones furnishing with an armed spiny proboscis (Conway Morris, 1977b, 1986, Malakhov and Adrianov, 1995)

Tardipolypodians could be either microphagous predators or scavengers because they lacked any jaws or teeth, but the majority of them were discovered in usual accumulations with other animals, such as sponges, worms, and eldonioideans (Conway Morris, 1977a, Whittington, 1978) *Microdictyon* was con-fined strictly to *Eldonia* (Chen et al , 1995a,b) None of 70 individuals of this tardipolypodian show an alimentary tract filled with mud, as is typical of detritivorous animals (Chen et al , 1995b).

After etching with a weak acetic acid, sponge spicules and particulated brachiopod shells have been distinguished in the middle part of the gut of a single trilobite individual ascribed to *Bergeroniellus spino-sus* There is some regularity in the arrangement of the scraps. broken spicules alternate with brachiopod shells along the gut One can suggest that the trilobite was able to select the food and macerate solid

particles Polymeric trilobites and *Phytophilaspis* from Algal Lens commonly preserve a narrow stripe following the rachis Its width is usually comprised of one fifth of the rachis width, but varies along the rachis bulging transversely up its entire width In some individuals, the stripe can be traced along the entire thorax to the pygidium The axial position, linearity, almost constant width, and sharp boundaries of this structure are typical of an arthropod alimentary tract

The hypostome and bases of anterior limbs in *Phytophilaspis* are anomalous in size (Ivantsov, 1999 Figs 3 and 4B) The presence of hypostome suggests an orientation of the mouth rearwards The trans-portation of the food to the mouth was maintained by a metachronic rhythm of appendages The data on other Cambrian arthropods with well-developed appendages and hypostome are evident for a predator propensity of *Phytophilaspis* (cf Briggs and Whit-tington, 1985, Fortey and Owens, 1999) Trilobite *Bergeroniaspis lenaica* preserves undissolved isomet-ric phosphatic structures under the glabella, some of which can be the stomach content Usually a straight gut and an absence of mud infill in it are typical of arthropods preferring fleshy food (Chatterton et al , 1994, How and Bergstrom, 1997) In such a case, the phosphatic nodules could originate from highly chemically reactive organic matter rich in phosphate (Briggs and Whittington, 1985, Butterfield, 2002)

Altogether, these features allow us to consider *Bergeroniellus*, *Phytophilaspis*, and to a certain extent *Bergeroniaspis* as possible predators and/or scav-engers, although in the absence of appendages, it is difficult to find out their exact trophic orientation

Many have suggested a planktic life style for eldonioideans and closely related paropsonemids (Clarke, 1900, Durham, 1974, Stanley, 1986), and Conway Morris (1986) emphasized the significance of *Eldonia* among nektobenthos of the Phyllopod Bed community Recently Dzik (1991) and Dzik et al (1997) judged that, by their anatomy (inflexible aboral cover, absence of any organs suitable for floating) and by the restriction of epibiont settlements to the oral surface margin of *Rotadiscus* from the Kaili For-mation, implied a passive position of this animal on organic-rich mud However, this interpretation does not match well to cosmopolitan distribution of eldonioideans without any restriction to certain facies

They are equally widespread as in relatively deep-water siltstones and mudstones as in coarse-grained shallow-water sandstones (van der Meer Mohr and Okulitch, 1967, Masiak and Zylińska, 1994, Dzik et al , 1997) Besides, they are concurrent only with other representatives of pelagic fauna (Conway Morris, 1979, Friend et al , 2002), while a distinct asymmetry of the catching organ does not fit well with the anatomy of a sedentary, radially symmetrical animal. Nektobenthic life style is preferable for eldonioideans They could live in the water column at the density interface where food was plentiful Under such conditions, epibionts would still fix upon margins of floating disc facing the upper water layer rich in oxygen whereas the rigid lower part maintained the animal buoyant. Miomeric trilobites would be a suitable food for *Eldonia* then

In summary, sedentary filter-feeders (53% of individuals), mobile epibenthic detritophages (19%), as well as nektobenthic eldonioideans (11%), vagrant epibenthic (9 4%) and infaunal (2 8%) predators/scavengers, and mobile filter-feeders (2 8%) formed the Algal Lens community in terms of individuals (Fig 5A) The same guilds comprised also the bulk of biovolume, though in a slightly different order (Fig 5B)· vagrant epibenthic (68 2% of biovolume) and nektobenthic predators/scavengers (21%), sessile filter-feeders (7 5%), and mobile epibenthic detritophages (1 2%) The data on the Phyllopod Bed palaeocommunity gives a similar figure (Conway Morris, 1986) Among individuals, vagrant (64%) and sessile epibenthos (30%, *Cambrorhytium* was incorrectly assigned to attached infauna) as well as mobile infauna (6%) dominated By biovolume, vagrant (51%) and sessile (22%) epibenthos, nektobenthos (18%, eldonioideans only), and mobile infauna (10%) prevailed

## 5 2 Trophic nucleus

The trophic nucleus concept was developed by Neyman (1967) and successfully applied to the analysis of the Phyllopod Bed oryctocoenosis by Conway Morris (1986) According to this concept, the trophic nucleus consists of species comprising over 80% of the community biomass In terms of biovolume of the Algal Lens oryctocoenosis, these species are *Phytophilaspis* (vagrant epibenthic pre-

dator/scavenger—38%), trilobite *Jakutus* (vagrant epibenthic predator/scavenger—28 5%), *Eldonia* (nektobenthic predator—21%), hexactinellid sponge *Lenica unica* (sessile epibenthic filter-feeder—3 7%), chancellorid *Archiasterella* (sessile epibenthic filter-feeder—1 6%), and trilobite *Bergeroniellus* (vagrant epibenthic predator/scavenger—1 3%) In terms of individuals, the bulk of the oryctocoenosis consists of lingulate brachiopods *Linnarssonia* (34%), *Botsfordia* (10 5%), and *Eoobolus* (3 4%) (sessile epibenthic filter-feeders), bradoriids *Sinskolutella* (vagrant epibenthic detritophag—18 1%), *Eldonia* (11%), and *Phytophilaspis* (3 1%) In both accounts, 6 species only, although different, represent the trophic nucleus of the palaeocommunity Individually the trophic nucleus consists of epibenthic filter-feeders while volumetrically it is formed by predators, mostly because of the inclusion of the largest fossils (*Phytophilaspis* and *Jakutus* up to 50 cm and 15 cm in length, respectively) However, the number of predators can be overestimated due to the presence of exuviae among fossils. Independently of this fact, the trophic nucleus was homogenous (filter-feeders and predators) and, therefore, the community was low structuralized and highly dominant (Fig 5A,B)

## 5 3 Trophic web

Two producent groups built the base of the trophic web of the Algal Lens community Among planktic producents, these were "net" phytoplankton (acritarchs) and, probably, bacterioplankton (free-living and attached bacteria) The sponge proliferation depends on the bacterioplankton abundance (Reiswig, 1971, Wilkinson et al , 1984) Abundant, although low diverse lingulate brachiopods and their planktic larvae, on the contrary seemingly preferred acritarchs representing large "net" phytoplankton, which is the main phosphate source for their shell growth (Chuang, 1959, Pan and Watabe, 1988). Other filter-feeders (chancellorids and miomeric trilobites) could like a catholic diet (Fig 7)

A reconstruction of trophic chain beginning from benthic producents is more difficult Single *Wiwaxia* and molluscs only are interpreted as grazers and browsers Possibly, the bulk of dead phytoplankton and phytobenthos was directed strictly to detritus because large accumulations of acritarchs and algae

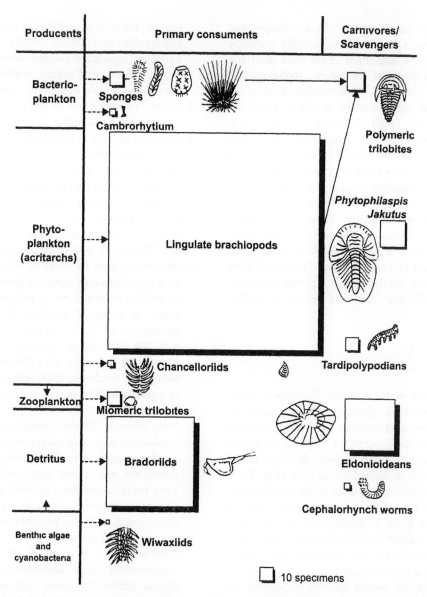

Fig 7 Tentative reconstruction of the trophic web in the Algal Lens (Sinsk Formation) palaeocommunity in terms of number of individuals alive at the time of burial Confirmed energy flows are shown by continuous arrows, inferred ones are indicated by dashed arrows Square area indicates the number of individuals in each group of animals

definitely was not required by any animals and preserved intact until nowadays Some bradoriids, miomeric and polymeric trilobites perhaps fed on the detritus In modern seas under oxygen shortage, the number of infaunal detritophages decreases and unutilised organic matter is accumulated (Kuznetsov, 1980) The abundance of anchoring in soft mud

epibenthic filter-feeders (sponges and juvenile brachiopods) proves indirectly the insignificance of detritophages in Sinsk communities Under favourable conditions, detritophages overcompete sedentary filter-feeders due to amensalism (Thayer, 1983)

Cephalorhynchs, tardipolypodians, possibly eldonioideans, and some polymeric trilobites were secondary consumers Large sizes of *Phytophilaspis* and trilobite *Jakutus* (14–15 cm in length) perfectly allowed them to be consumers of the third level and preyed on other predators, for instance cephalorhynch worms Trace fossils of trilobitomorph hunting on infaunal worms are described in many Cambrian localities (Jensen, 1990) Moreover, carapace segments of the largest trilobite *Jakutus* are ornamented by long axial spines that probably are evident for a predator pressure even on such sizeable animals

The study of lingulate brachiopod shells revealed three different groups of destructors at least in the Algal Lens palaeocommunity These were bacteria which, probably, destroying the organic matter incorporated within shell laminae, possible actinomycetes, and boring algae or cyanobacteria The borers were parasites of living animals which repaired injuries in the surface shell layer and rehabilitated its healthy development

## 6. Conclusions

A comparison of the number of primary (filter-feeders, detritophages, grazers, and browsers) and secondary consumers (counting out possible consumers of the third level) allow us to imply a relative amount of energetic transfer between these two levels It comprises 25% at the individual level (Fig 7) In modern communities the amount of energetic transfer approaches 10–12% This fact merely means that we have lost about a half of primary information rather than energetic transfer was approximately as twice as higher in the Cambrian This is not too much for an analysis of fossil communities

In summary we would like to emphasise the following features of the Algal Lens palaeocommunity

1 All organisms once upon a time comprising the community were adapted to a life under dysaerobic conditions.
2 The trophic web was firmly based on phytoplankton and bacterioplankton Possibly, a dynamic upwelling bringing onto the Siberian Platform by early Botoman transgression together with increased rates of nutrient cycling occurring within the water column at the aerobic/dysaerobic interface created conditions which were favourable for

the phytoplankton and bacterioplankton blooms (Zhuravlev and Wood, 1996) These blooms extracted significant oxygen volumes utilised on the oxygenation of organic necromass and, thus, maintained a wide dysaerobic zone
3 These conditions led to the development of low structuralized and highly dominate community
4 Nonetheless, the organisation of the Algal Lens palaeocommunity was not simple It comprises diverse feeding habits and various substrate relationships and the tiering among sessile suspension-feeders was well developed with the upper tier at the 50 cm level
5 A number of similar features of the Algal Lens and Phyllopod Bed (Burgess Shale) palaeocommunities studied by Conway Morris (1986), such as a significance of different groups of organisms in species composition, a relative percentage of fauna in terms of number of individuals and biovolumes of major groups, feeding habits, and substrate relationships, suggest a relative stability (during ca 25 my) of Cambrian communities occupying similar subtidal settings This stability is expressed also in the presence of the same genera and possibly even species (*Cambrorhytium, Wiwaxia, Eldonia*, many sponges and algae)
6 Cephalorhynchs, tardipolypodians, sponges, eldonioideans, and non-trilobite arthropods played a significant role in the organisation of the Algal Lens palaeocommunity, as well as in many other Cambrian marine communities This is well expressed in a relative percentages of these groups in species composition as well as in individual abundance and biovolumes In modern marine communities these animals either completely absent (tardipolypodians, eldonioideans, bradoriids) or their role is negligible (cephalorhynchs) Usually, they survive as meiobenthos, cryptobionts, or dysaerobic mud inhabitants The communities, where diverse trilobites and lingulate brachiopods predominated and which are commonly taken for the typical Cambrian communities, appeared during the late Middle Cambrian only.
7 Despite of an apparent difference in composition between inhabitants of shallow water well-oxygenated environments and those of dysaerobic zone, the basic features of the Early Cambrian

communities on the Siberian Platform were alike (Burzin et al , 2001) Both were dominated in terms of individuals and biovolume by filter-feeders: spicular sponges, tiny cnidarians, and lingulate brachiopods under dysaerobic conditions, calcified sponges (archaeocyaths and pharetronids), calcareous minute cnidarians, and different brachiopods as well as hyolithomorph hyoliths and stenothecoids within well-oxygenated bottom waters They were followed by detritophages bradoriids and miomeric trilobites and orthothecimorph hyoliths, respectively Finally, the same carnivores/scavengers represented by cephalorhynchs, tardipolypodians, and large polymeric trilobites were significant in both environments, although their sclerites and fragments only are preserved in limestones formed in well-agitated milieu

## Acknowledgements

The authors thank Loren Babcock for a kind invitation to participate in a special issue of Palaeogeography, Palaeoclimatology, Palaeoecology The field works were supported by "Quantas" and Russian Foundation for Basic Research projects 96-04-48372, 96-05-64224, and 98-04-49010 Mikhail Burzin is acknowledged for acritarch determinations

## References

Allison, P A , 1988 Konservat–Lagerstatten cause and classification Paleobiology 14, 331–344

Anderson, R F , Lyons, T W , Cowie, G L , 1994 Sedimentary record of a shoaling of the oxic/anoxic interface in the Black Sea Mar Geol 116, 373–384

Astashkin, V A , Varlamov, A I , Esakova, N V , Zhuravlev, A Yu , Repina, L N , Rozanov, A Yu , Fedorov, A B , Shabanov, Yu Ya , 1990 Guidebook for excursion on the Aldan and Lena rivers Siberian Platform 3d Internat Symp Cambrian System IGIG, Novosibirsk 115 pp

Babcock, L E , Zhang, W , 2001 Stratigraphy, paleontology, and depositional setting of the Chengjiang Lagerstatte (Lower Cambrian), Yunnan, China In Peng, S , Babcock, L E , Zhu, M (Eds ), Cambrian System of South China, Palaeoworld, vol 13 Univ Sci Tech China Press, Hefei, pp 66–86

Barskov, I S , Zhuravlev, A Yu , 1988 Myagkotelye organizmy kembriya Sibirskoy platformy (Soft-bodied organisms from the Cambrian of the Siberian Platform) [Paleontol J 1, 1–7] Palaeontol Ž 1988 (1), 3–9

Bengtson, S , Hou, Xianguang, 2001 The integument of Cambrian chancelloriids Acta Palaeontol Pol 46, 1–22

Bengtson, S , Missarzhevsky, V V , 1981 Coelosclenitophora—a major group of enigmatic Cambrian metazoans Open-File Rep (U S Geol Surv ) 81-743, 19–23

Briggs, D E G , 1978 A new trilobite-like arthropod from the Lower Cambrian Kinzers Formation, Pennsylvania J Paleontol 52, 132–140

Briggs, D E G , Kear, A J , 1993 Decay and preservation of polychaetes taphonomic thresholds in soft-bodied organisms Paleobiology 19, 107–135

Briggs, D E G , Nedin, C , 1997 The taphonomy and affinities of the problematic fossil Myoscolex from the Lower Cambrian Emu Bay Shale of South Australia J Paleontol 71, 22–32

Briggs, D E G , Whittington, H B , 1985 Modes of life of arthropods from the Burgess Shale, British Columbia Trans R Soc Edinb Earth Sci 76, 149–160

Budd, G E , 1998 Arthropod body-plan evolution in the Cambrian with an example from anomalocaridid muscle Lethaia 31, 197–210

Burzin, M B , Debrenne, F , Zhuravlev, A Yu , 2001 Evolution of shallow-water level-bottom communities In Zhuravlev, A Yu , Riding, R (Eds ), Ecology of the Cambrian Radiation Columbia University Press, New York, pp 217–237

Butterfield, N J , 1990 Organic preservation of non-mineralizing organisms and the taphonomy of the Burgess Shale Paleobiology 16, 272–286

Butterfield, N J , 2002 Leanchoilia guts and the interpretation of three-dimensional structures in Burgess Shale-type fossils Paleobiology 28, 155–171

Byers, C W , 1977 Biofacies patterns in euxinic basins a general model In Cook, H E , Enas, P (Eds ), Deep-water Carbonate Environments, Soc Econ Paleontols Minerals Spec Publ , vol 25, pp 5–17

Chatterton, B D E , Jahanson, Z , Sutherland, G , 1994 Form of the trilobite digestive system alimentary structures in Pterocephalia J Paleontol 68, 294–305

Chen, Junyuan, Zhou, Guiqing, 1997 Biology of the Chengjiang biota In Chen, Junyuan, Cheng, Yen-nien, Iten, H V (Eds ), The Cambrian Explosion and the Fossil Record Natl Mus Nat Sci , Taichung, Taiwan, Bull Natl Mus Nat Sci , vol 10, pp 11–105

Chen, Junyuan, Hou, Xianguang, Erdtmann, B D , 1989 New soft-bodied fossil fauna near the base of the Cambrian System at Chengjiang, eastern Yunnan, China In Chinese Acad Sci (Ed ), Developments in Geoscience, Contribution to 28th Internat Geol Congr , 1989, Washington DC, USA Science Press, Beijing, pp 265–277

Chen, Jun-Yuan, Zhou, Gui-Qqing, Ramskold, L , 1995a A new Early Cambrian onychophoran-like animal, Paucipodia gen nov , from the Chengjiang fauna, China Trans R Soc Edinb Earth Sci 85, 275–282 (for 1994)

Chen, Jun-Yuan, Zhou, Gui-Qing, Ramskold, L , 1995b The Cambrian lobopodian, Microdictyon sinicum Bull Natl Mus Nat Sci (Taichung, Taiwan) 5, 1–93

Chlupáč, I , 1995 Lower Cambrian arthropods from the Paseky Shale (Barrandian area, Czech Republic) J Czech Geol Soc 40, 9–36

Chuang, S H , 1959 The structure and function of the alimentary canal in *Lingula unguis* (L ) (Brachiopoda) Proc Zool Soc Lond 132, 283–311

Clarke, J M , 1900 *Paropsonema cryptophya*, a peculiar echinoderm from the intumescens zone (Portage beds) of western New York Bull N Y State Mus 39, 172–178

Clarkson, E N K , Ahlberg, P , Taylor, C M , 1998 Faunal dynamics and microevolutionary investigations in the Upper Cambrian *Olenus* Zone at Andrarum, Skane, Sweden Geol Foren Stockh Forh 120, 257–267

Coniglio, M , Dix, G R , 1992 Carbonate slopes In Walker, R G , James, N P (Eds ), Facies Models Response to Sea Level Change Geological Association of Canada, St John's, pp 349–374

Conway Morris, S , 1977a A new metazoan from the Cambrian Burgess Shale of British Columbia Palaeontology 20, 623–640

Conway Morris, S , 1977b Fossil priapulid worms Spec Pap Palaeontol 20 (iv+155 pp )

Conway Morris, S , 1979 The Burgess Shale (Middle Cambrian) fauna Ann Rev Ecolog Syst 10, 327–349

Conway Morris, S , 1985 The Middle Cambrian metazoan *Wiwaxia corrugata* (Matthew) from the Burgess Shale and *Ogygopsis* Shale, British Columbia, Canada Philos Trans R Soc Lond , B 307, 507–582

Conway Morris, S , 1986 The community structure of the Middle Cambrian Phyllopod Bed (Burgess Shale) Palaeontology 29, 423–467

Conway Morris, S , 1993 The fossil record and the early evolution of the Metazoa Nature 361, 219–225

Conway Morris, S , 1997 The cuticular structure of the 495-Myr-old type species of the fossil worm *Palaeoscolex, P piscatorum* (?Priapulida) Zool J Linn Soc 119, 69–82

Conway Morris, S , Peel, J S , 1995 Articulated halkieriids from the Lower Cambrian of North Greenland and their role in early protostome evolution Philos Trans R Soc Lond , B 347, 305–358

Conway Morris, S , Robison, R A , 1986 Middle Cambrian priapulids and other soft-bodied fossils from Utah and Spain Univ Kans Paleontol Contrib , Pap 117, 1–22

Conway Morris, S , Peel, J S , Higgins, A K , Soper, N J , Davis, N C , 1987 A Burgess Shale-like fauna from the Lower Cambrian of North Greenland Nature 326, 181–183

Durham, J W , 1974 Systematic position of *Eldonia ludwigi* Walcott J Paleontol 48 750–755

Dzik, J , 1991 Is fossil evidence consistent with traditional views of the early metazoan phylogeny? In Simonetta, A M , Conway Morris, S (Eds ), The Early Evolution of Metazoa and the Significance of Problematic Taxa Cambridge Univ Press, Cambridge, pp 47–56

Dzik, J , Zhao, Yuanlong, Zhu, Maoyan, 1997 Mode of life of the Middle Cambrian eldoniid Lophophorate *Rotadiscus* Palaeontology 40, 385–396

Eberli, G P , 1991 Calcareous turbidites and their relationship to sea-level fluctuations and tectonism In Einsele, G , Ricken, W , Seilacher, A (Eds ), Cycles and Events in Stratigraphy Springer, Berlin, pp 340–359

Elrick, M , Hinnov, L A , 1996 Millennial-scale climate origins for stratification in Cambrian and Devonian deep-water rhythmites, western USA Palaeogeogr Palaeoclimatol Palaeoecol 123, 353–372

Fatka, O , Konzalová, M , 1995 Microfossils of the Paseky Shale (Lower Cambrian, Czech Republic) J Czech Geol Soc 40, 55–66

Fortey, R A , Owens, R M , 1999 Feeding habits in trilobites Palaeontology 42, 429–465

Fortey, R A , Wilmot, N V , 1991 Trilobite cuticle thickness in relation to palaeoenvironment Palaontol Z 65, 141–151

Friend, D , Solov'ev, I A , Zhuravlev, A Yu , 2002 Latest *Eldonia* (Paropsonemidae, Echinodermata) from the Siberian Platform Palaeontol Ž (1), 22–26

Garcia-Bellido Capdevila, D , Conway Morris, S , 1999 New fossil worms from the Lower Cambrian of the Kinzers Formation, Pennsylvania, with some comments on Burgess Shale-type preservation J Paleontol 73 394–402

Glaessner, M , 1979 Lower Cambrian crustacea and annelid worms from Kangaroo Island, South Australia Alcheringa 3, 21–31

Hof, C H J , Briggs, D E G , 1997 Decay and mineralisation of mantis shrimps (Stomatopoda, Crustacea)—a key to their fossil record Palaios 12, 420–438

How, X -G , Bergstrom, J , 1997 Arthropods of the Lower Cambrian Chengjiang fauna, Southwest China Fossils Strata 45, 1–116

How, Xian-guang, Bergstrom, J , 1991 The arthropods of the Lower Cambrian Chengjiang fauna, with relationships and evolutionary significance In Simonetta, A M , Conway Morris, S (Eds ), The Early Evolution of Metazoa and the Significance of Problematic Taxa Cambridge Univ Press, Cambridge, pp 179–187

How, Xianguang, Bergstrom, J , 1995 Palaeoscolecid worms may be nematomorphs rather than annelids Lethaia 27, 11–17

Hou, Xianguang, Siveter, D J , Williams, M , Walossek, D , Bergstrom, J , 1996 Appendages of the arthropod *Kunmingella* from the early Cambrian of China its bearing on the systematic position of the Bradoriida and the fossil record of the Ostracoda Philos Trans R Soc Lond , B 351, 1131–1145

Huang, You-zhuang, Wang, Hua-yu, Zhao, Yuan-long, Dai, Xin-chun, 1994 Brachiopods from Early–Middle Cambrian Kaili Formation in Taijiang, Guizhou Acta Palaeontol Sin 33, 335–344 (in Chinese)

Huntley, M , Karl, D M , Niiler, P , Holm-Hansen, O , 1991 Research on Antarctic coastal ecosystem rates (RACER) an interdisciplinary field experiment Deep-Sea Res 38, 911–941

Ivantsov, A Yu , 1998 The richest of Sinsk Lagerstatten (Lower Cambrian, Siberian Platform) IV Field Conference of the Cambrian Stage Subdivision Working Group, Abstracts, Lund Publ Geol , vol 142, p 10

Ivantsov, A Yu , 1999 Trilobite-like arthropod from the Lower Cambrian of the Siberian Platform Acta Palaeontol Pol 44, 455–466

Ivantsov, A Yu , Zhuravlev, A Yu , Leonov, M V , Leguta, A V , 1996 New Lower Cambrian occurrence of Burgess Shale-type fossils in Siberia CSPG-SEPM Joint Convention, p 143

Jensen, S , 1990 Predation by early Cambrian trilobites on infaunal worms—evidence from the Swedish *Mickwitzia* Sandstone Lethaia 23, 29–42

Krautter, M , 1997 Aspekte zur Palaokologie postpalaozoischer Kieselschwamme Profil (Stuttg ) 11, 199–324

Kruse, P D , Zhuravlev, A Yu , James, N P , 1995 Primordial metazoan-calcimicrobial reefs Tommotian (Early Cambrian) of the Siberian Platform Palaios 10, 291–321

Kuznetsov, A P , 1980 Ekologiya donnykh soobshchestv Mirovogo okeana (The ecology of sea-bottom communities of the world ocean) Nauka, Moscow 243 pp (In Russian)

Lindstrom, M , 1995 The environment of the Early Cambrian Chengjiang fauna Intern Cambrian Explosion Symp (April, 1995, Nanjing) Progr Abstrs, p 17

Malakhov, V V , Adrianov, A V , 1995 Golovokhobotnye (Cephalorhyncha)—novyy tip zhivotnogo tsarstva (Cephalorhyncha—a new phylum of animal kingdom) KMK, Moscow 200 pp (in Russian)

Mao, Jia-ren, Zhao, Yuan-Long, Yu, Ping, Qian, Yi, 1992 Some Middle Cambrian hyolithids from Taijiang, Guizhou Acta Micropaleontol Sin 9, 257–265 (in Chinese)

Marss, T , 1988 Early Palaeozoic hadimopanellids of Estonia and Kirgizia (USSR) Eesti NSV Tead Akad Toim , Geol 37, 10–17

Masiak, M , Zylińska, A , 1994 Burgess Shale-type fossils in Cambrian sandstones of the Holy Cross Mountains Acta Palaeontol Pol 39, 329–340

Mehl, D , 1996 Organization and microstructure of the chancellorid skeleton implications for the biomineralization of the Chancelloriidae Bull Inst Océanogr , Monaco Numéro Spéc 14, 377–385

Melnikova, L M , 1998 Reviziya nekotorykh kembriyskikh bradoriid (Crustacea) Sibirskoy platformy (Revision of some Cambrian bradoriids (Crustacea) fiom the Siberian platform) [Paleontol J 32, 357–361] Paleontol Ž (4), 36–40

Melnikova, L M , 2000 Novyy rod bradoriid (Crustacea) iz kembriya Severnoy Evrazii (A new genus of Bradoriidae (Crustacea) from the Cambrian of Northern Eurasia) [Paleontol J 34, 180–185] Paleontol Ž (2), 65–68

Morri, C , Bavestrello, G , Bianchi, C N , 1991 Faunal and ecological notes on some benthic cnidarian species from the Tuscan Archipelago and eastern Ligurian Sea (western Mediterranean) Boll Mus Ist Biol Univ Genova 54–55, 27–47

Morris, K A , 1979 A classification of Jurassic marine shale sequences an example from Toarcian (Lower Jurassic) of Great Britain Palaeogeogr Palaeoclimatol Palaeoecol 26, 117–126

Muller, K J , Hinz-Schallreuter, I , 1993 Palaeoscolecid worms fiom the Middle Cambrian of Australia Palaeontology 36, 549–592

Nedin, C , 1995 The Emu Bay Shale, a Lower Cambrian fossil lagerstatten, Kangaroo Island, South Australia Mem Assoc Australas Palaeontol 18, 31–40

Nedin, C , 1997 Taphonomy of the Early Cambrian Emu Bay Shale Lagerstatte, Kangaroo Island, South Australia In Chen, Junyuan, Cheng, Yen-nien, Iten, H V (Eds ), The Cambrian

explosion and the fossil record Natl Mus Nat Sci , Taichung, Taiwan, Bull Natl Mus Nat Sci vol 10, pp 133–141

Neyman, A A , 1967 O granitsakh primeneniya nazvaniya "troficheskaya gruppirovka" pri izuchenii donnogo naseleniya (Limits to the application of the 'trophic group' concept in benthic studies (Oceanol Acad Sci USSR 7, 149–155) Okeanologia 7 (2), 195–207

Nikolaeva, I V , 1981 Fatsial'naya zonal'nost' khimicheskogo sostava mineralov gruppy glaukonita i opredelyayushchie ee faktory (Facies zonation of chemical composition of glauconite group minerals and the factors defining it) In Nikolaeva, I V (Ed ), Mineralogiya i geokhimiya glaukonita (Mineralogy and Geochemistry of Glauconite), Tr . Inst Geol Geofiz , Sibirskoe Otdelenie, Akad nauk SSSR vol 515, pp 4–41 (in Russian)

Pan, C -M , Watabe, N , 1988 Uptake and transport of shell minerals in *Glottidia pyramidata* Stimpson (Brachiopoda, Inarticulata) J Exp Mar Biol Ecol 118, 257–268

Pelman, Yu L , 1977 Ranne- i srednekembriyskie bezzamkovye biakhiopody Sibirskoy platformy (Early and middle Cambrian inarticulate brachiopods of the Siberian platform) Nauka, Novosibirsk 168 pp (in Russian)

Piper, D J W , 1972 Sediments of the Middle Cambrian Burgess Shale, Canada Lethaia 5, 169–175

Piper, D J W , Stow, D A V , 1991 Fine-grained turbidites In Einsele, G , Ricken, W , Seilacher, A (Eds ), Cycles and Events in Stratigraphy Springer, Berlin, pp 360–376

Reiswig, H M , 1971 In situ pumping activity of tropical Demospongiae Mar Biol 9, 38–50

Resser, C E , Howell, B F , 1938 Lower Cambrian *Olenellus* zone of the Appalachian Bull Geol Soc Am 49, 195–248

Rigby, J K , 1987 Early Cambrian sponges from Vermont and Pennsylvania, the only ones described from North America J Paleontol 61, 451–461

Rigby, J K , Gunther, L F , Gunther, F , 1997 The first occurrence of the Burgess Shale demosponge *Hazelia palmata* Walcott, 1920, in the Cambrian of Utah J Paleontol 71, 994–997

Robison, R A , 1990 Earliest-known uniramous arthropod Nature 343, 163–164

Robison, R A , 1991 Middle Cambrian biotic diversity examples from four Utah Lagerstatten In Simonetta, A M , Conway Morris, S (Eds ), The Early Evolution of Metazoa and the Significance of Problematic Taxa Cambridge Univ Press, Cambridge, pp 77–98

Robison, R A , Wiley, E O , 1995 A new arthropod, *Meristoma* more fallout from the Cambrian explosion J Paleontol 69, 447–459

Runnegar, B , Curry, G B , 1992 Amino acid sequences of hemerythrins from *Lingula* and a priapulid worm and the evolution of oxygen transport in the Metazoa 29th Internat Geol Congr Kyoto, Japan, 24 August–3 September 1992, Abstrs 2, p 346

Savdra, C E , Bottjer, D , 1986 Trace-fossil model for reconstruction of paleo-oxygenation in bottom waters Geology 14, 3–6

Savitskiy, V E , Astashkin, V A , 1979 Rol' i masshtaby rifoobrazovaniya v kembriyskoy istorii Sibirskoy platformy (The role and scales of reefbuilding in the Cambrian history of the

Siberian Platform) In Savitskiy, V E , Astashkin, V A (Eds ), Geologiya rifovykh sistem kembriya Zapadnoy Yakutii (Geology of Reef Systems in the Cambrian of Western Yakutia) Tr , Sibirskogo nauchno-issledovatel'skogo inst geol geofiz miner res vol 270, pp 5–18 (in Russian)

Savitskiy, V E , Evtushenko, V M , Egorova, L I , Kontorovich, A E , Shabanov, Yu Ya , 1972 Kembriy Sibirskoy platformy (Yudomo-Olenekskiy tip razreza) (The Cambrian of the Siberian Platform (Yudoma-Olenek section type)) Tr , Sibirskogo nauchno-issledovatel'skogo inst geol geofiz miner res vol 130, pp 1–200 (in Russian)

Shu, Degan, Vannier, J , Luo, Huilin, Chen, Ling, Zhang, Xingliang, Hu Shixue, 1999 Anatomy and lifestyle of *Kunmingella* (Arthropoda, Bradoriida) from the Chengjiang fossil Lagerstatte (lower Cambrian, Southwest China) Lethaia 32, 279–298

Smethurst, M A , Khramov, A N , Torsvik, T H , 1998 The Neoproterozoic and Palaeozoic palaeomagnetic data for the Siberian Platform from Rodinia to Pangea Earth-Sci Rev 43, 1–24

Sorokin, Yu I , 1990 Ekosistema korallovykh rifov (Ecosystem of Coral Reefs) Nauka, Moscow 503 pp (in Russian)

Sprinkle J , 1973 Morphology and evolution of blastozoan echinoderms Spec Publ -Mus Comp Zool Harv , pp 1–284

Stanley Jr , G D , 1986 Chondrophorine hydrozoans as problematic fossils In Hoffman, A , Nitecki, M H (Eds ), Problematic Fossil Taxa Oxford Univ Press, New York, Clarendon Press, Oxford Oxford Monogr Geol Geophys 5, 68–86

Steiner M , Fatka, O , 1996 Lower Cambrian tubular micro- to macrofossils from the Paseky Shale of the Barrandian area (Czech Republic) Palaontol Z 70, 275–299

Sumrall, C D , Sprinkle, J , 1999 *Ponticulocarpus*, a new cornutegrade stylophoran from the Middle Cambrian Spence Shale of Utah J Paleontol 73, 886–891

Thayer, C W , 1981 Ecology of living brachiopods review and preview Lophophorates Notes for a short course Univ Tennessee, Dept Geol Sci Stud Geol 5, 110–126

Thayer, C W , 1983 Sediment-mediated biological disturbance and the evolution of marine benthos In Tevesz, M J S , McCall P L (Eds ), Biotic Interactions in Recent and Fossil Benthic Communities Plenum Press, New York, pp 479–625

Tunncliffe, V , Wilson, K , 1988 Brachiopod populations distribution in fjords of British Columbia (Canada) and tolerance of low-oxygen concentrations Mar Ecol 47, 117–128

Ushatinskaya, G T , 2001 Brachiopods In Zhuravlev, A Yu , Riding, R (Eds ), Ecology of the Cambrian Radiation Columbia University Press, New York, pp 350–369

van der Land, J 1970 Systematics, zoogeography, and ecology of the Priapulida Zool Verh 112, 1–118 (Leiden)

van der Meer Mohr, C G , Okulitch, V J , 1967 On the occurrence of scyphomedusa in the Cambrian of the Cantabrian Mountains (NW Spain) Geol Mijnb 46, 361–362

Vannier, J , Williams, M , Siveter, D J , 1997 The Cambrian origin of the circulatory system of crustaceans Lethaia 30, 169–184

Vidal, G , Peel, J S , 1993 Acritarchs from the Lower Cambrian Buen Formation in North Greenland Bull -Grønl Geol Unders 164, 1–35

Wang, Hai-Feng, Zhang, Jun-Ming, Hu, Shi-Xue Li, Guo-Xiang, Zhu, Mao-Yan, Wang, Wei, Luo, Hui-Lin, 2001 Litho- and biostratigraphy of the Lower Cambrian Yu'anshan Formation near the village of Ercaicun, Haikou County, eastern Yunnan Province Acta Palaeontol Sin 40, 106–114 (Suppl )

Whittington, H B , 1978 The lobopod animal *Aysheaia pedunculata* Walcott, Middle Cambrian, Burgess Shale, British Columbia Philos Trans R Soc Lond , B 284, 165–197

Wilkinson, C R , Garrone, R , Vacelet, J , 1984 Marine sponges discriminate between food bacteria and bacterial symbionts electron microscope radiography and in situ evidence Proc R Soc Lond , B 220, 519–528

Wright J , Schrader, H , Holser, W T , 1987 Paleoredox variations in ancient oceans, recorded by rare earth elements in fossil apatite Geochim Cosmochim Acta 51, 631–644

Yin, Lei-Ming, Yang, Rui-Dong, 1999 Early-Middle Cambrian acritarchs in the Kaili Formation from Taijiang County, Guizhou, China Acta Palaeontol Sin 38, 58–78 (Suppl , in Chinese)

Zhang, Xi-Guang, Pratt, B R , 1996 Early Cambrian palaeoscolecid cuticles from Shaanxi, China J Paleontol 70, 275–279

Zhang, X , Shu, D , Li, Y , Han, J , 2001 New sites of Chengjiang fossils crucial windows of Cambrian explosion J Geol Soc (Lond ) 158, 211–218

Zhao, Yuanlong, Yuan, Jinliang, Zhang, Zhenghua, Huang, Youzhuang, Chen, Xiaoyuan, Zhou, Zhen, 1996 Composition and significance of the Middle Cambrian Kaili Lagerstatte in Taijiang County, Guizhou Province, China a new Burgess type Lagerstatte Geol Guizhou 13, 7–14 (in Chinese)

Zhao, Yuan-Long, Steiner, M , Yang, Rui-Dong, Erdtmann, B -D , Guo, Qin-Jun, Zhou, Zhen, Wallis, S , 1999a Discovery and significance of the early metazoan biotas from the Lower Cambrian Niutitang Formation, Zunyi, Guizhou, China Acta Palaeontol Sin 38, 132–144 (Suppl , in Chinese)

Zhao, Yuan-Long, Yuan, Jin-Liang, Zhu, Mao-Yan, Yang, Rui-Dong, Guo, Qing-Jun, Qian, Yi, 1999b A progress report on research on the early Middle Cambrian Kaili Biota, Guizhou, PRC Acta Palaeontol Sin 38, 1–14 (Suppl , in Chinese)

Zhuravlev A Yu , 2001 Paleoecology of Cambrian reef ecosystems In Stanley Jr , J D (Ed ), The History and Sedimentology of Ancient Reef Systems Plenum Press, New York, pp 121–157

Zhuravlev, A Yu , Debrenne, F , 1996 Pattern of evolution of Cambrian benthic communities environments of carbonate sedimentation Riv Ital Paleontol Stratigr 102, 333–340

Zhuravlev, A Yu , Wood, R A , 1996 Anoxia as the cause of the mid-Early Cambrian (Botomian) extinction event Geology 24, 311–314

Available online at www.sciencedirect.com

Palaeogeography, Palaeoclimatology, Palaeoecology 220 (2005) 89–117

www.elsevier.com/locate/palaeo

ELSEVIER

PALAEO

# Articulated sponges from the Lower Cambrian Hetang Formation in southern Anhui, South China: their age and implications for the early evolution of sponges

Shuhai Xiao[a,*], Jie Hu[b], Xunlai Yuan[b], Ronald L. Parsley[c], Ruiji Cao[b]

[a]*Department of Geological Sciences, Virginia Polytechnic Institute and State University, Blacksburg, VA 24061, USA*
[b]*Nanjing Institute of Geology and Palaeontology, Academia Sinica, Nanjing 210008, China*
[c]*Department of Earth and Environmental Sciences, Tulane University, New Orleans, LA 70118, USA*

Received 1 February 2002, accepted 15 February 2002

## Abstract

A well-preserved benthic, epifaunal assemblage of articulated sponges is described from the stone coal beds of the Lower Cambrian Hetang Formation at Lantian, southern Anhui Province, South China These sponges are Meishucunian–Qiongzhusian (=Diandongian–early Qiandongian) in age In Siberian terminology, they are probably Tommotian–Atdabanian, approximately 535–520 Ma The Hetang sponge fauna is taxonomically diverse and morphologically complex Eleven species of both demosponges and hexactinellids, including three new taxa (*Choia? striata* sp nov, *Protospongia gracilis* sp nov, and *Lantianospongia palifera* gen et sp nov), are described Two undetermined forms are also illustrated The Hetang and other Neoproterozoic–Cambrian sponge fossils, at their face value, indicate that that hexactinellids evolved no later than the Nemakit-Daldynian–Tommotian and perhaps in the late Neoproterozoic, and the demosponges and calcareans evolved no later than the Atdabanian The divergence of sponge classes therefore appears to be part of the Cambrian Radiation event In comparison with the eumetazoans which probably diverged at ca 600 Ma, however, the sponges (particularly the demosponges and calcareans) appear to have a missing fossil record in the late Neoproterozoic and earliest Cambrian The minimum implied gaps (MIGs) of the calcareans and demosponges are substantial (tens of Myrs to perhaps more than 100 Myrs), particularly if the calcareans constitute a sister group of the eumetazoans—a topology supported by currently available molecular evidence

*Keywords* Early Cambrian, Hetang Formation, South China, Sponges, Phylogeny

# 1. Introduction

The evolution of early animals at the Neoproterozoic–Cambrian transition (ca 600–520 Ma) is a key

\* Corresponding author Fax +1 540 231 3386
   *E-mail address* xiao@vt edu (S Xiao)

0031-0182/$ - see front matter © 2004 Elsevier B V All rights reserved
doi 10 1016/j palaeo 2002 02 001

event in the history of life Much of the current discussion has been centered on bilateran animals (Budd and Jensen, 2000) Bilateans are duly important because of their great diversity, but non-bilaterian animals (cnidarians and sponges) and protists are also part of the Cambrian radiation puzzle (Lipps, 2001, Moczydłowska, 2001) For example, phytoplankton was a crucial component of the Neoproterozoic–Cambrian food chain (Butterfield, 2001), and non-bilaterians were important players in ecological tiering in Neoproterozoic–Cambrian epifaunal communities (Yuan et al, 2002, Clapham and Narbonne, 2002)

The focus of this paper is the sponges Traditionally, sponges have been treated as a monophyletic group with three classes (Hexactinellida, Demospongea, and Calcarea), each of which is taken to be monophyletic (Reitner and Mehl, 1996) Molecular phylogenetic analyses in the recent years, however, appear to support sponge paraphyly with calcareans being a sister group to the eumetazoans, although the monophyly of the three individual classes has not been challenged Whether the molecular data support a demosponge+hexactinellid grouping to form a monophyletic Silicea is a matter of current controversy (Borchiellini et al, 2001, Medina et al, 2001)

Few have questioned and evaluated the quality of the sponge fossil record, despite considerable debate on a possible deep, but missing, bilaterian history in the Precambrian (Wray et al, 1996, Conway Morris, 1997, Budd and Jensen, 2000) The incongruence between sponge phylogeny and stratigraphy can be used to evaluate the quality of the sponge fossil record and to estimate the minimum implied gaps (MIGs)—minimum fossil gaps or range extension required for the congruence between stratigraphic and phylogenetic data (cf. Benton and

Storrs, 1996, Benton et al, 2000) Given the antiquity of eumetazoans, all sponge phylogenies (particularly those based on molecular data) require long MIGs in the early record of calcareans and demosponges, both of which first appeared certainly in the Atdabanian (and possibly in the Tommotian for demosponges described in this paper) This simple analysis of stratigraphy against phylogeny highlights the incompleteness of early sponge fossils

## 2. Geologic background and stratigraphic correlation

The sponge fossils were collected from the Early Cambrian Hetang Formation at Lantian (29°55′N, 118°05′E, Fig 1), southern Anhui Province in South China Because the age of these fossils is critical to our stratigraphic–phylogenetic analysis, we present in this section detailed discussion on the correlation between the Hetang Formation in southern Anhui and better known Early Cambrian sections in eastern Yunnan (regional type sections), Siberia, and Newfoundland (Neoproterozoic–Cambrian GSSP)

### 2 1 Eastern Yunnan

In South China, the Early Cambrian successions typically begin with a few to a few tens of meters of phosphatic dolostones that contain basal Cambrian small shelly fossils (SSFs) The phosphatic dolostones are succeeded by black shales that are widespread in South China (Erdtmann and Steiner, 2001; Steiner et al, 2001a) The Shiyantou and Hetang Formations, in eastern Yunnan and southern Anhui, respectively (see below), are examples of such black shales On weathered landscape, the black shales contrast sharply

---

Fig 1 Neoproterozoic–Lower Cambrian lithostratigraphy at Lantian (star on map), southern Anhui, and proposed correlations with eastern Yunnan and Siberian stages Column (1) Lower Cambrian stages of Siberia Column (2) Recently proposed Cambrian series and stages of south China (Peng and Babcock, 2001) Column (3) Traditional Lower Cambrian stages of south China Column (4) Lower Cambrian lithostratigraphic units and SSF assemblages (I–IV) in eastern Yunnan SSF I (*Anabarites trisulcatus–Protohertzina anabarica* assemblage), SSF II (*Siphogonuchites triangularis–Paragloborilus subglobosus* assemblage), SSF III (*Heraultipegma yunnanensis=Watsonella crosbyi* assemblage), SSF IV (*Sinosachites flabelliformis–Tannuolina zhangwentangi* assemblage) An ash bed from the SSF I assemblage is dated from 538 2±1 5 Ma (Jenkins et al, 2002) Dated ash beds from Siberia (534 6±0 5 Ma, Bowring et al, 1993) and Avalonia (530 7±0 9 Ma, Isachsen et al, 1994) are correlated to eastern Yunnan using the *Watsonella crosbyi* assemblage as a time indicator Column (5) Stratocolumn at Lantian Sponge fossils reported in this paper are from lower Hetang Formation, probably Meishucunian–Qiongzhusian in age ND—Nemakit-Daldynian, JN—Jinningian, LGW—Leigongwu diamictite

with thick-bedded dolostones of the terminal Neoproterozoic Dengying Formation and its equivalents
(for example, the Liuchapo Formation in central
Hunan and the Piyuancun Formation in southern
Anhui)

The most intensively studied Neoproterozoic–
Cambrian successions in South China are probably
those in Yunnan Province The Meishucun section in
eastern Yunnan was once a strong contender for the
Neoproterozoic–Cambrian boundary global stratotype

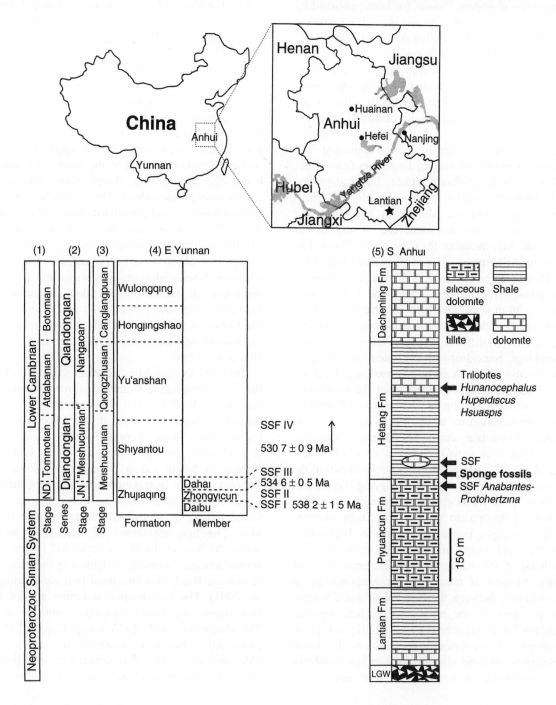

(Luo et al, 1982, Luo et al, 1984) The Meishucun and many other sections in eastern Yunnan are still used as yardsticks for Early Cambrian correlation in South China (Luo et al, 1994) The litho-, bio-, chemo-, and chronostratigraphy of Early Cambrian successions in eastern Yunnan has been published by others (Jiang et al, 1988, Brasier et al, 1990, Zhou et al, 1997, Shen and Schidlowski, 2000, Zhu et al, 2001) and is summarized below

In eastern Yunnan, Neoproterozoic–Early Cambrian successions are divided into the Zhujiaqing, Shiyantou, and Yu'anshan formations (Zhu et al, 2001, Fig 1) The Heilinpu Formation (Luo et al, 1994, Babcock et al, 2001) is equivalent to the Shiyantou and Yu'anshan formations combined The recently proposed Zhujiaqing Formation (Zhu et al, 2001) avoids the confusion between lithostratigraphic and chronostratigraphic terms—the old lithostratigraphic term was the Meishucun Formation but Meishucun is also the namesake of the Lower Cambrian Meishucunian Stage in South China The Zhujiaqing Formation consists of, in ascending order, the Daibu (cherty dolostone, 0–56 m thick), Zhongyicun (phosphorite, 1–90 m thick), and Dahai (dolostone, 1–70 m thick) members The Neoproterozoic–Cambrian boundary is placed at the base of the SSF *Anabarites trisulcatus–Protohertzina anabarica* assemblage, coincident with the base of the Zhongyicun Member of the Zhujiaqing Formation (Zhu et al, 2001) Earlier reports of SSFs from the underlying Xiaowaitoushan Member of the Meishucun Formation near Meishucun (Luo et al, 1982) have not been confirmed by subsequent studies (Qian and Bengtson, 1989, Qian et al, 1996) Zhu et al (2001) therefore considered the Xiaowaitoushan Member as part of the Baiyanshao Member of the Dengying Formation, because it is lithologically inseparable from the underlying Baiyanshao Member The Daibu Member of the Zhujiaqing Formation overlies the Baiyanshao Member and crops out in areas some 200 km northeast of Meishucun, in the Meishucun area, the Daibu Member is missing and corresponds to an unconformity between the Baiyanshao and Zhongyicun members So far, no SSFs have been reported from the Daibu Member (Li et al, 2001) and the *A trisulcatus–P anabarica* assemblage in the lower Zhongyicun Member represents the earliest Cambrian SSF assemblage in South China The second SSF

assemblage, the *Siphogonuchites triangularis–Paragloborilus subglobosus* assemblage, occurs in the upper Zhongyicun Member, most SSFs in this assemblage also extend upward into the Dahai Member (Qian and Bengtson, 1989, Qian et al, 2001) The Dahai Member contains a third and newly named SSF assemblage, the *Heraultipegma yunnanensis* assemblage (Qian et al, 1996, Qian et al, 2001), *H yunnanensis* is probably a junior synonym of *Watsonella crosbyi* (Landing, 1989)

The Shiyantou Formation (formerly Shiyantou Member of the Qiongzhusi Formation, Chen et al., 1996) consists of black siltstones and shales with phosphatic nodules Few fossils occur in the lower part of this formation, but the upper Shiyantou contains fossils belonging to the fourth SSF assemblage, the *Sinosachites flabelliformis–Tannuolina zhangwentangi* assemblage This assemblage extends upward into the basal phosphorite bed (<50 cm thick) of the Yu'anshan Formation (formerly Yu'anshan Member of the Qiongzhusi Formation; Chen et al, 1996) This phosphorite bed is succeeded by black siltstones, black shales, and gray shales and siltstones that contain the earliest Chinese trilobites including, in ascending order, the *Parabadiella* and *Wutingaspis–Eoredlichia* zones (Steiner et al, 2001b), as well as the astounding Chengjiang Biota (Chen et al, 1996, Hou et al, 1999)

Traditionally, the Meishucunian Stage represents the first, pre-trilobite Cambrian chronostratigraphic unit in South China The first occurrence of *Parabadiella* defines the base of the trilobite-bearing Qiongzhusian Stage Recently, a new chronostratigraphic scale has been proposed for the Cambrian System in South China (Peng, 1999; Peng and Babcock, 2001). In this new scheme, the Cambrian is divided into four series, in ascending order, the Diandongian, Qiandongian, Wulingian, and Hunanian The first appearance of *Treptichnus pedum* marks the base of the Diandongian and the Precambrian–Cambrian boundary as defined by the stratotype at Fortune Head, Newfoundland (but see Gehling et al, 2001) The Diandongian is further divided into two stages, the lower Jinningian and the upper "Meishucunian", with the first appearance of *Paragloborilus subglobosus* marking the base of the "Meishucunian" The first appearance of trilobites defines the base of the Qiandongian Series This series includes two stages, the Nangaoan and the Duyunian It should be noted that Peng's definition of the "Meishucunian" differs from the traditional Meishucunian (Qian, 1977), hence the use of quotation marks The Meishucunian Stage in the conventional sense includes the four SSF assemblages discussed above, roughly equivalent to the Jinningian plus the "Meishucunian" sensu Peng (1999) Both schemes are presented in Fig 1

### 2 2 Southern Anhui and correlation with eastern Yunnan

In southern Anhui and neighboring western Zhejiang and northern Jiangxi provinces, the Neoproterozoic Piyuancun Formation (siliceous dolostones) is overlain by the Cambrian Hetang Formation Regionally, the Hetang Formation can be divided into four lithostratigraphic units—in ascending order, a phosphorite unit typically <1 m thick, a "stone coal" (flammable, organic-rich mudrock and shale) unit <50 m thick, a siliceous–carbonaceous shale unit <30 m thick, and a carbonaceous shale unit (with carbonate nodules) <100 m thick (Xue and Yu, 1979) Articulated sponges and millimeter-size orthothecid hyoliths were collected from the second lithostratigraphic unit, about 30 m above the base of the Hetang Formation, in a stone coal quarry near the village of Lantian (29°55′N, 118°05′E), Xiuning County, southern Anhui Province

Correlation between the Lower Cambrian successions in southern Anhui and those in eastern Yunnan depends on SSFs, acritarchs, and trilobites from the Hetang Formation in the neighboring western Zhejiang province This is justified because the Hetang Formation can be traced with confidence from southern Anhui to western Zhejiang (Xue and Yu, 1979). Key elements of the *Anabarites trisulcatus–Protohertzina anabarica* assemblage, including *A trisulcatus* and *P anabarica,* as well as other SSFs such as *Kaiyangites jianshanensis* and *Zhejiangorhabdion comptum,* have been recovered from basal phosphorite beds of the Hetang Formation in western Zhejiang (Zhao and Yue, 1987, Yue and He, 1989, He and Yu, 1992, Yue and Zhao, 1993), although these phosphorite beds sometimes are mapped as the uppermost Dengying Formation These SSFs indicate that the sponge fossils from the stone coal unit cannot be older

than the *A trisulcatus–P anabarica* assemblage of the Meishucunian Stage *Kaiyangites* has been reported from phosphorite beds of the lowermost Yangjiaping Formation in northwestern Hunan (Ding and Qian, 1988) and the lowermost Niutitang and uppermost Liuchapo formations in central Guizhou (Qian and Yin, 1984) These phosphorites beds are considered Meishucunian in age (Qian, 1999), but the exact range of *Kaiyangites* is unknown

Carbonate nodules from the stone coal unit in western Zhejiang yield a unique SSF assemblage Some elements in this assemblage have been identified as species of *Siphogonuchites* and *Lopochites* (He and Yu, 1992) If these SSFs are correctly identified, they suggest equivalency to the second SSF assemblage, the *Siphogonuchites triangularis–Paragloborilus subglobosus* assemblage Qian (1999), however, questioned the identification of *Siphogonuchites* and *Lopochites* from the stone coal beds in western Zhejiang Other SSFs such as *Jiangshanodus triangularis, Kijacus kijanicus, Hagionella cultrate,* and *Allonia tripodophora,* as well as bivalved arthropods (bradoriids), also occur in carbonate nodules of the stone coal beds (Yue and He, 1989, Yue and Zhao, 1993, Qian, 1999) The biostratigraphic significance of these small shelly fossils is not clear Bradoriids, however, usually occur in Qiongzhusian or younger deposits in Yunnan, indicating that the stone coal beds may be Qiongzhusian

Further upsection in the carbonaceous shale unit, trilobites such as *Hunanocephalus, Hupeidiscus,* and *Hsuaspis* have been discovered in carbonate nodules (Li et al, 1990, He and Yu, 1992), suggesting a Qiongzhusian or early Canglangpuan age because these trilobites occur in Qiongzhusian to early Canglangpuan deposits in eastern Yunnan and eastern Guizhou (Zhou and Yuan, 1980; Steiner et al., 2001b) Taken together, SSFs and trilobites restrict the sponge-bearing stone coal beds to be no older than the Meishucunian and no younger than the Qiongzhusian Stage, equivalent to the Diandongian–early Qiandongian in Peng's chronostratigraphic system (Peng and Babcock, 2001) This is consistent with the fact that no trilobites have been recovered from the stone coal beds at Lantian

Further stratigraphic constraints based on acritarchs and sponges can also be considered First, abundant occurrence of *Micrhystridium*-like acritarchs seems to coincide with the SSF *Anabarites trisulcatus–Protohertzina anabarica* assemblage on the Yangtze platform (Yin, 1995) Abundant occurrence of *Micrhystridium*-like acritarchs in the stone coal beds of the Hetang Formation therefore indicates a Meishucunian age (Xue and Yu, 1979, Yin, 1995) Second, an unusual type of pentact sponge spicule is found in the Hetang sponge *Sanshapentella dapingi*, similar although smaller pentacts co-occur with SSFs of the *A trisulcatus–P anabarica* assemblage in Hunan (Ding and Qian, 1988, Qian and Bengtson, 1989) Furthermore, pre-trilobite stone coal beds are distributed widely across the Yangtze platform in South China Indeed, the Shiyantou Formation in eastern Yunnan, consisting of organic-rich siltstones, may be broadly correlated with the Hetang stone coal beds. If correct, the sponge-bearing Hetang stone coal beds may be older than the Qiongzhusian Chengjiang Biota that occurs in the trilobite-bearing Yu'anshan Formation However, these stratigraphic constraints are weak, as purely lithostratigraphic correlations may be misleading, and sponge spicules as biostratigraphic tools have not been independently tested Furthermore, the identification of *Micrhystridium*-like acritarchs is problematic in many publications (see Moczydłowska, 1991, 1998, for discussion) Indeed, *Micrhystridium*-like acritarchs may have a longer range, from the Nemakit-Daldynian to the Atdabanian (Sergeev, 1989; Moczydłowska, 1991), than suggested by Yin (1995) In addition, a Meishucunian age is contradictory to the occurrence of bradoriids in the stone coal beds. Therefore, the stone coal beds can only be confidently constrained to be Meishucunian–Qiongzhusian in age

### 2 3 Correlation with Siberia

More challenging is the correlation between Early Cambrian successions in South China and Siberia (Brasier et al, 1990; Landing, 1994, Zhuravlev, 1996, Qian et al, 2001, Jenkins et al, 2002) The difficulty lies in the strong provincialism of SSFs and facies dependence of archaeocyathans Siberian SSF assemblages are markedly different from those in South China, making interregional correlation on the basis of SSFs a difficult exercise (Qian and Bengtson, 1989) For example, archae-

ocyathans first appeared in the pre-trilobite Tommotian in Siberia (Khomentovsky and Karlova, 1993), whereas their first appearance in the Qiongzhusian in South China postdates that of trilobites (Yuan et al, 2001) As a consequence, diverging opinions exist regarding the correlation between the Tommotian and the Meishucunian stages Some think that the Tommotian is equivalent to the upper part of the Meishucunian (Brasier et al, 1990, Landing, 1994, Qian et al, 2001; Zhu et al, 2001) Others think that the Tommotian is largely missing in South China (Zhuravlev, 1996, Zhuravlev and Wood, 1996) Despite these uncertainties, opinion seems to converge on correlating the Meishucunian *Anabarites trisulcatus–Protohertzina anabarica* assemblage with either the entire Nemakit-Daldynian (Khomentovsky and Karlova, 1993, Landing, 1994, Geyer and Shergold, 2000) or only the *A trisulcatus* Zone of the Nemakit-Daldynian Stage in Siberia (Brasier et al, 1990, Qian et al, 2001) Regardless, the sponge-bearing stone coal beds of the Hetang Formation overlie the *A trisulcatus–P anabarica* assemblage and thus can be no older than the Nemakit-Daldynian and probably no older than the Tommotian The upper age limit for the stone coal beds depends on correlation between the trilobite-bearing Qiongzhusian and Atdabanian stages (Palmer, 1998) The exact correlation of trilobite zones between South China and Siberia is problematic, but most agree that the first appearance of the oldest Chinese trilobite *Parabadiella* in the Qiongzhusian Yu'anshan Formation in eastern Yunnan (Zhang, 1987; Steiner et al, 2001b) cannot be much younger than Atdabanian (Palmer, 1998) Therefore, the age of the stone coal beds of the Hetang Formation is probably Tommotian or Atdabanian

Carbon isotope chemostratigraphy is broadly consistent with these correlations A prominent positive $\delta^{13}C$ excursion has been reported from the Dahai Member of the Zhujiaqing Formation and its equivalents at multiple sections in South China (Brasier et al, 1990, Zhou et al, 1997) The Dahai positive is probably equivalent to the I (upper Nemakit-Daldynian) or the I' (lower Tommotian) positive excursion in Siberia (Brasier et al, 1994b, Knoll et al, 1995, Kouchinsky et al, 2001) Unfortunately, chemostratigraphic data from the Qiongzhusian and Canglangpuan successions are not sufficient to allow a reliable correlation between South China and Siberia Future chemostratigraphic research on the Lantian section may provide useful $\delta^{13}C_{carb}$ and $\delta^{13}C_{org}$ data for more precise correlation

### 2 4 Correlation with Newfoundland

Still more challenging is the correlation between Lower Cambrian beds in South China and the chosen Neoproterozoic–Cambrian boundary GSSP at Fortune Head, southeastern Newfoundland (Landing, 1994) The first occurrence of *Treptichnus pedum* in the primarily siliciclastic Fortune Head section defines the Neoproterozoic–Cambrian boundary, although it has been recently discovered that the range of *T pedum* extends a few meters below the designated boundary (Gehling et al, 2001) The base of the Nemakit-Daldynian and the Meishucunian is usually taken to approximate the Neoproterozoic–Cambrian boundary in carbonate successions (Narbonne et al, 1987, Brasier et al, 1994a, Landing, 1994, Rowland et al, 1998) This seems consistent with rare occurrences of *T pedum* within the Meishucunian *Anabarites trisulcatus–Protohertzina anabarica* assemblage in eastern Yunnan (Zhu et al, 2001) Given the unknown (and presumably long) range of *T pedum*, *A trisulcatus*, and *P anabarica* (Nowlan et al, 1985, Qian and Bengtson, 1989, Gehling et al, 2001), precise interfacies correlation cannot be achieved on the basis of scattered occurrences of these index fossils Unfortunately, the siliciclastic nature of the Fortune Head stratotype does not allow an independent chemostratigraphic test of hypothesized interfacies correlations (Rozanov et al, 1997).

### 2 5 Numerical age

It can be concluded from the preceding discussion that the stone coal beds of the Hetang Formation were likely deposited during the Meishucunian–Qiongzhusian, or, in Siberian terminology, the Tommotian–Atdabanian stages An ash bed from the (presumably Nemakit-Daldynian) *Anabarites trisulcatus–Protohertzina anabarica* assemblage of the lower Zhongyicun Member in eastern Yunan gives an age of 538 2±1 5 Ma ($\sigma$, Jenkins et al, 2002) This age

can be regarded as a maximum age for the stone coal beds in southern Anhui

The numerical age of the Tommotian Stage critically depends on the definition of the Nemakit-Daldynian–Tommotian boundary and inter-regional correlation (Rowland et al, 1998; Khomentovsky and Karlova, 2002) An ash bed (530 7±0 9 Ma) in the lower Placentian Series in New Brunswick was interpreted as upper Nemakit-Daldynian (Isachsen et al, 1994) In northern Siberia, an ash bed (534 6±0 5 Ma) in the Tyuser Formation was interpreted as lowest Tommotian (Bowring et al, 1993) These interpretations are contradictory and both ashes have later been reinterpreted as middle Tommotian (Vidal et al, 1995, Jenkins et al., 2002) It appears that the New Brunswick ash (530 7±0 9 Ma) postdates and the Siberia ash (534.6±0 5 Ma) predates the *Watsonella crosbyi* zone (Bowring et al, 1993, Jenkins et al, 2002) Because the stone coal beds of the Hetang Formation can be probably correlated with the black shales of the lower Shiyantuo Formation, which overlies the *Watsonella crosbyi* (=*Heraultipegma yunnanensis*) assemblage in the Dahai Member of the Zhujiaqing Formation, the radiometric date of 534 6±0 5 from Siberia can be taken as a maximum age of the stone coal beds A Re–Os date of 542±11 Ma from Lower Cambrian black shales in Hunan and Guizhou provinces in South China (Li et al, 2002), which are correlative to the stone coal beds in southern Anhui, is marginally consistent with our estimate

The minimum age of the stone coal beds is constrained by the Atdabanian–Botomian boundary, approximately 520 Ma (Bowring and Erwin, 1998, Landing et al, 1998, Jenkins et al, 2002) The stone coal beds are therefore broadly constrained to be between 535 and 520 Ma, making the sponge fossils reported in this paper among the earliest known articulated sponges (see also Steiner et al, 1993)

## 3. Description of sponge fossils

All specimens described in this paper are from the stone coal bed of the Hetang Formation at Lantian (29°55'N, 118°05'E), southern Anhui Province Most articulated sponges are preserved as compressed body fossils Siliceous spicules are typically replaced by diagenetic pyrite On the weathered outcrop, the pyrite is typically oxidized to form rusty limonite

The illustrated sponge fossils are reposited at Nanjing Institute of Geology and Palaeontology under the catalog numbers NIGPAS-134519 to 134543

Class Demospongea Sollas, 1875
Order Monaxonida Sollas, 1883
Family Choiidae DeLaubenfels, 1955
Genus *Choia* Walcott, 1920
*Type species* —*Choia carteri* Walcott, 1920, from the Middle Cambrian Burgess Shale in British Columbia

*Remarks* —The genus *Choia* is characterized as a disc-shaped sponge consisting of two types of radiating monaxonal spicules fine monaxons thatched in a central disc and larger coronal monaxons that extend beyond the central disc Species of *Choia* are differentiated by the radius of the central disc and the length of the coronal spicules (Rigby, 1986)

*Choia utahensis* Walcott, 1920
(Fig 2)

*Choia utahensis* Walcott, 1920, p 25, pl 75, fig 1, Rigby, 1978, p 1331, pl 2, fig 2, Rigby, 1983, p 252, fig 5A,C
*Choia*-like demosponge, Yuan et al , 2002, p 364, fig 3D

*Description* —Hemispherical demosponge that consists of a well-defined central disc and two types of monaxonal spicules, both radiating from the center of the central disc. The central semicircular disc is about 20 mm in radius and thatched with fine monaxons, they are typically 0 01 mm in diameter but their full length is difficult to determine because of they are densely matted in the central disc Large coronal oxeas are 0 3–2 mm in maximum width and can extend 20–30 mm beyond the margin of the central disc

*Discussion* —Coronal spicules of *Choia utahensis* from the Middle Cambrian Marjum Limestone and Wheeler Shale in Utah are typically less than 0 5 mm in maximum diameter (Walcott, 1920, Rigby, 1978, 1983) The Hetang specimens have larger coronal oxeas but are otherwise similar to the North American specimens The semicircular central disc of the Hetang specimens is perhaps a result of preservation, similarly preserved semicircular *Choia* is also known from Middle Cambrian in Utah (Rigby, 1983)

Rigby (1986) reconstructed *Choia carteri* as a conical sponge with its apex pointing upward This is in contrast to Walcott's (1920) original interpretation in which *Choia* was envisioned as a conical sponge with the apex pointing downward The Rigby reconstruction is hydrodynamically more stable, but such a sponge is less likely to have been preserved as a semicircular body fossil upon compression The

Fig 2 *Choia utahensis* Walcott, 1920 Fine spicules in the central disc are not discernible at this magnification NIGPAS-134519 Specimen is about 100 mm in height

semicircular specimen from the Hetang Formation is more consistent with Walcott's reconstruction

*Choia xiaolantianensis* from the Lower Cambrian Chengjiang Biota is minimally described (Hou et al, 1999) Published photographs suggest that it is conspecific with either *Ch utahensis* or the smaller *Ch carteri* (Rigby, 1986)

*Material* —Four specimens

*Choia? striata* sp nov
(Fig 3A–D)

*Diagnosis* —A possible species of *Choia* with a poorly defined central disc and abundant, densely packed coronal oxeas Coronal oxeas large (up to 2 5 mm in maximum width and 50–100 mm in length), speckled with subcircular patches of organic matter Fine monaxons about 0 07–0 1 mm in diameter are distributed in the central disc area Coronal oxeas and some fine monaxons bear longitudinal striae

*Description* —An incomplete, fan-shaped specimen about 110×240 mm in size The poorly defined central disc is roughly 30–40 mm in radius Numerous fine monaxons (0 07–0.1 mm in diameter and about 1–10 mm in length) occur in the central disc, but the central disc is not thatched Coronal oxeas, 1 0–2 5 mm in maximum width and 50–100 mm in length, taper to pointed ends at both termini. Coronal oxeas are flattened in their middle part where the width is the greatest, and become cylindrical toward tapering termini Subcircular patches (about 0 1 mm in diameter) of amber-colored organic matter are found on most coronal oxeas, but not in rock matrix Striae occur in both types of spicules with a consistent spacing of 0 03–0 04 mm, but some spicules have no striae probably because of poor preservation

*Discussion* —This new species can be differentiated from all other *Choia* species (Walcott, 1920, Rigby, 1978, 1983, 1986) by its poorly defined central disc, large and striated coronal spicules, and organic speckles on the coronal spicules These features depart from the original diagnosis of *Choia*, thus this species is placed in the genus *Choia* with uncertainty Indeed, even the sponge affinity of this species can be questioned, but more specimens are needed for further study

*Etymology*—*Striatus*, Latin, referring to the striae on the large blade-like coronal oxeas

*Holotype* —NIGPAS-134520, Fig 3A–D
*Material* —A single specimen

Class Hexactinellida Schmidt, 1870
Subclass Amphidiscophora Schulze, 1887
Order Reticulosa Reid, 1958
Superfamily Protospongioidea Finks, 1960
Family Protospongiidae Hinde, 1887
Genus *Diagoniella* Rauff, 1894
*Type species* —*Diagoniella coronata* (Dawson and Hinde, 1889) from Cambro-Silurian strata in Little Métis, Quebec

*Remarks*.—The genus *Diagoniella* is characterized by its diagonally oriented stauracts Stauracts in the genus *Protospongia* are more or less parallel to the principal axis of the sponge body The goblet-shaped *Gabelia* consists mostly of regularly oriented hexacts rather than stauracts (Rigby and Murphy, 1983)

*Diagoniella cyathiformis* (Dawson and Hinde, 1889)
(Fig 3E–F)

*Diagoniella cyathiformis* (Dawson and Hinde, 1889), p 23, Rigby, 1978, p 1336, pl 1, fig 2–3, pl 2, fig 1, Rigby, 1983, p 255, fig 6F–H

*Description* —Subconical to oval, thin-walled, protosponge with diagonally oriented stauracts Sponge has a somewhat rounded base and wide osculum Sponge body 25–90 mm in height and 15–45 mm in maximum width Stauracts are regularly spaced 1 5–3 mm apart to form rhombic quadrules Each quadrule appears to be formed by two (rather than four) first-order stauracts First-order stauracts have ray diameters of 0 03–0 05 mm and ray lengths about 2 5 mm Second-order stauracts occur in quadrules. No marginalia or prostalia are preserved The Hetang specimens are closely similar to *Diagoniella cyathiformis* from the Middle Cambrian Wheeler Shale and Marjum Limestone (Rigby, 1978, 1983)

*Material* —Three specimens

Genus *Protospongia* Salter, 1864
*Type species* —*Protospongia fenestrata* Salter, 1864, from the Cambrian of Wales, Great Britain
*Remarks* —*Protospongia* is a common sponge genus in the Cambrian It can be distinguished from

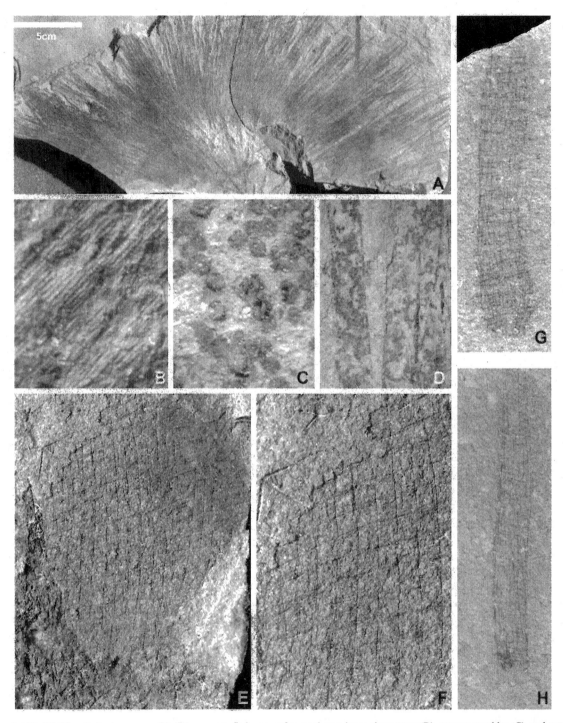

Fig 3 (A)–(D) *Choia? striata* sp nov (B)–(D) are magnified views of coronal spicules to show striae (B), organic speckles (C), and general morphology of coronal spicules (D) NIGPAS-134520, holotype (E)–(F) *Diagoniella cyathiformis* (Dawson and Hinde, 1889) (F) is close-up of the upper left part of (E) NIGPAS-134521 (G)–(H) *Protospongia gracilis* sp nov G, NIGPAS-134522, holotype (H) NIGPAS-134523 Both specimens are incompletely preserved Scale bar in (A) is 50 mm for (A), 0 5 mm for (B)–(C), 2 5 mm for (D), 10 mm for (E), 6 mm for (F), 8 mm for (G) and (H)

other genera (for example *Diagoniella*, *Gabelia*, *Phormosella*, *Cyathophycus*, *Plectoderma*) in this family by its thin-walled skeleton consisting of regularly arranged stauracts parallel to the principal body axis (Rigby, 1978, 1986, Rigby and Murphy, 1983)

*Protospongia gracilis* sp nov
(Fig 3G–H)

Unnamed sponge, Yuan et al , 2002, p 364, fig 2F
*Diagnosis* —A thin-walled protosponge with slender tubular skeletal net Skeletal net composed principally of stauracts, but with some hexacts Stauracts with elongate vertical rays that are parallel to the long axis of the skeletal net Quadrules are irregularly arranged Spicular ranks are not apparent Longitudinal rays of stauracts converge at oscular margin Some horizontal rays protrude from sponge surface
*Description* —Sponges about 3–5 mm in diameter and 30–70 mm in height. Stauracts have ray diameters of approximately 0 03–0 05 mm and ray lengths of 2–3 mm (vertical rays) and 0 8 mm (horizontal rays) Quadrules vary in size, about 0 5–1 mm in maximum dimensions Stauracts in the holotype (Fig 3G) appear to be arranged such that the horizontal rays are somewhat bundled, but this may be a preservational artifact
*Discussion* —This new species differs from other Cambrian *Protospongia* species (Walcott, 1920, Rigby, 1986) in its slender tubular form, irregular quadrules, and vertically elongate stauracts It is somewhat similar to the Silurian protosponge *Gabelia* (Rigby and Murphy, 1983, Rigby and Maher, 1995), but its slender form is distinct from the goblet-shaped sponge body of *Gabelia* Furthermore, *Gabelia* is primarily composed of hexacts rather than stauracts
*Etymology* —*Gracilis*, Latin, referring to the slender tubular form of this new species
*Holotype* —NIGPAS-134522, Fig 3G
*Material* —Three specimens

*Protospongia* cf *conica* Rigby and Harris, 1979
(Fig 4A)

*Protospongia conica* Rigby and Harris, 1979, p 974, pl 1, fig 1, pl 2, fig 1, 5, text-fig 2, 3

*Description* —A single specimen with spindle-shaped skeletal net that is composed primarily of stauracts with some hexacts Stauracts slightly elongate vertically Rectangular quadrules occur between stauracts Spicular ranks are not apparent Longitudinal rays of stauracts converge at both ends of sponge body Some stauract rays protrude from sponge surface
The specimen is about 20 mm in height and 9 mm in width Stauracts have ray diameters of approximately 0 03–0 05 mm and ray lengths of 1 2 mm (vertical rays) and 0 8 mm (horizontal rays) Quadrules vary in size, between 0 6 and 0 8 mm in maximum dimensions
*Discussion* —*Protospongia conica* was first described from the Silurian in northern British Columbia (Rigby and Harris, 1979). The specimen from the Hetang Formation resembles the Silurian specimens with its pointed base and somewhat irregularly arranged spicules The Silurian specimens, however, are larger (up to 44 mm in height) and have ranked stauracts (0 04–0 1 mm in ray diameter and 0 4–1 0 mm in ray length) Their oscular end does not narrow as much as the Hetang specimen Considering these differences and the Cambrian age of the Hetang specimen, it is possible that the Hetang specimen represents a new protosponge species More specimens are needed to differentiate the Hetang and the Silurian populations At present, we tentatively place the single specimen from the Hetang Formation in open nomenclature, *Protospongia* cf *conica*
*Material* —A single specimen

Genus *Gabelia* Rigby and Murphy, 1983
*Type species* —*Gabelia pedunculus* Rigby and Murphy, 1983, from Devonian shales in the northern Roberts Mountains, Neveda

*Gabelia* sp indet
(Fig 4B–E)

*Description* —A fragmented specimen with more or less regularly arranged hexacts Putative stauracts may be present too First-order hexacts have ray diameters of 0 16 mm and ray lengths of 3–5 mm Smaller hexacts of several ranks divide, although unevenly, quadrules between larger spi-

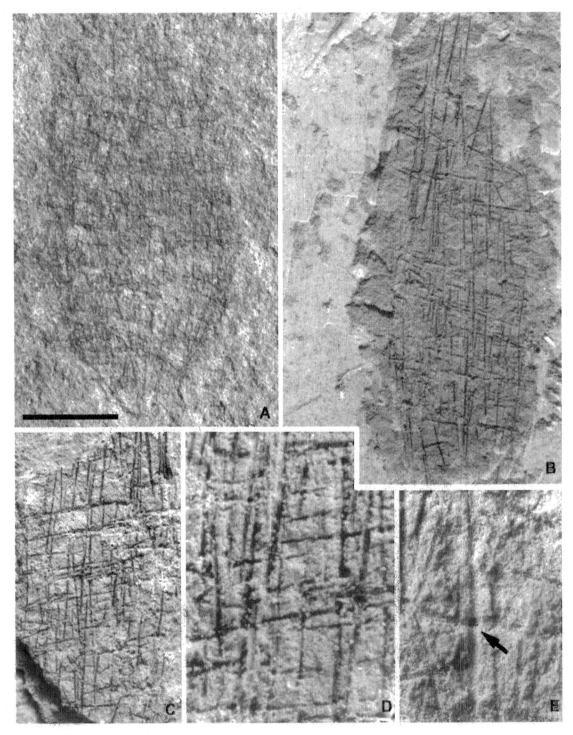

Fig 4 (A) *Protospongia* cf *conica* Rigby and Harris, 1979 NIGPAS-134524 (B)–(E) *Gabelia* sp indet (C) is counterpart of (B) (D) and (E) are close-ups of (B) and (C), respectively, to show molds of hexacts, two of the six rays are perpendicular to the bedding plane, making a hole (arrow) on the bedding plane NIGPAS-134525 Scale bar in (A) is 5 mm for (A), 4 mm for (B), 5 mm for (C), 2 mm for (D), 1 mm for (E)

cules Because of its incomplete preservation, identification at species level is impossible, but its skeleton composed primarily of hexacts places this form in the genus *Gabelia* (Rigby and Murphy, 1983, Rigby et al, 1991, Rigby and Maher, 1995)

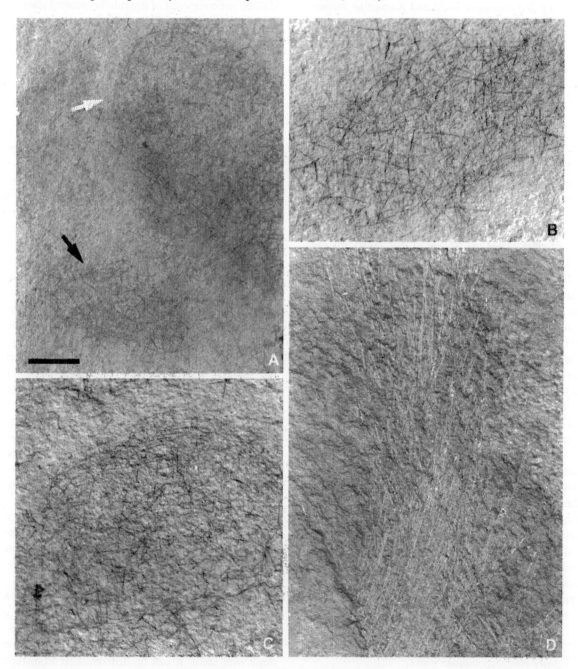

Fig 5 (A)–(C) *Triticispongia diagonata* Mehl and Reitner *in* Steiner et al , 1993 (A) shows two specimens—one in the upper right, the other lower left (B) is close-up view of the lower left specimen in (A) (black arrow) NIGPAS-134526 (C) is a close-up view of the upper right specimen in (A) (white arrow) NIGPAS-134527 (D) *Hyalosinica archaica* Mehl and Reitner *in* Steiner et al , 1993 NIGPAS-134544 Scale bar in (A) is 10 mm for (A), 4 mm for (B) and (C), 7 mm for (D)

*Material* —A single specimen with part and counterpart

Genus *Triticispongia* Mehl and Reitner *in* Steiner et al , 1993

*Type species* —*Triticispongia diagonata* Mehl and Reitner *in* Steiner et al , 1993

*Diagnosis* —This monospecific genus is diagnosed as "sponge body hardly exceeds 10 mm total size Spicules are small triaxons, mainly stauracts with their paratangentialia in a diagonal arrangement" (Mehl and Reitner *in* Steiner et al., 1993)

*Triticispongia diagonata* Mehl and Reitner *in* Steiner et al , 1993
(Fig 5A–C)

*Triticispongia diagonata* Mehl and Reitner *in* Steiner et al , 1993, p 307, pl 3, fig 3, Rigby and Hou, 1995, p 1011, fig 4 1–4 2

*Description* —Compressed sponge oval to subcircular Two specimens measure 13×27 mm and 30×55 mm in size, respectively Spiculation made mainly of stauracts that typically have a ray diameter about 0 03–0 05 mm and ray length about 3–5 mm Second-order, shorter stauracts and hexacts (about 0 03–0 05 mm in ray diameter and 1–2 mm in ray length) are also present Because of compression, quadrular spicule organization is not obvious

*Discussion* —Rigby and Hou (1995) observed marginalia and root tufts, as well as crude quadrules of at least three ranks, in *Triticispongia diagonata* from the Lower Cambrian Chengjiang Biota These features are poorly preserved in the Hetang specimens and those from the coeval Niutitang Formation in northwestern Hunan (Steiner et al , 1993) Additionally, the Hetang specimens are larger than those from the Niutitang Formation in northwestern Hunan (Steiner et al , 1993) and the Chengjiang Biota in eastern Yunnan (Rigby and Hou, 1995), which are typically less than 10 mm in height

*Material* —Two specimens

Superfamily Hintzespongioidea Finks, 1983
Family Hintzespongiidae Finks, 1983
Genus *Lantianospongia* gen nov
*Type species* —*Lantianospongia palifera* gen et sp nov

*Diagnosis* —A genus of the Hintzespongiidae with ovoidal skeleton. Sponge walls thin and perforated by numerous, large, elliptical to circular parietal gaps. Skeletal net composed of irregularly oriented stauracts of several sizes Basal part of sponge body reinforced by long, diagonally oriented, probably bundled monacts or diacts Oscular margins with regularly spaced serrations and indentations Long, bundled monact or diact supports each of the serrations; the bundled spicules fan out beneath the apex of serrations (Fig 6E)

*Discussion* —Several hexactinellids have parietal gaps in their skeletal nets *Stephenospongia* Rigby, 1986, from the Middle Cambrian Burgess Shale, is more than 44 mm high and has elliptical parietal gaps 10–14 mm in maximum diameter Its skeleton is made of hexactine-based spicules with uniform diameter (ca 0 06 mm) but variable ray length (from less than 1 mm to more than 3 mm). Three other hexactinellid genera, *Ratcliffespongia* (Rigby, 1969), *Hintzespongia* (Rigby and Gutschick, 1976; Rigby, 1983), and *Valospongia* (Rigby, 1983), all from the Middle Cambrian Marjum Formation in western Utah, also bear parietal gaps in their skeletal nets *Ratcliffespongia* (up to 70 mm high) has somewhat smaller parietal gaps (3–5 mm in maximum diameter) and loosely woven stauracts of several sizes (ray diameter between 0.02 and 0.06 mm, ray length between 0 5 and 3.5 mm) *Hintzespongia* (22–32 mm) has double-layered walls, with parietal gaps (0 5–0 7 mm in diameter) occurring on the inner wall, and stauracts of several sizes (ray diameter between 0 02 and 0 06 mm, ray length between 0.3 and 2 mm) Parietal gaps (or "mounds" in the terminology of Rigby, 1983) of two sizes (2 mm and 6–8 mm in diameter) occur in the endosomal net of *Valospongia* (193 mm high, 90 mm wide). *Valospongia* has several sizes of hexactine-based spicules, ranging from 0 3 mm to several centimeters in ray length and from 0 04 to 0 3 mm in ray diameter

*Lantianospongia* shares some characters, such as the presence of parietal gaps and several sizes of stauracts, with the Middle Cambrian genera mentioned above, indicating close evolutionary relationships with them It can be differentiated from those Middle Cambrian sponges, however, by its basal diagonal monacts/diacts and oscular supporting monacts/diacts as well as its serrated oscular margin

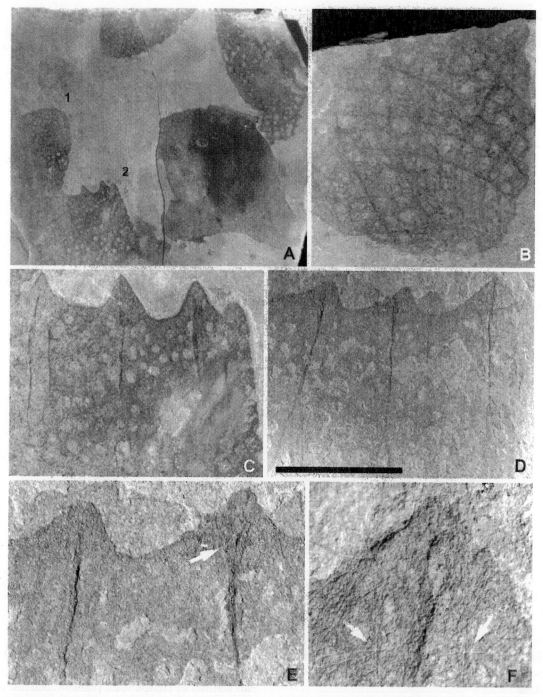

Fig 6 *Lantianospongia palifera* gen et sp nov (A) A slab with five incompletely preserved specimens (B) Close-up view of the counterpart of specimen 1 in (A), showing diagonally arranged spicule bundles near the base of the sponge body NIGPAS-134528 (C) Close-up view of specimen 2 in (A), showing serrated oscular margin, parietal gaps, and vertically oriented spicule bundles Holotype, NIGPAS-134529 (D)–(F) Another specimen with serrated oscular margin and vertically oriented spicule bundles NIGPAS-134530 (E) and (F) are closer views of serrated oscular margin Bundled supporting spicules appear to fan out below the right serration in (E) (arrow) Notice stauracts in (F) (arrows) Scale bar in (D) is 300 mm for (A), 60 mm for (B), 80 mm for (C), 60 mm for (D), 30 mm for (E), 15 mm for (F)

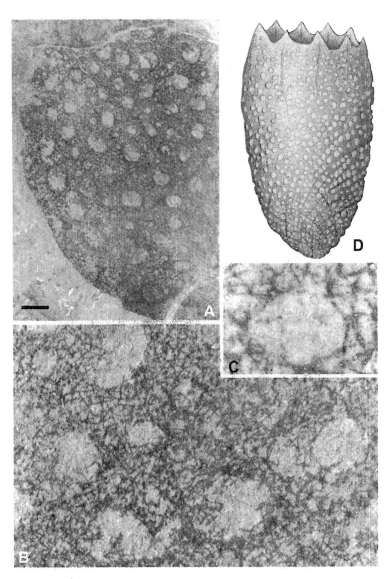

Fig 7 *Lantianospongia palifera* gen et sp nov (A)–(C) A specimen with parietal gaps of variable sizes Neither serrated oscular margin nor bundled supporting monaxons are preserved, probably due to the fragmentary nature of this specimen (B) and (C) are close-ups of (A), showing variable sizes of parietal gaps (B) and spicule arrangement around parietal gaps (C) NIGPAS-134531 (D) Conceptual reconstruction Scale bar in (A) is 10 mm for (A), 2 5 mm for (B). 1 5 mm for (C)

*Etymology.*—The generic name refers to the locality (Lantian of southern Anhui Province, South China) where the type species *Lantianospongia palifera* was collected

*Lantianospongia palifera* gen et sp nov
(Figs. 6 and 7)
Unnamed sponge, Yuan et al , 2002, p 364, fig 3A–B

*Diagnosis* —Same as for genus

*Description* —Ovoidal skeletal net 100–300 mm high and 80–150 mm wide Basal end rounded Lower part consists of conspicuous, probably bundled monacts/diacts that are diagonally oriented Oscular margin serrated There are at least 4 serrations, separated by indentations, on oscular margin Neighboring serration apices are about 50 mm apart Serration apices stand about 15 mm above the base of indentations Each serration is subtended by a group of bundled monacts or diacts that extend downward about one-third of sponge height Parietal gaps circular or elliptical, between 5 and 15 mm in maximum diameter (mean=8 4 mm, N=20) and spaced at 15–20 mm apart They tend to be concentrated in the upper part of the sponge Basal diagonal monacts/diacts and oscular supporting monacts/diacts can be 50–100 mm in length and are organized in bundles of 0 8–1 5 mm in diameter Much smaller diacts, about 0 15 mm in diameter and 15 mm in length, occur sporadically on the skeletal net Most stauracts in tracts have ray diameters of 0 04–0 08 mm and ray length of 0 5–1 5 mm Larger stauracts (about 0 2–0 25 mm in ray diameter and 1 5–2 mm in ray length) sporadically occur in tracts Most parietal gaps are void of spicules, but a few stauracts can occur in some parietal gaps although their density is much lower than in the tracts This is probably because of the same sponge wall superimposed on itself as a result of compression

*Discussion* —Comparison with other parietal-gap-bearing hexactinellids has been given in discussion of the genus.

*Etymology* —*Palus*, Latin, pole, referring to the supporting monacts/diacts beneath each serration on the oscular margin *Fero*, Latin, carry

*Holotype* —NIGPAS-134529, Fig 6C

*Material* —Eight specimens, five of which preserved on a single slab

Class Hexactinellida Schmidt, 1870
Subclass, Order, Superfamily, and Family indet
Genus *Sanshapentella* Mehl and Erdtmann, 1994
*Type species* —*Sanshapentella dapingi* Mehl and Erdtmann, 1994

*Remarks* —This monospecific genus is characterized by its dermal pentacts with their four paratangential rays bent toward the spongocoel Stauracts and hexacts also occur in this genus It is clearly a hexactinellid genus, but its systematic relationships with established hexactinellid orders and families (Finks, 1983) cannot be determined

Isolated pentacts from the Lower Cambrian Yangjiaping Formation (probably coeval to the Hetang stone coal beds), identified as *Hunanospongia delicata* (Ding and Qian, 1988), are broadly similar to pentacts of *Sanshapentella dapingi* Spicules in *H delicata*,

however, are about an order of magnitude smaller than pentacts of *S dapingi* We agree with Mehl and Erdtmann (1994) that these two taxa should be separated, but *Hunanospongia* and *Sanshapentella* are probably closely related

*Sanshapentella dapingi* Mehl and Erdtmann, 1994
(Fig 8)

?*Hunanospongia* sp Mehl and Reitner *in* Steiner et al, 1993, p 4, fig 2
*Sanshapentella dapingi* Mehl and Erdtmann, 1994, p 316, pl 1, fig 1–3, Yuan et al, 2002, p 364, fig 2A–D

*Description* —An incomplete specimen measures about 500 mm in height (Fig 8A); a complete individual may have been over 1 m high Skeletal net appears to consist of two layers of spicules The dermal spicules are large pentacts (about 0 4 mm in ray diameter) Such pentacts have one short ray (about 1 mm in ray length) pointing outward and four longer paratangential rays (about 3–5 mm in ray length) sharply pointing toward the spongocoel The short ray may be greatly reduced or absent Sponge body has sharp protrusions where these large pentacts are present The large pentacts are found mostly along the periphery and become rare toward the center of the compressed sponge body, this may have to do with the way the pentacts are compressed—laterally compressed pentacts are more easily recognized Subjacent to the dermal pentacts are smaller (0 10–0 15 mm in ray diameter and 1–2 mm in ray length) stauracts and occasionally hexacts A lateral bud is present in one specimen (Yuan et al, 2002, fig 2D), dermal spiculation can be traced from the bud to the stem

*Discussion* —This species was established on the basis of fragmented specimens (Mehl and Erdtmann, 1994) Better-preserved specimens from the Hetang Formation indicate this is a very large, cylindrical sponge

*Material* —Eight specimens

Genus *Hyalosinica* Mehl and Reitner *in* Steiner et al, 1993

*Type species* —*Hyalosinica archaica* Mehl and Reitner *in* Steiner et al, 1993

*Remarks* —The monospecific genus *Hyalosinica* was established on the basis of incomplete fragments

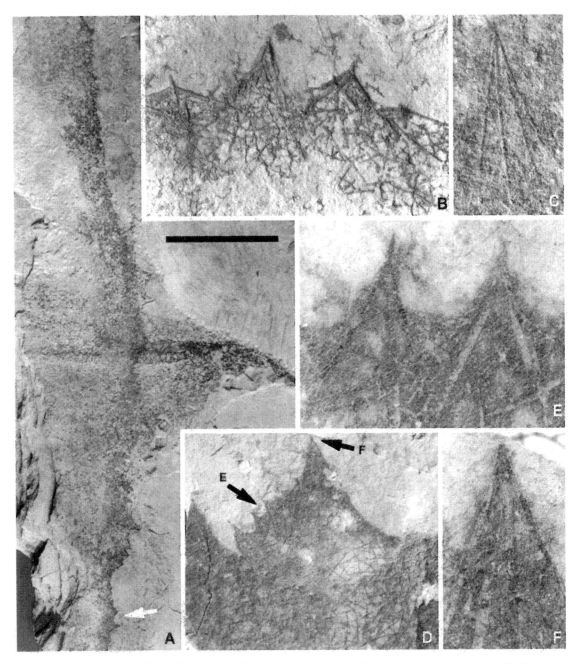

Fig 8 *Sanshapentella dapingi* Mehl and Erdtmann, 1994  (A) Slab showing possibly two overlapping specimens  (B) Close-up of the counterpart of (A) (corresponding to arrowed area in A) showing details of pentact spicules  NIGPAS-134532  (C) An isolated pentact spicule NIGPAS-134533  (D)–(F) A specimen showing marginally arranged pentacts  (E) and (F) are close-ups of (D) (arrows in D)  NIGPAS-134534  Scale bar in (A) is 100 mm for (A), 10 mm for (B), 13 mm for (C), 24 mm for (D), 5 mm for (E) and (F)

interpreted as root tufts (Steiner et al , 1993)  The root tufts are associated with triaxonal spicules and thus *Hyalosinica* was interpreted as a hexactinellid (Steiner et al , 1993)  Its systematic relationships with the established hexactinellid orders and families (Finks, 1983), however, are uncertain

*Hyalosinica archaica* Mehl and Reitner *in* Steiner et al, 1993

(Fig 5D)

*Hyalosinica archaica* Mehl and Reitner *in* Steiner et al , 1993, p 305, pl 4, fig 1a–b

Unnamed sponge, Yuan et al, 2002, p 364, fig 3G

*Discussion* —A single fragmented specimen is composed of twisted tufts of long (about 50 mm) monacts or diacts Individual spicules are about 0 1– 0 2 mm in maximum diameter The Hetang specimen is similar to those described from the Niutitang black shales (equivalent to the Hetang stone coal beds) in northwestern Hunan Province (Steiner et al , 1993) *Hyalosinica archaica* is considered to be

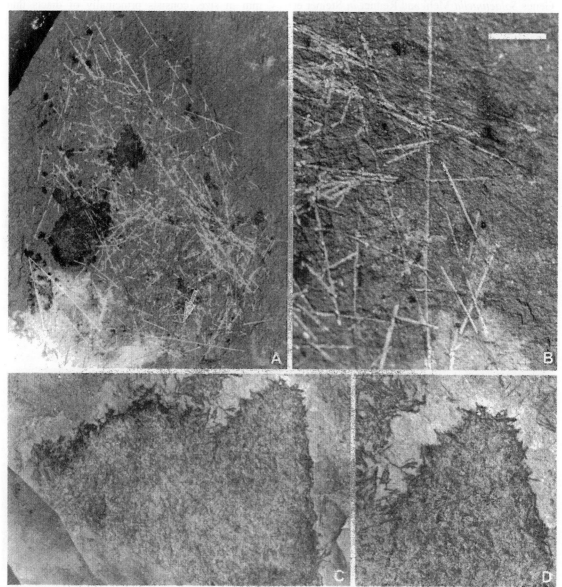

Fig 9 (A)–(B) *Solactiniella plumata* Mehl and Reitner *in* Steiner et al . 1993 (B) is magnified view of (A) (arrow) NIGPAS-134535 (C)–(D) Undetermined form 1 (D) is magnified view of the upper left of (C) NIGPAS-134536 Scale bar in (B) is 20 mm for (A), 5 5 mm for (B), 10 mm for (C), 5 mm for (D)

a hexactinellid because some spicules appear to have four paratangential rays (Steiner et al, 1993) No paratangential rays have been observed in this specimen

*Material* —A single specimen

Phylum Porifera

Class, Order, and Family undet

Genus *Solactiniella* Mehl and Reitner *in* Steiner et al, 1993

*Type species* —*Solactiniella plumata* Mehl and Reitner *in* Steiner et al, 1993

*Solactiniella plumata* Mehl and Reitner in Steiner et al, 1993

(Fig 9A–B)

*Solactiniella plumata* Mehl and Reitner in Steiner et al, 1993, p 309, pl 2, fig 1

Unnamed sponge, Yuan et al, 2002, p 364, fig 3E

*Description* —Compressed sponge body 40×60 mm in size, consists principally of diacts that are up to 40 mm in length and 0 1–0 2 mm in diameter These diacts seem to be tangentially oriented along sponge margin, but a few also

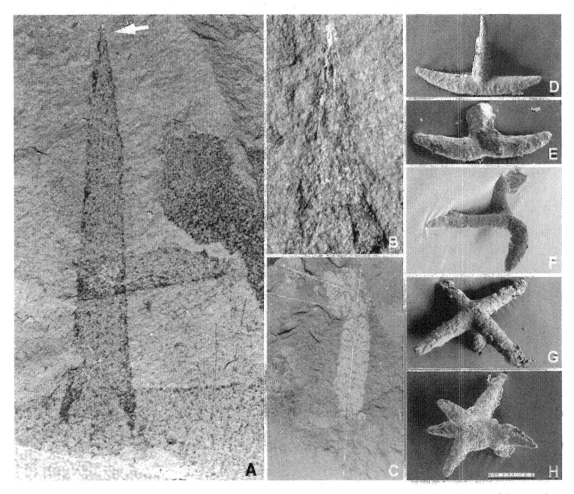

Fig 10 (A)–(B) Undetermined form 2 (B) is magnified view of (A) (arrow), showing a single pentact spicule at the apex of the conical sponge NIGPAS-134537 (C) Pyritized monaxonal spicules NIGPAS-134538 (D)–(H) Triaxonal spicules including triacts (D–F), pentacts (G), and hexacts (H) (D) NIGPAS-134539 (E) NIGPAS-134540, (F) NIGPAS-134541, (G) NIGPAS-134542, (H) NIGPAS-134543 Scale bar in (H) is 10 mm for (A), 4 mm for (B), 2 5 mm for (C), 0 25 mm for (D), 0 1 mm for (E), (F), (H), 0 15 mm for (G)

protrude beyond the margin Shorter (0 2–1 mm) and slightly thinner (0 05–0 1 mm in diameter) diacts are present preferentially in the center of compressed sponge body Probable stauracts are rarely present, but these may be superimposed diacts—the poor preservation of the single available specimen does not allow an unambiguous determination of stauracts

*Discussion* —Because of the inability to identify unambiguous triaxonal spicules, we cannot be certain whether this sponge is a hexactinellid or demosponge The Hetang specimen differs from those from the Niutitang population in its spicule orientation, the Niutitang specimens tend to have plumose arrangement of diacts (Steiner et al , 1993) At present, the Hetang specimen is tentatively described under *Solactiniella plumata*

*Material* —A single specimen

Undetermined form 1
(Fig 9C–D)

Unnamed sponge, Yuan et al , 2002, p 364, fig 3C

*Description* —Poorly preserved sponge consisting of short spicules, possibly oxeas Shape of sponge body unknown because of the fragmentary nature of the only known specimen. The preserved fragment is about 50×70 mm in size The specimen does not appear to be spicule mat, because it has clearly defined boundaries Spicules about 3–4 mm in length and 0 1–0 5 mm in maximum diameter, and randomly oriented in the compressed specimen, they seem to be centrally dilated and distally tapering It is difficult to determine whether the spicules are monaxons or modified triaxons with shortened paratangential rays

*Discussion* —Spicules of this Hetang sponge superficially resemble those of the Silurian hexactinellid *Divaricospongia dilata* (Rigby and Maher, 1995), but the latter is a cylindrical sponge with unambiguously hexact-based spicules Because only one specimen is available, it is described here as an unnamed form

*Material* —A single specimen

Undetermined form 2
(Fig 10A–B)

*Description* —A conical sponge body that appears to be a branch of a larger specimen The branching, however, could be an artifact because of two specimens overlapping each other, if so, this conical sponge may have its apex attached to the substrate The conical sponge body is about 15 mm in maximum width and gradually tapers toward a pointed apex along its 78 mm height There appear to be spicule-like objects, about 0 5–1 mm in length and 0.1–0 2 mm in diameter, on the sponge body, but the nature of these spicules (possibly stauracts or hexacts) cannot be determined with confidence There is one large pentact spicule, about 0 1 mm in ray diameter and 6 mm in ray length, present at the apex of this sponge (Fig. 10B) This pentact is similar to those in *Sanshapentella dapingi* and may indicate a close affinity with *S dapingi*

*Material* —A single specimen

Dispersed sponge spicules
(Fig 10C–H)

Abundant dispersed siliceous and secondarily pyritized sponge spicules, including monacts, diacts, triacts, stauracts, pentacts, and hexacts, occur in the stone coal beds of the Hetang Formation Some spicules can be as long as 15 cm (Fig 10C) A thin (3–20 μm), organic or pyritic axial filament is present in the center of some spicules (Yuan et al , 2002, fig 3K) Concentric layers of silica were deposited around the axial filaments (Yuan et al , 2002, fig 3L), indicating incremental spicule growth similar to modern sponge spicules (Simpson, 1984)

## 4. Discussion

*4 1 Sponge fossil record in the Neoproterozoic–Cambrian transition are sponges part of the Cambrian radiation?*

Animal phylogeny indicates that sponges (or stem-group sponges) must be among the earliest animals Indeed, sponge biomarkers have been reported in lower Neoproterozoic rocks (McCaffrey et al , 1994), suggesting the presence of at least stem-group sponges The fossil record of crown-group sponges, however, occurs much later in the geologic history Spicule-like objects have been reported from the late Neoproterozoic Doushantuo and Dengying formations in South China (Tang et al, 1978, Zhao et al, 1988, Steiner et al, 1993) and the uppermost Neoproterozoic rocks in Mongolia (Brasier et al, 1997), these are often interpreted as sponge spicules, but alternative interpretations (such as casts and molds of cyanobacterial filaments, volcanic shards, twined arsenopyrite crystals) have been proposed (Steiner et al, 1993, Zhou et al, 1998) Possible sponge body fossils occur in the late Neoproterozoic Doushantuo Formation (Li et al, 1998), but these have been questioned by Zhang et al (1998) and Yin et al (2001) The Ediacaran fossil *Palaeophragmodictya reticulata* has been interpreted as a hexactinellid by Gehling and Rigby (1996) but alternatively as a stem-group sponge by Mehl (1998) Therefore, Neoproterozoic sponge fossils are scanty and their phylogenetic interpretations are problematic Currently, no unquestionable demosponges or calcareans are known from the Neoproterozoic

Lower Cambrian sponge fossils are diverse and abundant Dispersed hexactinellid spicules are common in Nemakit-Daldynian–Tommotian deposits (Ding and Qian, 1988, Rozanov and Zhuravlev, 1992, Qian, 1999) Archaeocyathans, now accepted as a sponge group (Wood, 1999; Rowland, 2001), also appeared in the Tommotian Demosponge spicules made their first appearance in the fossil record during the Lower Cambrian Atdabanian (Bengtson et al, 1990, Gruber and Reitner, 1991, Rozanov and Zhuravlev, 1992, Zhang and Pratt, 1994, Reitner and Mehl, 1995) Calcarean spicules, such as *Dodecaactinella cynodonota* and *Eiffelia araniformis* (Bengtson et al, 1990, Reitner and Mehl, 1995), also occur in Atdabanian deposits, although Steiner et al (1993, p 302) considered *D cynodonota* demosponge spicules

Articulated sponge fossils are not known from Nemakit-Daldynian deposits. The sponge body fossils described here, along with those from the Niutitang Formation in northwestern Hunan (Steiner et al, 1993, Mehl and Erdtmann, 1994, Table 1), are probably the earliest known sponge body fossils They are broadly constrained to be Tommotian–Atdabanian in age The Hetang and Niutitang assemblages contain both hexactinellids and demo-

Table 1

List of Hetang sponges described in this paper and their occurrences in the Niutitang Formation, Chengjiang Biota, and Middle Cambrian biotas (BMW, including the Burgess Shale, Marjum Limestone, and Wheeler Shale)

| Taxa | Niutitang | Chengjiang | BMW |
|---|---|---|---|
| Demosponges | | | |
| *Choia utahensis* | | Yes | |
| *Choia? striata* sp nov | | | |
| Hexactinellids | | | |
| *Diagoniella cyathiformis* | | | Yes |
| *Protospongia gracilis* sp nov | | | |
| *Protospongia* cf *conica* | | | |
| *Gabelia* sp indet | | | |
| *Triticispongia diagonata* | Yes | Yes | |
| *Lantianospongia palifera* gen et sp nov | | | |
| *Sanshapentella dapingi* | Yes | | |
| *Hyalosinica archaica* | Yes | | |
| Porifera *incertae sedis* | | | |
| *Solactiniella plumata* | Yes | | |
| Undetermined form 1 | | | |
| Undetermined form 2 | | | |

sponges, but articulated calcareans are lacking (Table 1) In the Atdabanian-age Chengjiang Biota, articulated hexactinellids and demosponges are diverse—at least 17 species have been described to date (Chen et al, 1989, 1990, 1996, Rigby and Hou, 1995, Hou et al, 1999) Interpretation of some Chengjiang sponge fossils (such as *Quadrolaminiella*, Chen et al, 1990) as demosponges has been questioned (Reitner and Mehl, 1995, Mehl, 1998), but true demosponges such as *Choia, Choiella*, and *Allantospongia*, do occur in the Chengjiang Biota (Rigby and Hou, 1995; Chen et al, 1996; Hou et al, 1999) Articulated calcareans are absent from the Chengjiang Biota The earliest known, articulated calcareans are probably the heteractinids (including *Eiffelia globosa* and *Canistrumella alternata*) from the Burgess Shale (Rigby, 1986)

The paleontological evidence collectively suggests that, although sponges may have diverged in the early Neoproterozoic, the hexactinellids did not evolve until near the Neoproterozoic–Cambrian transition, followed by archaeocyaths in the Tommotian, and demosponges and calcareans no later than the Atdabanian The paleontological data seem to suggest that much of the sponge diversification at the class-level occurred between the Nemakit-Daldynian (or terminal Neoproterozoic if *Palaeophrag-modictya reticulata* or Doushantuo spicules are accepted as hexactinellids) and the Atdabanian Therefore, just like the bilaterians, sponges are part of the Cambrian Radiation, but only at the class level

## 4 2 Stratigraphy meets phylogeny are there significant gaps in the fossil record of early sponges?

The incompleteness of the early sponge fossil record has not attracted much attention among paleontologists In the following paragraphs, we

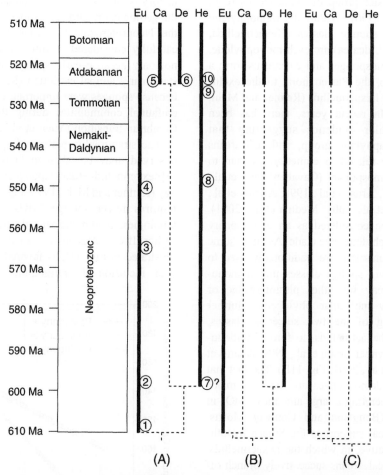

Fig 11 Sponge phylogeny, fossil record, and minimal inferred gaps (MIGs) (A) Sponges are a monophyletic group and the demosponges and calcareans form the clade of Pinacophora (Reitner and Mehl, 1996) (B) Sponges are a paraphyletic group that represents an evolutionary grade, with calcareans being closely related to eumetazoans and hexactinellids and demosponges forming the clade of Silicea (Cavalier-Smith et al, 1996, Adams et al, 1999, Medina et al, 2001) (C) Hexactinellids, demosponges, and calcareans successively branch off from the metazoan tree (Kruse et al, 1998, Borchiellini et al, 2001) Solid bars represent known fossil record, and dashed lines represent ghost lineages and gap extension Early Cambrian time scale after Vidal et al, 1995, Bowring and Erwin (1998), Landing et al (1998), and Jenkins et al (2002) Paleontological and geochronological information 1—Hofmann et al, 1990, 2—Xiao et al, 2000, Barfod et al, 2002, 3—Benus, 1988, 4—Fedonkin, 1994, Fedonkin and Waggoner, 1997, Martin et al, 2000, 5—Bengtson et al, 1990, Reitner and Mehl, 1995, 6—Chen et al, 1989, Bengtson et al, 1990, Gruber and Reitner, 1991, Rozanov and Zhuravlev, 1992, Zhang and Pratt, 1994, Reitner and Mehl, 1995, Rigby and Hou, 1995, Chen et al, 1996, Hou et al, 1999, this paper, 7—Tang et al, 1978, Zhao et al, 1988, but see Steiner et al, 1993, 8—Gehling and Rigby, 1996, 9—Ding and Qian, 1988, Rozanov and Zhuravlev, 1992, Qian, 1999, 10—Bengtson et al, 1990, Zhang and Pratt, 1994, Rigby and Hou, 1995, this paper Demosponges and calcareans reported from the Doushantuo Formation by Li et al (1998) are questioned by Zhang et al (1998) Eu eumetazoans, Ca calcareans, De demosponges, He hexactinellids

discuss the sponge fossil record in the context of sponge phylogeny and argue that, although sponges are part of the Cambrian Radiation, there are significant gaps in the fossil record of early sponges, particularly the calcareans and demosponges

On the basis of morphological data, the Porifera has been traditionally treated both as a monophyletic clade and as the sister group of eumetazoans (e g, Reitner and Mehl, 1996) Within the monophyletic Porifera, the three sponge classes (Hexactinellida, Demospongea, and Calcarea) are each monophyletic, and the cellular demosponges and calcareans form a monophyletic group, the Pinacophora, to the exclusion of the syncytial hexactinellids (Reitner and Mehl, 1996, Fig 11A) In recent years, there has been increasing molecular evidence suggesting that sponges are a paraphyletic group, with calcareans being more closely related to eumetazoans than to hexactinellids or demosponges (Cavalier-Smith et al., 1996, Collins, 1998, Kruse et al, 1998; Adams et al, 1999; Borchiellini et al, 2001; Medina et al, 2001) The Porifera can be considered as an evolutionary grade rather than a phylogenetic clade As such, some morphological features that are traditionally used to unite the three living sponge classes into a monophyletic phylum may be either plesiomorphic or convergent A few molecular phylogenies further support a grouping of the two siliceous classes, demosponges and hexactinellids, to form the monophyletic Silicea (Cavalier-Smith et al, 1996; Adams et al, 1999, Medina et al, 2001, Fig 11B) The fact that hexactinellids and demosponges are the only major metazoan groups secreting a large amount of $SiO_2$ in biomineralization (Bengtson and Conway Morris, 1992) adds additional support to the Silicea hypothesis A third alternative, in which the hexactinellids, demosponges, and calcareans successively branch off from the metazoan tree (Fig 11C), cannot be rejected with confidence (Kruse et al, 1998, Borchiellini et al., 2001)

Eumetazoans existed at least 550 Ma (Fedonkin, 1994; Fedonkin and Waggoner, 1997, Martin et al, 2000) and most likely earlier Probable stem group cnidarians have been described from the ca 600 Ma Doushantuo Formation and earlier deposits (Hofmann et al, 1990, Xiao et al, 2000, Barfod et al, 2002) Accepting each of the three sponge classes as a monophyletic clade (Adams et al, 1999, Borchiellini

et al, 2001), it becomes apparent that significant gaps exist in the sponge fossil record, particularly the demosponges and calcareans Allowing the uncertainties of sponge phylogeny and the earliest known eumetazoans, the minimum implied gaps (MIGs; Benton and Storrs, 1996) range from 50 to 180 Ma (Fig 12) This implies that much of the early sponge history is missing from or remains to be discovered in the geological record The sponge fossil record across the Neoproterozoic–Cambrian boundary should be carefully considered in future paleontological work, not only because early sponges are a key to understand early animal evolution (Muller, 2001) but also because sponges were important ecological players in epifaunal communities during the Neoproterozoic–Cambrian transition (Yuan et al, 2002)

The stratigraphic pattern of sponges, as currently observed, may be taken to favor the {eumetazoans+[hexactinellids+(demosponges+calcareans)]} topology (Reitner and Mehl, 1996; Fig 11A) This topology requires fewer sponge MIGs than the other two topologies, if eumetazoans diverged ca 560 Ma or earlier If eumetazoans diverged later, the phylogeny illustrated in Fig 11C is favored by the stratigraphic data Inclusion of stratigraphic data in phylogenetic

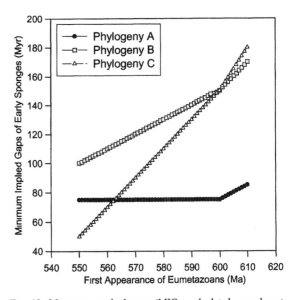

Fig 12 Minimum implied gaps (MIGs, calculated according to Benton and Storrs, 1996) of the sponge fossil record assuming the three phylogenies illustrated in Fig 11 and the first appearance of eumetazoans between 550 and 600 Ma MIGs of the early sponge fossil record ranges from 50 to 180 Ma

reconstruction and optimization, however, is controversial (Smith, 2000) Until the completeness of the sponge fossil record is fully investigated and molecular, morphological, and paleontological data converge to a consensus, the current debate about Porifera, Pinacophora, and Silicea monophyly is likely to continue

## 5. Conclusions

Both demosponges and hexactinellids, but no calcareans, occur in the Hetang assemblage Eleven articulated sponge species (including three new taxa) and two undetermined forms are present in this assemblage These sponge fossils are Meishucunian–Qiongzhusian in age Paleontological evidence allows us to correlle Meishucunian–Qiongzhusian deposits in South China with Nemakit-Daldynian–Atdabanian deposits in Siberia Radiometric dates from eastern Yunnan, Siberia, and Avalonia constrain the sponge fossils to be between ca 535 and 520 Ma

The Hetang and other sponge fossils suggest that hexactinellids evolved no later than the Nemakit-Daldynian–Tommotian and probably did so in the Neoproterozoic Demosponges and calcareans evolved no later than the Atdabanian The occurrence of eumetazoans at 550 Ma and probably 600 Ma implies substantial gaps in the fossil record of early sponges The minimum implied gaps (MIGs) are particularly evident if the sponges are a paraphyletic group and the calcareans are a sister group of the eumetazoans, a topology that is supported by current molecular data

## Acknowledgements

We thank Loren Babcock for inviting us to contribute to this special volume, and Zhou Chuanming, Chen Zhe, Yan Kui, Li Jun, and Wang Jinlong for field assistance Constructive comments by J K Rigby, Guoxiang Li, and Brian Pratt have greatly improved this manuscript Financial support for this research is provided by the Petroleum Research Fund, the Chinese Academy of Sciences (KZCX2-116), the National Natural Science Foundation of China, and the Chinese Ministry of Science and Technology (G2000077700)

## References

Adams, C L, McInerney, J O, Kelly, M, 1999 Indications of relationships between poriferan classes using full-length 18s rRNA gene sequences Proceedings of the 5th international sponge symposium, Mem Queensl Mus, vol 44, pp 33–43

Babcock, L E, Zhang, W, Leslie, S A, 2001 The Chengjiang Biota record of the Early Cambrian diversification of life and clues to exceptional preservation of fossils GSA Today 11 (2), 4–9

Barfod, G H, Albaréde, F, Knoll, A H, Xiao, S, Télouk, P, Frei, R, Baker, J, 2002 New Lu–Hf and Pb–Pb age constraints on the earliest animal fossils Earth Planet Sci Lett 201, 203–212

Bengtson, S, Conway Morris, S, 1992 Early radiation of biomineralizing phyla In Lipps, J H, Signor, P W (Eds), Origin and Early Evolution of Metazoa Plenum Press, New York, pp 447–481

Bengtson, S, Conway Morris, S, Cooper, B J, Jell, P A, Runnegar, B N, 1990 Early Cambrian fossils from south Australia Memoir 9 of the association of Australian palaeontologists Association of Australian Palaeontologists, Brisbane (1–364 pp)

Benton, M J, Storrs, G W, 1996 Testing the quality of the fossil record paleontological knowledge is improving Geology 22, 111–114

Benton, M J, Willis, M A, Hitchin, R, 2000 Quality of the fossil record through time Nature 403, 534–537

Benus, A P, 1988 Sedimentological context of a deep-water Ediacaran fauna (Mistaken Point Formation, Avalon Zone, eastern Newfoundland) In Landing, E, Narbonne, G M, Myrow, P (Eds), Trace Fossils, Small Shelly Fossils and the Precambrian–Cambrian Boundary Bulletin of the New York State Museum, pp 8–9

Borchiellini, C, Manuel, M, Alivon, E, Boury-Esnault, N, Vacelet, J, Le Parco, Y, 2001 Sponge paraphyly and the origin of Metazoa J Evol Biol 14, 171–179

Bowring, S A, Erwin, D H, 1998 A new look at evolutionary rates in deep time uniting paleontology and high-precision geochronology GSA Today 8 (9), 1–8

Bowring, S A, Grotzinger, J P, Isachsen, C E, Knoll, A H, Pelechaty, S M, Kolosov, P, 1993 Calibrating rates of Early Cambrian evolution Science 261, 1293–1298

Brasier, M D, Magaritz, M, Corfield, R, Luo, Huilin, Wu, Xiche, Jiang, Zhiwen, Hamdi, B, He, Tinggui, Fraser, A G, 1990 The carbon- and oxygen-isotope record of the Precambrian–Cambrian boundary interval in China and Iran and their correlation Geol Mag 127, 319–332

Brasier, M, Cowie, J, Taylor, M, 1994a Decision on the Precambrian–Cambrian boundary stratotype Episodes 17, 3–8

Brasier, M D, Corfield, R M, Derry, L A, Rozanov, A Y, Zhuravlev, A Y, 1994b Multiple $\delta^{13}C$ excursions spanning the Cambrian explosion to the Botomian crisis in Siberia Geology 22, 455–458

Brasier, M, Green, O, Shields, G, 1997 Ediacarian sponge spicule clusters from southwestern Mongolia and the origins of the Cambrian fauna Geology 25, 303–306

Budd, G E , Jensen, S , 2000 A critical reappraisal of the fossil record of the bilaterian phyla Biol Rev Camb Philos Soc 75, 253–295

Butterfield, N J , 2001 Ecology and evolution of Cambrian plankton In Zhuravlev, A Y, Riding, R (Eds ), The Ecology of the Cambrian Radiation Columbia University Press, New York, pp 200–216

Cavalier-Smith, T, Allsopp M T E P, Chao, E E , Boury-Esnault, N , Vavelet, J , 1996 Sponge phylogeny, animal monophyly, and the origin of the nervous system 18s rRNS evidence Can J Zool 74, 2031–2045

Chen, J , Hou, X , Lu, H , 1989 Lower Cambrian leptomitids (Demospongea), Chengjiang, Yunnan Acta Palaeontol Sin 28, 17–31

Chen, J , Hou, X Li, G , 1990 New Lower Cambrian demosponges—*Quadrolaminiella* gen nov from Chengjiang, Yunnan Acta Palaeontol Sin 29, 402–414

Chen, J Y, Zhou, G Q Zhu, M Y, Yeh, K Y, 1996 The Chengjiang biota a unique window of the Cambrian explosion National Museum of Natural History, Taichung (1–222 pp )

Clapham, M E , Narbonne, G M , 2002 Ediacaran epifaunal tiering Geology 30, 627–630

Collins, A G , 1998 Evaluating multiple alternative hypotheses for the origin of Bilateria an analysis of 18s rRNA molecular evidence Proc Natl Acad Sci U S A 95, 15458–15463

Conway Morris, S , 1997 Molecular clocks defusing the Cambrian "explosion"? Curr Biol 7, R71–R74

Dawson, J W, Hinde, G J , 1889 New species of fossil sponges from the Siluro–Cambrian at Little Métis on the lower St Lawrence Trans R Soc Can 7, 31–55

Ding, W, Qian, Y, 1988 Late Sinian to Early Cambrian small shelly fossils from Yangjiaping, Shimen, Hunan Acta Micropalaeontol Sin 5, 39–55

Erdtmann, B -D , Steiner, M , 2001 Special observations concerning the Sinian–Cambrian transition and its stratigraphic implications on the central and SW Yangtze Platform, China In Peng, S , Babcock, L E , Zhu, M (Eds ), Palaeoworld 13 Cambrian system of South China University of Science and Technology of China Press, Hefei, pp 52–65

Fedonkin, M A , 1994 Vendian body fossils and trace fossils In Bengtson, S (Ed ), Early Life on Earth Columbia University Press, New York, pp 370–388

Fedonkin, M A , Waggoner, B M 1997 The late Precambrian fossil *Kimberella* is a mollusc-like bilaterian organism Nature 388 868–871

Finks, R M , 1983 Fossil Hexactinellida In Rigby, J K , Stearn, C W (Ed ), Sponges and Spongiomorphs University of Tennessee and the Paleontological Society, Knoxville, Tennessee, pp 101–115

Gehling, J G , Rigby, J K , 1996 Long expected sponges from the Neoproterozoic Ediacara fauna of South Australia J Paleontol 70, 185–195

Gehling, J G , Jensen, S , Droser, M L Myrow, P M , Narbonne, G M , 2001 Burrowing below the basal Cambrian GSSP, Fortune Head, Newfoundland Geol Mag 138, 213–218

Geyer, G , Shergold, J , 2000 The quest for internationally recognized divisions of Cambrian time Episodes 23, 188–195

Gruber, G , Reitner, J , 1991 Isolierte Mikro- und Megaskleren von Poriferen aus dem Untercampan von Hover (Norddeutschland) und Bemerkungen zur Phylogenie der Geodiidae (Demospongiae) Berl Geowiss Abh , Reihe A, Geol Palaontol 134, 107–117

He, S , Yu, G , 1992 The small shelly fossils from the Palaeocambrian Meishucunian Stage in western Zhejiang Zhejiang Geol 8, 1–7

Hofmann, H J , Narbonne, G M , Aitken, J D , 1990 Ediacaran remains from intertillite beds in northwestern Canada Geology 18, 1199–1202

Hou, X Bergstrom, J , Wang, H , Feng, X , Chen, A , 1999 The Chengjiang Fauna Exceptionally Well-Preserved Animals from 530 Million Years Ago Yunnan Science and Technology Press, Kunming, China (170 pp )

Isachsen, C E , Bowring, S A , Landing, E , Samson, S D , 1994 New constraint on the division of Cambrian time Geology 22, 496–498

Jenkins, R J F , Cooper, J A , Compston, W , 2002 Age and biostratigraphy of Early Cambrian tuffs from SE Australia and southern China J Geol Soc (Lond ) 159, 645–658

Jiang, Z , Brasier M D , Hamdi, B , 1988 Correlation of the Meishucun Stage in South Asia Acta Geol Sin 60, 191–199

Khomentovsky, V V, Karlova, G A , 1993 Biostratigraphy of the Vendian–Cambrian beds and the Lower Cambrian boundary in Siberia Geol Mag 130, 29–45

Khomentovsky, V V, Karlova, G A , 2002 The boundary between Nemakit-Daldynian and Tommotian Stages (Vendian–Cambrian systems) of Siberia Stratigr Geol Correl 10, 217–238

Knoll, A H Kaufman, A J , Semikhatov, M A , Grotzinger, J P , Adams, W , 1995 Sizing up the sub-Tommotian unconformity in Siberia Geology 23, 1139–1143

Kouchinsky, A , Bengtson, S , Missarzhevsky, V V Pelechaty, S , Torssander, P , Val'kov, A K , 2001 Carbon isotope stratigraphy and the problem of a pre-Tommotian stage in Siberia Geol Mag 138, 387–396

Kruse, M , Leys, S P , Muller, I M , Muller, W E G , 1998 Phylogenetic position of the Hexactinellida within the Phylum Porifera based on the amino acid sequence of the protein kinase C from *Rhabdocalyptus dawsoni* J Mol Evol 46, 721–728

Landing, E , 1989 Paleoecology and distribution of the Early Cambrian rostroconch *Watsonella crosbyi* Grabau J Paleontol 63, 566–573

Landing, E , 1994 Precambrian–Cambrian boundary global stratotype ratified and a new perspective of Cambrian time Geology 22, 179–182

Landing, E , Bowring, S A , Davidek, K L , Westrop, S R , Geyer, G , Heldmaier, W , 1998 Duration of the Early Cambrian U–Pb ages of volcanic ashes from Avalon and Gondwana Can J Earth Sci 35, 329–338

Li, C , He, J , Ye, H , 1990 Discovery of Early Cambrian trilobites in Guichi of Anhui Province J Stratigr 14, 159–160

Li, C -W, Chen, J -Y, Hua, T -E 1998 Precambrian sponges with cellular structures Science 279, 879–882

Li, G, Zhang, J, Zhu, M, 2001 Litho- and biostratigraphy of the Lower Cambrian Meishucunian stage in the Xiaotan section, eastern Yunnan Acta Palaeontol Sin 40, 40–53 (Suppl)

Li, S, Xiao, Q, Shen, J, Sun, L, Liu, B, Yan, B, 2002 Re–Os isotopic constraints on the source and age of Lower Cambrian platinum group element ores in Hunan and Guizhou Sci China Ser D 32, 568–575

Lipps, J H, 2001 Protists and the Precambrian–Cambrian skeletonization event In Peng, S, Babcock, L E, Zhu, M (Eds), Cambrian System of South China, Palaeoworld, vol 13 University of Science and Technology of China Press, Hefei, p 280

Luo, H, Jiang, Z, Wu, X, Song, X, Ouyang, L, 1982 The Sinian–Cambrian Boundary in Eastern Yunnan, China People's Publishing House of Yunnan, Kunming, Yunnan (265 pp)

Luo, H, Jiang, Z, Wu, X, Song, X, Ouyang, L, Xing, Y, Liu, G, Zhang, S, Tao, Y, 1984 Sinian–Cambrian Boundary Stratotype Section at Meishucun, Jinning, Yunnan, China People's Publishing House of Yunnan, Kunming, Yunnan (154 pp)

Luo, H, Jiang, Z, Tang, L, 1994 Stratotype section for Lower Cambrian stages in China Yunnan Science and Technology Press, Kunming, Yunnan (183 pp)

Martin, M W, Grazhdankin, D V, Bowring, S A, Evans, D A D, Fedonkin, M A, Kirschvink, J L, 2000 Age of Neoproterozoic bilaterian body and trace fossils, White Sea, Russia implications for metazoan evolution Science 288, 841–845

McCaffrey, M A, Moldowan, J M, Lipton, P A, Summons, R E, Peters, K E, Jeganathan, A, Watt, D S, 1994 Palaeoenvironmental implications of novel C30 steranes in Precambrian to Cenozoic age petroleum and bitumen Geochim Cosmochim Acta 58, 529–532

Medina, M, Collins, A G, Silberman, J D, Sogin, M L, 2001 Evaluating hypotheses of basal animal phylogeny using complete sequences of large and small subunit rRNA Proc Natl Acad Sci U S A 98, 9707–9712

Mehl, D, 1998 Porifera and Chancelloriidae from the Middle Cambrian of the Georgina Basin, Australia Palaeontology 41, 1153–1182

Mehl, D, Erdtmann, B -D, 1994 *Sanshapentella dapingi* n gen n sp —a new hexactinellid sponge from the Early Cambrian (Tommotian) of China Berl Geowiss Abh 13, 315–319

Moczydłowska, M, 1991 Acritarch biostratigraphy of the Lower Cambrian and the Precambrian–Cambrian boundary in southeastern Poland Fossils Strata 29, 1–127

Moczydłowska, M, 1998 Cambrian acritarchs from Upper Silesia, Poland biochronology and tectonic implications Fossils Strata 46, 1–121

Moczydłowska, M, 2001 Early Cambrian phytoplankton radiations and appearance of metazoans In Peng, S, Babcock, L E, Zhu, M (Eds), Cambrian system of South China, Palaeoworld, vol 13 University of Science and Technology of China Press, Hefei, pp 293–296

Muller, W E G, 2001 Review how was metazoan threshold crossed? The hypothetical Urmetazoa Comp Biochem Physiol, Part A Mol Integr Physiol 129, 433–460

Narbonne, G M, Myrow, P M, Landing, E, Anderson, M M, 1987 A candidate stratotype for the Precambrian–Cambrian boundary, Fortune Head, Burin Peninsula, southeastern Newfoundland Can J Earth Sci 24, 1277–1293

Nowlan, G S, Narbonne, G M, Fritz, W H, 1985 Small shelly fossils and trace fossils near the Precambrian–Cambrian boundary in the Yukon Territory Canada Lethaia 18, 233–256

Palmer, A R, 1998 Why is intercontinental correlation within the Lower Cambrian so difficult? Rev Esp Paleontol, 17–21 (Special Issue In Memory of Prof Gonzalo)

Peng, S, 1999 A proposal of the Cambrian chronostratigraphic scale in China Geoscience 13, 242

Peng, S, Babcock, L E, 2001 Cambrian of the Hunan–Guizhou region, South China In Peng, S, Babcock, L E, Zhu, M (Eds), Cambrian System of South China, Palaeoworld vol 13 University of Science and Technology of China Press, Hefei, pp 3–51

Qian, Y, 1977 Hyolitha and some problematica from the Lower Cambrian Meishucun Stage in central and southwestern China Acta Palaeontol Sin 16, 255–275

Qian, Y, 1999 Taxonomy and Biostratigraphy of Small Shelly Fossils in China Science Press, Beijing (247 pp)

Qian, Y, Bengtson, S, 1989 Palaeontology and biostratigraphy of the Early Cambrian Meishucunian stage in Yunnan Province, South China Fossils Strata 24, 1–156

Qian, Y, Yin, G, 1984 Small shelly fossils from the lowerest Cambrian in Guizhou Prof Pap Stratigr Palaeontol 13, 91–124

Qian, Y, Zhu, M, He, T, Jiang, Z, 1996 New investigation of Precambrian–Cambrian boundary sections in eastern Yunnan Acta Micropalaeontol Sin 13, 225–240

Qian, Y, Li, G, Zhu, M, 2001 The Meishucunian Stage and its small shelly fossil sequence in China Acta Palaeontol Sin 40, 54–62 (supplement)

Reitner, J, Mehl, D, 1995 Early paleozoic diversification of sponges new data and evidence Geol -Palaontol Mitt Innsbruck 20, 335–347

Reitner, J, Mehl, D, 1996 Monophyly of the Porifera Verh Nat wiss Ver Hamb (Neue Folge) 36, 5–32

Rigby, J K, 1978 Porifera of the Middle Cambrian Wheeler Shale, from the Wheeler Amphitheater, House Range, in western Utah J Paleontol 52, 1325–1345

Rigby, J K, 1983 Sponges of the Middle Cambrian Marjum limestone from the House Range and Drum Mountains of western Millard County, Utah J Paleontol 57, 240–270

Rigby, J K, 1986 Sponges of the Burgess Shale (Middle Cambrian), British Columbia Palaeontogr Can 2, 1–105

Rigby, J K, Harris, D R, 1979 A new Silurian sponge fauna from northern British Columbia Can J Paleontol 53, 968–980

Rigby, J K, Hou, X, 1995 Lower Cambrian demosponges and hexactinellid sponges from Yunnan, China J Paleontol 69, 1009–1019

Rigby, J K, Maher, B J, 1995 Age of the hexactinellid beds of the Roberts Mountains Formation, Snake Mountains, Nevada, and additions to the Silurian sponge fauna J Paleontol 69, 1020–1029

Rigby, J K , Murphy, M A , 1983 *Gabelia*, a new Late Devonian lyssakid protosponge from the Roberts Mountains, Neveda J Paleontol 57, 797–803

Rigby, J K , Maher, B , Browne, Q , 1991 New hexactinellids from the Siluro–Devonian of the Snake Mountains, Elko County, Nevada, and a new locality for *Gabelia* J Paleontol 65, 709–714

Rowland, S M , 2001 Archaeocyaths—a history of phylogenetic interpretation J Paleontol 75, 1065–1078

Rowland, S M , Luchinina, V A , Korovnikov, I V , Sipin, D P , Tarletskov, A I , Fedoseev, A V , 1998 Biostratigraphy of the Vendian–Cambrian Sukharikha River section, northwestern Siberia Platform Can J Earth Sci 35, 339–352

Rozanov, A Y , Zhuravlev, A Y , 1992 The Lower Cambrian fossil record of the Soviet Union In Lipps, J H , Signor, P W (Eds ), Origin and early evolution of the Metazoa Plenum Press, New York, pp 205–282

Rozanov, A Y , Semikhatov, M A , Sokolov, B S , Fedonkin, M A , Khomentovskii, V V , 1997 The decision on the Precambrian–Cambrian boundary stratotype a breakthrough or misleading action? Stratigr Geol Corr 5, 19–28

Sergeev, V N , 1989 Microfossils from transitional Precambrian–Phanerozoic strata of central Asia Himal Geol 13, 269–278

Shen, Y , Schidlowski, M , 2000 New C isotope stratigraphy from southwest China implications for the placement of the Precambrian–Cambrian boundary on the Yangtze Platform and global correlations Geology 28, 623–626

Simpson, T L , 1984 The cell biology of sponges Springer-Verlag, Berlin (662 pp )

Smith, A B , 2000 Stratigraphy in phylogenetic reconstruction J Paleontol 74, 763–766

Steiner, M , Mehl, D , Reitner, J , Erdtmann, B -D , 1993 Oldest entirely preserved sponges and other fossils from the lowermost Cambrian and a new facies reconstruction of the Yangtze Platform (China) Berl Geowiss Abh 9, 293–329

Steiner, M , Wallis, E , Erdtmann, B -D , Zhao, Y , Yang, R , 2001a Submarine-hydrothermal exhalative ore layers in black shales from South China and associated fossils—insights into a Lower Cambrian facies and bio-evolution Palaeogeogr Palaeoclimatol Palaeoecol 169, 165–191

Steiner, M , Zhu, M , Weber, B , Geyer, G , 2001b The Lower Cambrian of eastern Yunnan trilobite-based biostratigraphy and related faunas Acta Palaeontol Sin 40, 63–79 (Suppl )

Tang, T , Zhang, J , Jiang, X , 1978 Discovery and significance of the Late Sinian fauna from western Hunan and Hubei Acta Stratig Sin 2, 32–45

Vidal, G , Moczydlowska, M , Rudavskaya, V R , 1995 Constraints on the early Cambrian radiation and correlation of the Tommotian and Nemakit-Daldynian regional stages of eastern Siberia J Geol Soc (Lond ) 152, 499–510

Walcott, C D , 1920 Cambrian Geology and Paleontology IV middle Cambrian Spongiae Smithson Misc Collect 67, 261–364

Wood, R , 1999 Reef Evolution Oxford University Press, Oxford (414 pp )

Wray, G A , Levinton, J S , Shapiro, L H , 1996 Molecular evidence for deep Precambrian divergences among metazoan phyla Science 274, 568–573

Xiao, S , Yuan, X , Knoll, A H , 2000 Eumetazoan fossils in terminal Proterozoic phosphorites? Proc Natl Acad Sci U S A 97, 13684–13689

Xue, Y , Yu, C , 1979 Lithological characteristics and sedimentary environments of the Lower Cambrian Hetang Formation in western Zhejiang and northern Jiangxi J Stratigr 3, 283–293

Yin, L , 1995 Microflora from the Precambrian–Cambrian boundary strata in the Yangtze Platform J Stratigr 19, 299–307

Yin, L , Xiao, S , Yuan, Y , 2001 New observations on spicule-like structures from Doushantuo phosphorites at Weng'an, Guizhou Province Chin Sci Bull 46, 1031–1036

Yuan, K , Zhu, M , Zhang, J , van Iten, H , 2001 Biostratigraphy of archaeocyathan horizons in the Lower Cambrian Fucheng section, south Shaanxi Province implications for regional correlations and archaeocyanthan evolution Acta Palaeontol Sin 40, 115–129 (Suppl )

Yuan, X , Xiao, S , Parsley, R L , Zhou, C , Chen, Z , Hu, J , 2002 Towering sponges in an Early Cambrian Lagerstatte disparity between non-bilaterian and bilaterian epifaunal tiers during the Neoproterozoic–Cambrian transition Geology 30, 363–366

Yue, Z , He, S , 1989 Early Cambrian conodonts and bradoriids from Zhejiang Acta Micropalaeontol Sin 6, 289–300

Yue, Z , Zhao, J , 1993 Meishucunian (Early Cambrian) rod-like fossils from western Zhejiang Acta Micropalaeontol Sin 10, 89–97

Zhang, W , 1987 World's oldest Cambrian trilobites from eastern Yunnan In Nanjing Institute of Geology and Palaeontology, A S (Ed ), Stratigraphy and Palaeontology of Systemic Boundaries in China Precambrian–Cambrian Boundary (1) Nanjing University Press, Nanjing, pp 1–18

Zhang, X , Pratt, B R , 1994 New and extraordinary Early Cambrian sponge spicule assemblage from China Geology 22, 43–44

Zhang, Y , Yuan, X , Yin, L , 1998 Interpreting Late Precambrian microfossils Science 282, 1783

Zhao, J , Yue, Z , 1987 New discovery of Early Cambrian Meishucunian small shelly fossils in western Zhejiang province and the Sinian–Cambrian boundary Chin Sci Bull 32, 1168–1170

Zhao, Z , Xing, Y , Ding Q , Liu, G , Zhao, Y , Zhang, S , Meng, X , Yin, C , Ning, B , Han, P , 1988 The Sinian System of Hubei China University of Geosciences Press, Wuhan (205 pp )

Zhou, Z , Yuan, J , 1980 Lower Cambrian trilobite succession in southwest China Acta Palaeontol Sin 19, 331–339

Zhou, C , Zhang, J , Li, G , Yu, Z , 1997 Carbon and oxygen isotopic record of the Early Cambrian from the Xiaotan Section, Yunnan, South China Sci Geol Sin 32, 201–211

Zhou, C , Yuan, X , Xue, Y , 1998 Sponge spicule-like pseudofossils frm the Neoproterozoic Doushantuo Formation in Weng'an, Guizhou, China Acta Micropalaeontol Sin 15, 380–384

Zhu, M , Li, G , Zhang, J , Steiner, M , Qian, Y , Jiang, Z , 2001 Early Cambrian stratigraphy of east Yunnan, southwestern China a synthesis Acta Palaeontol Sin 40, 4–39 (supplement)

Zhuravlev, A Y , 1996 Preliminary suggestions on the global Early Cambrian zonation Beringeria 2, 147–160 (special issue)

Zhuravlev, A Y , Wood, R A , 1996 Anoxia as the cause of the mid-Early Cambrian (Botomian) extinction event Geology 24, 311–314

Available online at www.sciencedirect.com

Palaeogeography, Palaeoclimatology, Palaeoecology 220 (2005) 119–127

www elsevier com/locate/palaeo

# Cambrian *Sphenothallus* from Guizhou Province, China: early sessile predators

Jin Peng[a,*], Loren E. Babcock[b], Yuanlong Zhao[a], Pingli Wang[a], Rongjun Yang[a]

[a]*College of Resource and Environment, Guizhou University, Guiyang 550003, China*
[b]*Department of Geological Sciences, 125 South Oval Mall, The Ohio State University, Columbus, Ohio 43210, USA*

Received 14 June 2004, accepted 6 September 2004

## Abstract

Two species of *Sphenothallus* (Cnidaria) from Cambrian rocks of eastern Guizhou, China, are reported One species, *Sphenothallus songlinensis* n sp , from the Niutitang Biota (Lower Cambrian Nangaoan Stage), represents the oldest reported example of the genus Other specimens, referred to *Sphenothallus taijiangensis?*, are from the Kaili Formation (Lower Cambrian Duyuanian Stage) The new material adds to the record of sessile cnidarians, an important group of predators, from the early part of the Cambrian Occurrences of the genus in strata traditionally assigned to the Lower Cambrian show that *Sphenothallus* was among the earliest animals to produce a biomineralized (phosphatic) skeleton
© 2004 Elsevier B V All rights reserved

*Keywords* Cnidaria, Cambrian, China, predation, biomineralization

## 1. Introduction

Elongate, tubular, phosphatic fossils assigned to the genus *Sphenothallus* are well known from Ordovician and overlying strata but few examples are known from the Cambrian Here we document the genus from two Cambrian occurrences in China: the Niutitang Biota and the somewhat younger Taijiang Biota of the Kaili Formation *Sphenothallus* was reported earlier from the Kaili Formation (Zhao et al , 1999, Zhu et al., 2000), but this is the first report from the Niutitang Biota

Together, specimens from these two biotas provide important information concerning the early Paleozoic record of biomineralized animals

The Niutitang Biota, which occurs in dark gray to black shales (weathering dark gray) of the lower part of the Niutitang Formation (Lower Cambrian Nangaoan Stage), Zunyi County, Guizhou Province, China, includes trilobites and exceptionally preserved sponges, non-biomineralized arthropods, cnidarians, mollusks, rhabdopleuroids, and algae (Zhao et al , 1999; Erdtmann and Steiner, 2001, Steiner and Erdtmann, 2001, Steiner et al , 2001, Zhao et al , 2001, 2002, Fig 1) Abundant specimens of *Sphenothallus* from the Niutitang Biota all belong to a new

---

\* Corresponding author
*E-mail address* gzpengjin@sina com (J Peng)

0031-0182/$ - see front matter © 2004 Elsevier B V All rights reserved
doi 10 1016/j palaeo 2004 09 014

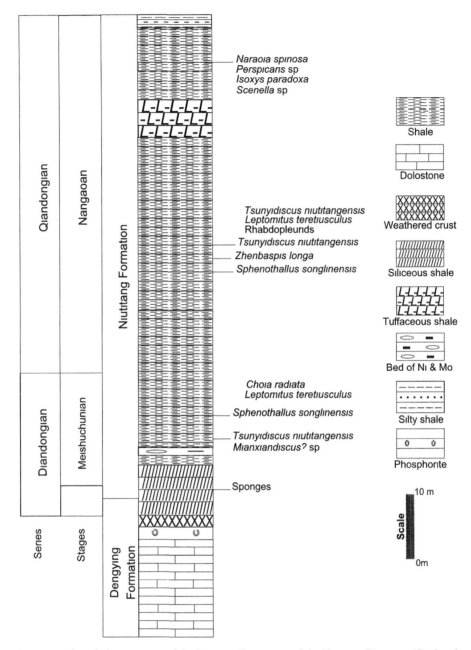

Fig 1 Stratigraphic section through the upper part of the Dengying Formation and the Niutitang Formation (Cambrian) in the village of Zhongnan, Songlin County, Guizhou Province, China Positions of fossils collected from the section are indicated

species, *Sphenothallus songlinensis*, and represent the oldest known examples of the genus For comparative purposes, specimens of a species left in open nomenclature are illustrated from the Taijiang Biota (upper Lower Cambrian. Duyuanian Stage) of Guizhou Zhu

et al (2000) described one species, *Sphenothallus taijiangensis*, from the Kaili Formation

*Sphenothallus* specimens from the Cambrian, reported here and elsewhere (Zhao et al , 1999; Zhu et al , 2000, Van Iten et al , 2002) add to a growing list of cnidarians recognized from the Cambrian Similar to their modern relatives, these ancient cnidarians are inferred to have been predators Examples of Cambrian cnidarians other than *Sphenothallus* include anemones (Chen and Erdtmann, 1991; Briggs et al, 1994; Chen et al, 1996, Hou et al, 2004), pennatulaceans (Conway Morris, 1993, Briggs et al, 1994; Zhang and Babcock, 2001), a probable octacoral (Ausich and Babcock, 1998, 2000), conulariids (Hughes et al, 2000), *Byronia* (e g, Bischoff, 1989, Zhu et al, 2000), and *Cambrorhytium* (Conway Morris and Robison, 1988, Chen et al, 1996) All inferred cnidarians from the Cambrian that are known at present represent the sessile polypoid (or strobilus) life phase In addition to Cambrian cnidarians, some putative Neoproterozoic cnidarians have been reported (e g, Glaessner, 1984, Fedonkin, 1982, 1985, 1994; Hahn et al, 1982, Jenkins, 1984, Babcock et al, 2004), and they include both sessile polypoid and free-living medusoid life phases

The presence of body fossils of cnidarian predators in upper Neoproterozoic (Ediacaran) and Cambrian strata serves to reinforce arguments (Lipps et al, 2001, Vannier and Chen, 2002, Babcock, 2003; Zhu et al, 2004) that predation played an important role in animal evolution across the Neoproterozoic– Cambrian transition Previously, much of this argument was based on body fossils (e g, Whittington, 1980; Whittington and Briggs, 1985, Briggs et al, 1994, Chen et al, 1994, Vannier and Chen, 2002, Babcock, 2003), broken or disarticulated skeletal remains (e g, Pratt, 1998, Babcock, 2003), gut tracts containing skeletal elements (e g, Conway Morris, 1977; Bruton, 1981; Conway Morris and Robison, 1988; Babcock, 2003; Zhu et al, 2004), trace fossils including coprolites (Sprinkle, 1973, Conway Morris and Jenkins, 1985; Conway Morris and Robison, 1988, Conway Morris and Bengtson, 1994, Babcock, 2003), and interpretations of life-history patterns (Babcock, 2003) Most of these records of carnivores or carnivorous activity represent vagile predators or the activity of vagile carnivores Lipps et al (2001) noted that sessile predators, notably cnidarians, rarely leave clear traces of their trophic behavior to the fossil record even though they were likely to have been important carnivores in marine ecosystems Trace fossils of predaceous activity involving cnidarians as predators have been reported from

Cambrian strata (Alpert and Moore, 1975) but are unconvincing and have been reinterpreted as the result of carnivory involving vagile arthropods (Babcock, 1993)

## 2. Affinities and occurrences of *Sphenothallus*

*Sphenothallus* Hall, 1847 is a cosmopolitan fossil distributed through marine strata of Paleozoic age The elongate, conical tube is composed of calcium phosphate and reinforced by a pair of long, thin, hollow, longitudinal thickenings extending the length of the skeleton (Van Iten et al, 1992, 2002, Zhu et al, 2000, Fig 2A, B) Attachment disks have been reported from some species assigned to the genus (Babcock et al, 1987, Van Iten et al, 1992, 1996, 2002) although Bolton (1994) suggested that many *Sphenothallus* species lacked attachment disks Because of a simple morphology, the phylogenetic affinities of *Sphenothallus* have been difficult to assess Although once commonly regarded as an annelid or other "worm" (e g, Mason and Yochelson, 1985, Fauchald et al, 1986; Pashin and Ettensohn, 1987; Neal and Hannbal, 2000), arguments have been posed to suggest that *Sphenothallus* has affinities with cnidarians (e g, Bodenbender et al, 1989, Van Iten et al, 1992, Van Iten, 1994; Zhu et al, 2000) A cnidarian affinity is supported by the presence of a sucker disc, an asexual reproductive habit (budding), and a laminar skeletal microstructure resembling that of conulariids Lateral branching of a few individuals, indicative of budding, is present on some specimens reported here from Songlin, Guizhou (Fig 2 3)

This report of *Sphenothallus* from the Niutitang Formation of China extends the stratigraphic range of the genus downward from the Middle Cambrian of traditional usage into the Lower Cambrian of traditional usage This, the first known occurrence of *Sphenothallus*, approximates the first occurrence of trilobites in the fossil record and succeeds the first occurrence of small shelly fossils As a calcium phosphate-secreting organism, *Sphenothallus* should now be regarded as one of the earliest Paleozoic animals to have developed a biomineralized skeleton

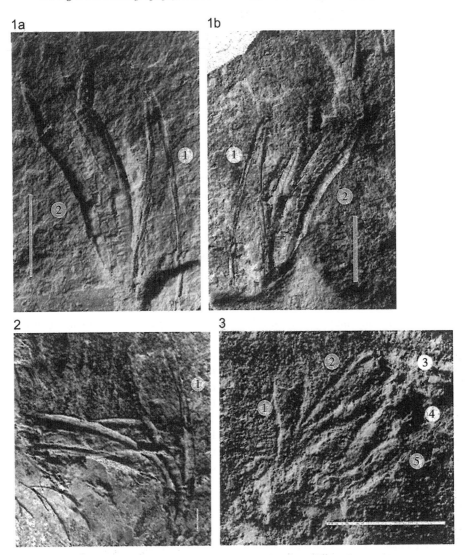

Fig. 2. *Sphenothallus songlinensis* n. sp. from the lowermost part of the Niutitang Formation (Cambrian), Heishapo section, Songlin Town, Zunyi County, Guizhou Province, China. 1, Colony of paratype individuals; 1a, part, Sh-6-1928a-②; 1b, counterpart, Sh-6-1928b-②. 2, Holotype, (identified with number 1), Sh-6-1712-①. 3, Paratype showing branching of small individual from a larger one, Sh-c-133. Scale bar in 2 is 2 mm; other scale bars are 5 mm.

*Sphenothallus* remains are commonly reported from strata inferred to represent dysaerobic environments (e.g., Feldmann et al., 1986; Van Iten et al., 1992, 1996; Bolton, 1994; Neal and Hannbal, 2000), although occurrences in strata inferred to represent aerobic environments are also known (Zhu et al., 2000). Specimens from both the Kaili Formation and the lowermost Niutitang Formation are preserved in dark gray shales that were probably deposited under dysaerobic or exaerobic conditions.

## 3. Locations and stratigraphy

New specimens of *Sphenothallus* illustrated here are from two localities in eastern Guizhou Province, China: (1) the lower part of the Niutitang Formation in a section close to the town of Songlin Town, Zunyi County (Fig. 1); and (2) the lower part of the Kaili Formation in the Wuliu–Zengjiaya section, Balang Village, Taijiang County (see Zhao et al., 2001). Specimens from the Niutitang Formation occur in association with fossils indicative of the *Mianxiandiscus* Zone and are correlated with the lower part of the *Hupeidiscus-Sinodiscus* Zone (Nangaoan Stage) according to the South China standard (see Peng and Babcock, 2001) Specimens from the Kaili Formation occur in association with fossils indicative of the *Bathynotus holopygus-Ovatoryctocara granulata* Zone and are correlated with the *Bathynotus* Zone (Duyuanian Stage) according to the South China standard (see Peng and Babcock, 2001) Global chronostratigraphic subdivision of the Cambrian is

unresolved below the uppermost series (Furongian Series, Peng et al , 2004), but subdivision of the strata in South China (Peng and Babcock, 2001) approximates a growing consensus as to where series and stage boundaries should be delineated Ultimately, four series are likely to be established (Peng et al , 2004) but those below the Furongian are as yet unnamed The Niutitang Formation occurrence is in the lowermost part of the second series (equivalent to the Qiandongian Series of South China), and the Kaili Formation occurrence is in the uppermost part of the

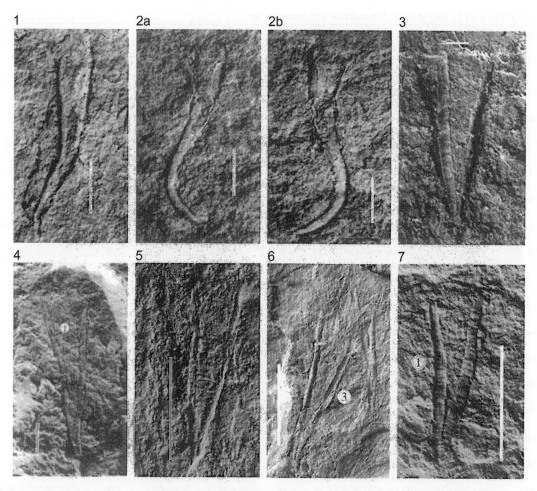

Fig 3 *Spenothallus* spp from the Cambrian of China 1, 2, *Sphenothallus taijiangensis* Zhu et al , 2000 from the lower part of the Kaili Formation (Cambrian), Wuliu-Zengjiaya section, Taijiang County, Guizhou Province. China 1, GTB-7-101 2a, part, GTB-2-189a, and 2b, counterpart GTB-2-189b of same specimen 3–7, *Sphenothallus songlinensis* n sp from the lowermost part of the Niutitang Formation (Cambrian). Heishapo section, Songlin Town, Zunyi County. Guizhou Province, China 3, Paratype, Sh-c-50, right wall tube is inner mold, left one is outer mold 4, Paratype. Sh-6-17, with one small individual inside of larger one 5, Paratype, Sh-c-135, showing transverse wall, 6, Paratype, Sh-6-1928-③, right wall tube is inner mold, left one is outer mold 7, Paratype, Sh-c-26-①, two wall tubes with transverse striae and ridges Scale bars for panels 1–4 are 2 mm Scale bars for panels 5–7 are 5 mm

second series (equivalent to the Qiandongian Series of South China) as currently envisioned

## 4. Systematic paleontology

All described specimens are reposited in the Palaeontological Museum of the Guizhou University, Guiyang, China (GU) Specimens having the prefix Sh are in the collection from the Niutitang Formation, and specimens having the prefix GTBM are in the collection from the Kaili Formation

Phylum Cnidaria Hatscheck, 1888
Class, order, and family uncertain
Genus *Sphenothallus* Hall, 1847

Table 1
Measurements (in mm) for individuals of *Sphenothallus songlinensis* n sp

| Specimen no | Thecal wall | Preservation | Length | Width of wall tube (max) | Width of wall tube (min) | Width of apical end | Width of aperture | Apetural expanding angle (degree) | Remarks |
|---|---|---|---|---|---|---|---|---|---|
| Sh-c-98 | tubular wall | right inner mold, left outer mold | 11 90 | 0 80 | 0 55 | ? | 2 90 | 25 | |
| Sh-c-34 | tubular wall | right inner mold, left outer mold | 9 10 | 0 65 | 0 53 | ? | 2 20 | 21 | |
| Sh-c-50 | tubular wall | right inner mold, left outer mold | 8 50 | 0 68 | 0 55 | 0 40 | 3 10 | 28 | holdfast? |
| Sh-c-1712-① | tubular wall | right inner mold, left outer mold | 14 20 | 0 75 | 0 54 | 0 60 | 2 50 | 3 | Holotype |
| Sh-c-1712-② | tubular wall | right outer mold, left inner mold | 14 10 | 0 74 | 0 50 | ? | 2 50 | 8 | |
| Sh-c-26-① | tubular wall | right inner mold, left outer mold | 8 60 | 0 78 | 0 45 | 0 20 | 2 80 | 26 | Paratype1 |
| Sh-c-26-② | tubular wall | right inner mold, left outer mold | 10 00 | 0 78 | 0 54 | ? | 2 90 | 26 | |
| Sh-c-1703 | tubular wall | right outer mold, left inner mold | 10 00 | 0 68 | 0 48 | | | 22 | |
| Sh-c-20-① | tubular wall | right outer mold, left inner mold | 6 80 | 0 50 | 0 35 | 0 40 | 2 20 | 23 | |
| Sh-c-24-① | tubular wall | right outer mold, left inner mold | 10 00 | 0 70 | 0 40 | 0 30 | 2 80 | 19 | |
| Sh-c-29-① | tubular wall | right inner mold, left outer mold | 11 00 | 0 45 | 0 38 | ? | 2 50 | 18 | |
| Sh-c-29-② | tubular wall | right outer mold, left inner mold | 17 30 | 0 48 | 0 38 | 0 50 | 3 10 | 9 | |
| Sh-c-29-③ | tubular wall | right outer mold, left inner mold | 11 00 | 0 45 | 0 38 | 0 35 | 2 50 | 21 | |
| Sh-6-1928a-① | tubular wall | right inner mold, left outer mold | 13 20 | 0 50 | 0 40 | ? | 3 90 | | |
| Sh-6-1928a-② | tubular wall | right outer mold, left inner mold | 14 20 | 1 15 | 0 50 | ? | 5 00 | 15 | |
| Sh-6-17 | tubular wall | right inner mold, left outer mold | 13 20 | 0 58 | 0 34 | 0 45 | 4 00 | 22 | holdfast? Paratype2 |
| Sh-6-1928a-③ | tubular wall | right outer mold, left inner mold | 10 20 | 0 50 | 0 40 | 0 70 | 3 00 | 18 | |
| Sh-c-135 | tubular wall | right outer mold, left inner mold | 9 20 | 0 54 | 0 40 | 0 30 | 2 10 | 12 | transverse wall |

Specimen numbers are field collection numbers

*Remarks* In general, skeletons of *Sphenothallus* have two principal shapes· a tubal type and a conical–tubal type Species having tubal skeletons exhibit either long thin tubes (e g , *Sphenothallus angustifolius* Hall, 1847, *Sphenothallus taijiangensis* Zhu et al , 2000) or short thin tubes (e g , *Sphenothallus bicarinatus* (Girty, 1911)) Species having conical-tubal skeletons exhibit either long conical tubes (e g , *Sphenothallus ruedemanni* (Kobayashi, 1934, see also Choi, 1990)) or short conical tubes (e g , *Sphenothallus songlinensis* n sp )

*Sphenothallus songlinensis* n sp
Figs 2 1– 3 and 3 3– 7, Table 1

*Etymology·* Named for Songlin, the town where the species was collected
*Material* Forty individuals on 15 slabs
*Types* Holotype, GUT Sh-6-1712-①, paratypes GUT Sh-c-26-①, GUT Sh-6-17, GUT Sh-c-133
*Diagnosis*. Skeleton short, conical-tubular, angle of expansion from the aperture 15°–26°
*Description* *Sphenothallus* having short (up to 17 5 mm), conical–tubular skeleton, with pair of narrow, elongate longitudinal thickenings (0 2 to 1 4 mm wide) Apical end, 0 2–0 6 mm wide, aperture, 2 5–5 0 mm wide, angle of expansion from apertural end, 15°–26° Transverse ridges (25–45/cm), transverse striae (3–5/mm), and longitudinal lines present on tubes, transverse walls present on some individuals Branches (budded individuals) present on some specimens Holdfast not observed
*Remarks* *Sphenothallus songlinensis* is unique among Cambrian species assigned to the genus in having a short conical tube Other Cambrian examples of *Sphenothallus* include a species from the Lower–Middle Cambrian Kaili Formation (*Sphenothallus taijiangensis*, Zhu et al , 2000, herein, questionably assigned) and a species from the Middle Cambrian Burgess Shale (*Sphenothallus* sp , Van Iten et al , 2002) Both species have tubes that are narrower and more elongate than *S songlinensis*
One large specimen assigned to this species (Fig 2 1) shows a slightly constricted apertural region It is uncertain whether this constriction is related to individual variation, ontogenetic variation, or taphonomic factors

*Occurrence* Dark gray shale (weathering light gray) in the lower part of the Niutitang Formation (*Mianxiandiscus* Zone, Lower Cambrian· Nangaoan Stage) at Heshanpi, Zhongnan Village, Songlin Town, Zunyi County, Guizhou, China

*Sphenothallus taijiangensis?* Zhu et al , 2000
Fig 3 1– 2

*Material* 5 specimens, including GUT GTB-7-101, GUT GTB-9-2-189a, b, GUT GTB-6-2-16, GUT GTB-9-1-6, GUT GTB-6-2-16
*Remarks* Five specimens are referred with question to *Sphenothallus taijiangensis* The specimens range in length from 7 5 to 11 0 mm, have angles of expansion ranging from 20° to 22° and have transverse walls The new specimens seem to be less elongate than specimens referred earlier (Zhu et al , 2000) to *S taijiangensis*, although this characteristic may be a reflection of intraspecific variation The new specimens also exhibit a bend in the apical half of the tube
*Occurrence*. Gray–green shale of the lower part of the Kaili Formation (*Bathynotus holopygus–Ovatoryctocara granulata* Zone; upper Lower Cambrian Duyunian Stage, see Yuan et al , 2002), Wuliu–Zengjiaya section, Balang, Taijiang County, Guizhou Province, China

## Acknowledgements

Assistance in collecting the material discussed here was provided by Michael Steiner and Bernd-D Erdtmann (Technische Universitat Berlin); Qingjun Guo (Institute of Geochemistry, Chinese Academy of Sciences, Beijing); Ruidong Yang and Yue Wang, Yunming Chen, Peirong Bai, Jian Wang, Guangyuan Mo, Kaikun Liu (Guizhou University of Technology), and Demei Wang (Zhongnan Village) Xingdong Deng (Nanjing Institute of Geology and Palaeontology, Chinese Academy of Sciences, Nanjing) photographed the specimens This research was supported in parts by grants from the National Natural Sciences Foundation of China (40162002, 40372023, 40232020), from the Foundation of the Key and Basic Project of Science and Technology of Guizhou (Gui No 2002-309), and from the Early and Special Project of the Key and Basic Project of the Ministry of Technology and Science of

China (2002-456), to Zhao, and by grants from the U S National Science Foundation (EAR 0106883, EAR-OPP 0229757) to Babcock

# References

Alpert, S P, Moore, J N, 1975 Lower Cambrian trace fossil evidence for predation on trilobites Lethaia 8, 223–230

Ausich, W I, Babcock, L E, 1998 The phylogenetic position of *Echmatocrinus brachiatus*, a probable octocoral from the Burgess Shale Palaeontology 41, 269–279

Ausich, W I, Babcock, L E, 2000 *Echmatocrinus*, a Burgess Shale animal reconsidered Lethaia 33, 92–94

Babcock, L E, 1993 Trilobite malformations and the fossil record of behavioral asymmetry J Paleontol 67, 217–229

Babcock, L E, 2003 Trilobites in Paleozoic predator–prey systems, and their role in reorganization of early Paleozoic ecosystems In Kelley, P A, Kowalewski, M, Hansen, T A (Eds), Predator–prey interactions in the fossil record Kluwer Academic/Plenum Publishers, New York, pp 55–92

Babcock, L E, Feldmann, R M, Wilson, M T, Suárez-Riglos, M, 1987 Devonian conulariids of Bolivia Nat Geogr Res 3, 210–231

Babcock, L E, Leslie, S A, Grunow, A M, Sadowski, G R, 2004 *Corumbella*, an Ediacaran-grade organism from the Late Neoproterozoic of Brazil Palaeogeogr Palaeoclim Palaeoecol 220, 210–231

Bischoff, G C O, 1989 Byroniida new order from early Paleozoic strata of eastern Australia (Cnidaria, thecate scyphopolyps) Senckenb Lethaea 69, 467–521

Bodenbender, B E, Wilson, M A, Palmer, T J, 1989 Paleoecology of *Sphenothallus* on an Upper Ordovician hardground Lethaia 22, 217–225

Bolton, T E, 1994 *Sphenothallus angustifolius* Hall, 1847 from the lower Upper Ordovician of Ontario and Quebec Bull Geol Surv Can 479, 1–11

Briggs, D E G, Erwin, D H, Collier, F J, 1994 The fossils of the Burgess Shale Smithsonian Institution Press, Washington and London 238 pp

Bruton, D L, 1981 The arthropod *Sidneyia inexpectans*, Middle Cambrian, Burgess Shale, British Columbia Philos Trans R Soc Lond, B 295, 619–656

Chen, J Y, Erdtmann, B -D, 1991 Lower Cambrian Lagerstatte from Chengjiang, Yunnan, China insights for reconstructing early metazoan life In Simonetta, A M, Conway Morris, S (Eds), The Early Evolution of Metazoa and the Significance of Problematic Taxa Cambridge University Press, Cambridge, pp 57–76

Chen, J Y, Ramskold, L, Zhou, G Q, 1994 Evidence for monophyly and arthropod affinity of Cambrian giant predators Science 264, 1304–1308

Chen, J Y, Zhou, G Q, Zhu, M Y, Yeh, K Y, 1996 The Chengjiang biota a unique window of the Cambrian explosion National Museum of Natural Science, Taichung, Taiwan 222 pp

Choi, D K, 1990 *Sphenothallus* ("Vermes") from the Tremadocian Dumugol Formation, Korea J Paleontol 64, 403–408

Conway Morris, S, 1977 Fossil priapulid worms Spec Pap Palaeontol 20, 1–95

Conway Morris, S, 1993 Ediacaran-like fossils in Cambrian Burgess Shale-type faunas of North America Palaeontology 36, 593–635

Conway Morris, S, Bengtson, S, 1994 Cambrian predators possible evidence from boreholes J Paleontol 68, 1–23

Conway Morris, S, Jenkins, R J F, 1985 Healed injuries in Early Cambrian trilobites from South Australia Alcheringa 9, 167–177

Conway Morris, S, Robison, R A, 1988 More soft-bodied animals and algae from the Middle Cambrian of Utah and British Columbia Univ Kans Paleontol Contrib, Pap 122, 1–48

Erdtmann, B -D, Steiner, M, 2001 Special observations concerning the Sinian–Cambrian transition and its stratigraphic implications on the central and SW Yangtze Platform, China In Peng, S C, Babcock, L E, Zhu, M Y (Eds), Cambrian System of South China University of Science and Technology of China Press, pp 52–65

Fauchald, K, Sturmer, W, Yochelson, E L, 1986 *Sphenothallus* "Vermes" in the Early Devonian Hunsruck Slate, West Germany Paleontol Z 60, 57–64

Fedonkin, M A, 1982 Vendian faunas and the early evolution of Metazoa In Lipps, J H, Signor, P W (Eds), Origin and Early Evolution of the Metazoa Kluwer Avademic/Plenum Publishers, New York, pp 87–129

Fedonkin, M A, 1985 Precambrian metazoans the problems of preservation, systematics and evolution Philos Trans R Soc Lond, B 311, 27–45

Fedonkin, M A, 1994 Early multicellular fossils In Bengtson, S (Ed) Early Life on Earth Columbia Univ Press, New York, pp 370–388

Feldmann, R M, Hannibal, J T, Babcock, L E, 1986 Fossil worms from the Devonian of North America (*Sphenothallus*) and Burma ("Vermes") previously identified as phyllocarid arthropods J Paleontol 60, 341–346

Girty, G H, 1911 The fauna of the Moorefield Shale of Arkansas U S Geol Surv Bull vol 439 338 pp

Glaessner, M F, 1984 The dawn of animal life Cambridge University Press, Cambridge 244 pp

Hahn, G, Hahn, R, Leonardos, O H, Pflug, H D, Walde, D H G, 1982 Korperlich erhaltene Scyphozoen-Reste aus dem Jungprakambrium Brasiliens Geol Palaeontol 16, 1–18

Hall, J, 1847 Palaeontology of New York Volume I Containing descriptions of the organic remains of the lower division of the New York System C Van Benthuysen, Albany, New York 338 pp

Hatscheck, B, 1888 Lehrbuch der Zoologie, eine Morphologische Ubersicht des Thierreiches zur Einführung in das Studium Dieser Wissenschaft Gustav Fischer, Jena 432 pp

Hou, X G, Aldridge, R J, Bergstrom, J, Siveter, D J, Siveter, D J, Feng, X H, 2004 The Cambrian Fossils of Chengjiang, China the flowering of animal life Blackwell Publishing, Malden, Massachusetts 233 pp

Hughes, N C , Gunderson, G O , Weedon, M , 2000 Late Cambrian conulariids from Wisconsin and Minnesota J Paleontol 74, 828–838

Jenkins, R J F , 1984 Interpreting the oldest fossil cnidarians Palaeontogr Am 54, 95–104

Kobayashi, T , 1934 The Cambro-Ordovician formations and faunas of South Chosen Palaeontology Part II Lower Ordovician faunas J Fac Sci , Imp Univ Tokyo, Sect II 3, 521–585

Lipps, J H , Gershwin, L A , Fedonkin, M A , 2001 Trophic styles in Neoproterozoic and Cambrian marine communities In Peng, S C , Babcock, L E , Zhu, M Y (Eds ), Cambrian System of South China University of Science and Technology of China Press, pp 281–282

Mason, C , Yochelson, E L , 1985 Some tubular fossils (Sphenothallus "Vermes") from the middle and late Paleozoic of the United States J Paleontol 59, 85–95

Neal, M , Hannbal, J T , 2000 Paleoecologic and taxonomic implications of Sphenothallus and Sphenothallus-like specimens from Ohio and areas adjacent to Ohio J Paleontol 74, 369–380

Pashin, J C , Ettensohn, F R , 1987 An epeiric shelf-to-basin transition Bedford–Berea sequence, northeastern Kentucky and south-central Ohio Am J Sci 287, 893–926

Peng, S C , Babcock, L E , 2001 Cambrian of the Hunan–Guizhou region, South China In Peng, S C , Babcock, L E , Zhu, M Y (Eds ), Cambrian System of South China University of Science and Technology of China Press, pp 3–51

Peng, S C , Babcock, L E , Robison, R A , Lin, H L , Rees, M N , Saltzman, M R , 2004 Global Standard Stratotype-section and Point (GSSP) of the Furongian Series and Paibian Stage (Cambrian) Lethaia 37, 365–379

Pratt, B R , 1998 Probable predation on upper Cambrian trilobites and its relevance for the extinction of soft-bodied Burgess Shale-type animals Lethaia 31, 73–88

Sprinkle, J , 1973 Morphology and evolution of blastozoan echinoderms Mus Compar Zool , Harvard Univ Spec Publ 284 pp

Steiner, M , Erdtmann, B -D , 2001 The Sancha–Wangjiashan section, Hunan Province, China occurrence of syngenetic Fe–Ni–Mo ore layers near the base of the lowermost Cambrian Niutitang Formation In Peng, S C , Babcock, L E , Zhu, M Y (Eds ), Cambrian System of South China University of Science and Technology of China Press, pp 125–131

Steiner, M , Wallis, E , Erdtmann, B -D , Zhao, Y L , Yang, R D , 2001 Submarine–hydrothermal exhalative ore layers in black shales from south China and associated fossils—insight into a Lower Cambrian facies and bio-evolution Palaeogeogr Palaeoclimatol Palaeoecol 169, 165–191

Van Iten, H , 1994 Redescription of Glyptoconularia gracilis (Hall), an Ordovician conulariid from North America In

Landing, E (Ed ), Studies in Stratigraphy and Paleontology in Honor of Donald W Fisher Bulletin of the New York State Museum, vol 481, pp 363–366

Van Iten, H , Cox, R S , Mapes, R H , 1992 New data on the morphology of Sphenothallus Hall implications for its affinities Lethaia 25, 135–144

Van Iten, H , Fitzke, J A , Cox, R S , 1996 Problematical fossil cnidarians from the upper Ordovician of the north-central USA Acta Palaeontol Pol 12, 99–132

Van Iten, H , Zhu, M Y , Collins, D , 2002 First report of Sphenothallus Hall, 1847 in the Middle Cambrian J Paleontol 76, 902–905

Vannier, J , Chen, J Y , 2002 Digestive system and feeding mode in Cambrian naraoiid arthropods Lethaia 35, 107–120

Whittington, H B , 1980 Exoskeleton, moult stage, appendage morphology and habits of the Middle Cambrian trilobite Olenoides serratus Palaeontology 23, 171–204

Whittington, H B , Briggs, D E G , 1985 The largest Cambrian animal, Anomalocaris, Burgess Shale, British Columbia Philos Trans R Soc Lond , B 306, 569–609

Yuan, J L , Zhao, Y L , Li, Y , Huang, Y Z , 2002 Trilobite fauna of the Kaili Formation (uppermost Lower Cambrian–lower Middle Cambrian) from southeastern Guizhou, South China Shanghai Science and Technology Press, Shanghai 433 pp (in Chinese)

Zhang, W T , Babcock, L E , 2001 New extraordinarily preserved enigmatic fossils, possibly with Ediacaran affinities, from the Lower Cambrian of Yunnan, China Acta Palaeontol Sin 40(Suppl ), 201–213

Zhao, Y L , Steiner, M , Yang, R L , Erdtmann, B -D , Guo, Q J , Zhou, Z , Wallis, E , 1999 Discovery and significance of the early Metazoan biotas from the Lower Cambrian Niutitang Formation, Zunyi, Guizhou Acta Palaeontol Sin 38(Suppl ), 132–144

Zhao, Y L , Yu, Y Y , Yuan, J L , Yang, X L , Guo, Q J , 2001 In Peng, S C , Babcock, L E , Zhu, M Y (Eds ), Cambrian System of South China University of Science and Technology of China Press, pp 172–183

Zhao, Y L , Erdtmann, B -D , Steiner, M , Guo, Q J , 2002 Early Cambrian Niutitang Formation Biota In Zhao, Y L (Ed ), Guizhou–Palaeontological Kingdom Guizhou Science and Technology Press, Guiyang, pp 94–109 (in Chinese)

Zhu, M Y , Van Iten, H , Cox, R S , Zhao, Y L , Erdtmann, B -D , 2000 Occurrence of Byronia Matthew and Sphenothallus Hall in the Lower Cambrian of China Paleontol Z 74 (3), 227–238

Zhu, M Y , Vannier, J , Van Iten, H , Zhao, Y L , 2004 Direct evidence for predation in the Cambrian Proc R Soc Lond , B 271 (Suppl , online 10 1098), S277–S280

Available online at www.sciencedirect.com

Palaeogeography, Palaeoclimatology, Palaeoecology 220 (2005) 129–152

www.elsevier.com/locate/palaeo

# Lower Cambrian Burgess Shale-type fossil associations of South China

Michael Steiner[a,*], Maoyan Zhu[b], Yuanlong Zhao[c], Bernd-Dietrich Erdtmann[a]

[a]TU Berlin, Sekr. ACK 14, Ackerstrasse 71-76, 13355 Berlin, Germany
[b]Nanjing Institute of Geology and Palaeontology, Chinese Academy of Sciences, Nanjing 210008, China
[c]Guizhou University of Technology, Guiyang 550003, China

Received 15 September 2002, accepted 15 June 2003

## Abstract

Burgess Shale-type preservation has been reported from numerous Qiongzhusian and Canglangpuian sections of southern China during the last few decades. The Early Cambrian Chengjiang-type faunas of East Yunnan were of particular interest due to their excellent preservation and taxonomic diversity. A new definition and revision of the Chengjiang-type faunas is given, which indicates that distinct ecological variations occurred between the Chengjiang and Haikou faunas on a smaller scale, and between the arthropod-dominated faunas of shallow shelf and the sponge-dominated faunas of deeper shelf (e.g. Hunan) on a larger scale. Recent finds from the Yuanshan Formation of Deze, East Yunnan, and the Niutitang Formation of Zhongnan, Zunyi County, Guizhou, indicate a much wider distribution of Burgess Shale-type fauna during this period than was previously known. The present investigations of the latter fauna revealed the presence of *Tsunyidiscus niutitangensis*, *Naraoia spinosa*, *N.* cf. *longicaudata*, *Isoxys curvirostratus*, *I. paradoxus*, *Skioldia aldna*, *Amplectobelua trispinata*, cf. *Tsunyiella diandongensis*, *Scenella* sp., *Cambrorhytium elongatum*, *Hyalosinica archaica*, and *Choiaella radiata*. The occurrence of Burgess Shale-type preservation on the Yangtze Platform is not restricted to a small area as in the Burgess Shale, but known from a ca. 80 km wide geographic stripe extending for more than 500 km along the rim of Proterozoic platform core. This wide distribution of Burgess Shale-type preservation is interpreted to be triggered by the coincidence of specific palaeoecological conditions, including the rapid sedimentation of finest siliciclastic particles and the partial influence of suboxic water masses.

*Keywords:* Palaeontology; Yangtze Platform, China; Lower Cambrian; Chengjiang; Burgess Shale-type preservation

## 1. Introduction

Soft-bodied fossil preservation is of great importance in reconstructing the evolution of the early metazoan lineages. For a long time, the Middle Cambrian Burgess Shale was the earliest known

* Corresponding author.
 *E-mail address:* steishhb@mailbox.tu-berlin.de (M. Steiner).

0031-0182/$ - see front matter © 2004 Elsevier B.V. All rights reserved.
doi:10.1016/j.palaeo.2003.06.001

occurrence of metazoan fossils, with slightly sclero-tized exoskeletons and/or labile tissues preserved This kind of the preservation, first discovered by Charles Doolittle Walcott in the Burgess Shale in 1909 (Walcott, 1911), was initially regarded as extraordinary and unique However, at the beginning of the 20th century, the first Burgess Shale-type fossils of China were discovered by H Mansuy in Kebaocun, Yiliang County, Yunnan (Mansuy, 1912· pl 4, Figs 6–8) Walcott knew of these finds by communication with Mansuy and published a short remark on a Lower Cambrian arthropod from southern China (Walcott, 1911 19, 28) Therefore, the first published note of the Chengjiang-type biota, which was named after Chengjiang County because of the occurrence, also originated from Walcott In the following years, Lower Cambrian Burgess Shale-type fossils from Yunnan were described sporadically, but received little attention Pan (1957) described *Tuzoia* from the Canglangpu Formation of Qiongzhusi, Kunming Luo et al (1984 pl 16, Fig 13) figured a worm-like fossil from the Yuanshan Formation of Meishucun, Jinning County, as *Sabellidites* sp. Finds of the priapulid worm *Cricocosmia* from the Yuanshan Formation of Nazhang, Malong County (Luo et al, 1999· pl 19, Fig 1), which were misinterpreted as trace fossils, have been stored in the collection of the Geological Survey of Yunnan since 1972 (Luo et al, 1999, personal communication Hu Shixue, 2002)

Oblivion stole these earliest records, until Hou Xianguang, without knowledge of these former collections, discovered arthropods with soft-bodied preservation during investigations of bradoriids in the Yuanshan Formation at Mount Maotian in 1984 (Zhang and Hou, 1985) This discovery triggered systematic excavations by different Chinese research groups at several sites (classical outcrops Maotianshan, Maanshan, Dapotou) in the Yuanshan Formation of Chengjiang County during the 1980's and 1990's (overview in Chen et al, 1996, Chen and Zhou, 1997, Hou et al, 1999) The Burgess Shale-type fauna of this region became known as the "Chengjiang fauna" More occurrences of fossil associations with soft-bodied preservation were described from different counties of East Yunnan, which resembled the Chengjiang fauna in their composition or were quasi identical to it Here particularly, the discovery of different smaller out-

crops around Haikou, Xishan district of Kunming City, capital of Yunnan Province, is worthy of special emphasis (Luo et al, 1997, 1999, Chen et al, 2002) All the finds from the Yuanshan Formation of different counties of East Yunnan, within an area of approximately 120×120 km, were previously treated as "Chengjiang fauna" or "Chengjiang fossils" Here, we apply the term "Chengjiang fauna" in a more strict way, including only the fossil occurrences of the upper Yuanshan Formation in Chengjiang County (Jiucun, Maanshan, Maotianshan, Xiaolantian, Dapotou, Haikou sections), in order to allow differentiation from the associations of Haikou district (Ercaicun, Mafang, Dazicun sections—"Haikou fauna") and the Jinning (Meishucun/Badaowan section), Anning (Sankoucun, Haoyicun sections), Wuding, Malong, and Qujing counties (Fig 1), according to minor differences in taxic composition As a general term for the arthropod-dominated fossil associations with common occurrence of the genera *Naraoia* and *Isoxys* from the upper Yuanshan Formation (Maotianshan Shale unit and Upper Siltstone unit) and its strati-graphical equivalents of South China, we propose "Chengjiang-type fauna" The Chengjiang-type fauna represents a taxonomically, temporally, and spatially specific case of a Burgess Shale-type fauna In contrast to this taxonomic description, Burgess Shale-type preservation was designated as a particular type of taphonomic mode (Butterfield, 1990, 1995), also covering the taphonomic window of the Cheng-jiang-type faunas It was defined as an exceptional preservation of "non-mineralizing organisms pre-served as carbonaceous compressions in fully marine sediments" (Butterfield, 1995 2) Later taphonomical investigations, however, doubted the exclusively organic preservation of Burgess Shale-type fossils (Towe, 1996) and documented that authigenic miner-alizations played an important role in preserving selective tissues of these fossils, although partial organic preservation is retained (Orr et al, 1998) Hence, Burgess Shale-type preservation is considered here as a preservation of easily decomposable tissues in siliciclastic matrix in organic mode and by partly early diagenetic mineralization Although the pyritiz-ation and phosphatization of specific tissues may occur in Burgess Shale-type fossils, Burgess Shale-type preservation differs from predominant pyritiza-tion, as in the Devonian Hunsruck fossil *lagerstatte*

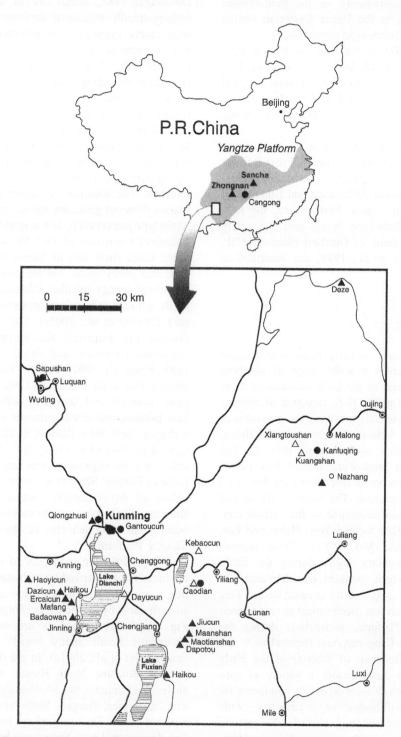

Fig 1 Map showing all previously known occurrences of Burgess Shale-type faunas from the Qiongzhusian (▲ described and figured fauna, △ mentioned occurrences) and Canglangpuian (●)

(Hunsruck-type preservation), or the predominant phosphatization, as in the Upper Cambrian Orsten fossil *lagerstatte* (Orsten-type preservation)

In the year 2000, a National Geological Park covering an area of 18 km² was established in Chengjiang County to protect the important classical fossil localities of the Yuanshan Formation and to coordinate scientific excavations (Zhao and Wang, 2002)

The aim of the present article is to give an overview on the distribution of the different Burgess Shale-type faunas of the Qiongzhusian and Canglangpuian (equivalent to the Atdabanian and Botoman of Siberia) in southern China Furthermore, the little known Burgess Shale-type faunas and the newly discovered Zunyi fauna of Guizhou (Steiner et al, 1998, 2001b, Zhao et al, 1999) are described in greater detail

## 2. Geologic setting

The Neoproterozoic to Early Palaeozoic Yangtze Paraplatform occupies a wide range of southern China, with an extent in the E–W direction of more than 2000 km and in the N–S -direction of approximately 900 km, essentially covering the provinces of East Yunnan, East Sichuan, Guizhou, Hunan, Hubei, Jiangxi, South Anhui, and Zhejiang (Fig. 1) The consolidation of the Yangtze craton basically occurred during the Jinning orogeny (approximately 850 m y ) during the Neoproterozoic The Sichuan Massif and the Kangdian Massif developed as the earliest continental nuclei in East Sichuan/West Hubei and East Yunnan (Wang and Mo, 1985) These regions represented sedimentary highs during the Early Cambrian, from which sediment was delivered into the nearby shallow seas and the oceanic basins Fine clastic sedimentation was predominant in vast regions of the Yangtze Platform, particularly during the Qiongzhusian and Canglangpuian (equivalent to the Atdabanian and Botoman of Siberia) of the Early Cambrian During these periods a variety of conditions prevailed which were especially conducive for preservation of soft-bodied organisms (e g wide distribution of metazoans, high rate of sedimentation of finest siliciclastic particles, comparable to contemporary nepheloid water bodies, Zoutendyk and

Duvenage, 1989, occurrence of a shelf basin, a bathymetrically structured depositional environment with partial occurrence of stratified water masses, local occurrence of anoxic or dysoxic water layers, compare Babcock et al, 2001) Due to the combination of the preceding palaeoenvironmental conditions in a large area, Burgess Shale-type preservation occurs in the Lower Cambrian of South China, particularly in the upper Yuanshan and Niutitang formations of Yunnan, Guizhou, and Hunan Provinces (Figs 1, 2) In contrast to the Burgess Shale, the Chengjiang-type fauna occurrences are not strictly punctual The presence of anoxic or dysoxic water masses does not guarantee the occurrence of Burgess Shale-type preservation, as it is evident from the lower Yuanshan Formation of East Yunnan Although the lower black shale unit of Yuanshan Formation was deposited partly under euxinic conditions, only disarticulated sponge spicules, trilobites, and brachiopod shells without soft-bodied preservation were recovered (Steiner et al, 2001a) Oxygen-depleted conditions are required for Burgess Shale-type preservation (Allison and Brett, 1995, Butterfield, 1995, Petrovich, 2001) For the Yangtze Platform, it appears that a fluctuation of OMZ possibly caused mass mortality and facilitated soft-bodied preservation, because anoxic water masses were distributed in a deeper shelf basin (Steiner et al, 2001b) A very rapid deposition of relatively thick beds of fine mud, entombing the organisms, was crucial for the occurrence of Burgess Shale-type preservation The combination of environmental characters necessary for Burgess Shale-type preservation probably only occurred in a limited area of shallow shelf on the Yangtze Platform

The pre-trilobitic sequences of the Meishucunian are mainly characterized by carbonates and phosphorites, representing shallow marine environments, and subordinately by clastic rocks (Zhujiaqing Formation, Fig 2) These sequences are especially rich in phosphatic Small Shelly Fossils (SSF) and trace fossils (Zhu et al, 2001a) In the deeper shelf basin (central Guizhou, North Hunan, etc ), organic-rich shales predominate the Meishucunian record However, no typical Burgess Shale-type fauna has been reported so far This can perhaps be explained by the low depositional rate Syngenetic hydrothermal exhalations were reconstructed for the deeper shelf

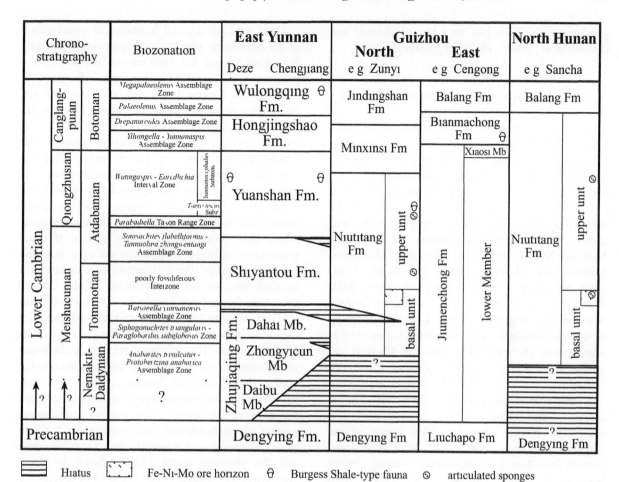

| Chrono-stratigraphy | | | Biozonation | East Yunnan | | Guizhou | | North Hunan |
|---|---|---|---|---|---|---|---|---|
| | | | | Deze | Chengjiang | North<br>e g Zunyi | East<br>e g Cengong | e g Sancha |
| Lower Cambrian | Qiongzhusian | Canglang-puian | Botoman | *Megapalaeolenus* Assemblage Zone | Wulongqing Fm. θ | Jindingshan Fm | Balang Fm | Balang Fm |
| | | | | *Palaeolenus* Assemblage Zone | | | | |
| | | | | *Drepanuroides* Assemblage Zone | Hongjingshao Fm. | Minxinsi Fm | Bianmachong Fm θ | |
| | | | | *Yiliangella - Yunnanaspis* Assemblage Zone | | | Xiaosi Mb | |
| | Atdabanian | | | *Wutingaspis - Eoredlichia* Interval Zone | Yuanshan Fm. θ θ | upper unit θθ | | upper unit ⊙ |
| | | | | *Tsunyiscus* Subz | | | | |
| | | | | *Parabadiella* Taxon Range Zone | | | Jiumenchong Fm lower Member | Niutitang Fm |
| | Meishucunian | Tommotian | | *Sinosachites flabelliformis - Tannuolina zhangwentangi* Assemblage Zone | Niutitang Fm | | | |
| | | | | poorly fossiliferous Interzone | Shiyantou Fm. | basal unit ⊙ | | |
| | | | | *Watsonella yunnanensis* Assemblage Zone | Zhujiaqing Fm. Dahai Mb. | | | basal unit ⊙ |
| | | Nemakit-Daldynian ? | | *Siphogonuchites triangularis - Paragloborilus subglobosus* Zone | Zhongyicun Mb | ? | | |
| | | | | *Anabarites trisulcatus - Protohertzina anabarica* Assemblage Zone | Daibu Mb. | | | ? |
| | ? | ? | | ? | | | | |
| Precambrian | | | | Dengying Fm. | | Dengying Fm | Liuchapo Fm | Dengying Fm |

▬▬ Hiatus    ⬚ Fe-Ni-Mo ore horizon    θ Burgess Shale-type fauna    ⊙ articulated sponges

Fig 2  Correlation table of the Meishucunian to Canglangpuian of selected regions in East Yunnan, Guizhou, and North Hunan showing the stratigraphic position of the described Burgess Shale-type faunas

environments of the Niutitang Formation in the Sancha (Hunan) and Zunyi (Guizhou) regions (Steiner et al., 2001b)  The unique Fe–Ni–Mo-ore horizon also contains mass occurrences of bivalved arthropods, sponges, and undetermined shelly fossils  Burgess Shale-type preservation could be detected in both regions in the younger sequences of the Niutitang Formation, which are not characterized by the influence of hydrothermal exhalations  However, these faunas cannot be compared in quality, composition, or quantity with those from East Yunnan

The chronostratigraphic subdivision and correlation of Early Cambrian strata is difficult, because there are few reliable radiometric ages available from South China  An international and regional correlation is mainly based on biostratigraphic and lithostrati-

graphic evidences (Fig 2)  Here, the distribution of trace fossils, acritarchs, SSF, and trilobites is of great importance  Archaeocyathids, which are of principal importance for the biostratigraphy of the Lower Cambrian, are largely unknown on the Yangtze Platform due to biofacies restrictions (Yuan et al, 2001)  Therefore, the exact age of the Chengjiang-type fauna has been controversial  Although published ages for the Chengjiang-type fauna vary over an absolute time span of more than 10 m y and include the chronostratigraphic units equivalent to the Siberian pre-Tommotian (Chen and Erdtmann, 1991), early Atdabanian (Chen and Zhou, 1997), late Atdabanian–early Botoman (Conway Morris, 1998), early Botoman (Zhuravlev, 1995, Shu et al, 1995), and Botoman (Landing, 1994), most recent biostratigraphic investigations indicate a late Atdabanian age (Steiner et al , 2001a, Zhu et al , 2001a)

## 3. Chengjiang-type fossil occurrences in Yunnan

Chengjiang-type fossil associations have been reported from a large area in East Yunnan, including Chengjiang, Jinning, Anning, Wuding, Yiliang, Malong counties, and the Kunming district (Fig 1) Here, we report a Chengjiang-type fauna from Deze of Qujing County for the first time However, the most productive outcrops in both quantity and quality for Chengjiang-type faunas are those near Haikou, Xishan district of Kunming (Ercaicun, Dazicun, and Mafang sections, Luo et al , 1997, 1999, Shu et al , 1999a,b, Chen et al , 1999, 2002, Zhang et al , 2000), and those of Chengjiang County Systematic descriptions of the Chengjiang fossils have been published, particularly from the sections at Maotianshan (Sun and Hou, 1987a,b, Chen et al , 1989a, Chen and Zhou, 1997; Hou et al , 1999), Maanshan (Shu et al , 1996a,b; Hou et al , 1999), Xiaolantian (Hou and Bergstrom, 1994, 1997, Hou et al , 1999), and Dapotou (Chen et al , 1989a; Hou et al , 1989) Poorer Chengjiang-type faunas of the comparable stratigraphical interval have been documented from Meishucun/Badaowan, Jinning County (Luo et al , 1984, Hou and Sun, 1988), from Nazhang, Malong County (Luo et al , 1999), from Sapushan, Wuding County (Luo et al , 1999), from Haoyicun, Anning County (Zhang et al , 2001), and Haikou, Chengjiang County (Steiner et al , 2001a) Further occurrences of Burgess Shale-type fossils were mentioned, but not described in detail or figured These include examples from Dayucun, Chenggong County, Kebaocun and Caodian, Yiliang County (Luo et al , 1999), as well as from Kuangshan, Malong County (Hou et al , 1999)

The classical localities for Lower Cambrian Burgess Shale-type fossils are situated in the so-called "Maotianshan Shale unit", though rarely, fossils with soft-bodied preservation are also found in the "Upper Siltstone unit" of the upper Yuanshan Formation, both of which are of late Qiongzhusian age (Fig 2) The sedimentology of the Yuanshan Formation of the sections at Maotianshan and Dapotou was described in detail by Zhu et al (2001b)

Until now, no detailed biostatistical investigation of the community structure of Chengjiang-type fauna from East Yunnan has been carried out, and only a preliminary counting of fossil specimens from collections of the Haikou fauna and a Chengjiang-type fauna from Anning has been published (Zhang et al., 2001) A subsequent counting of individuals and species from the extensive collection material from Chengjiang outcrops may be misleading, because previous collecting campaigns have always been selective and preferred exceptionally preserved and exotic fossils Our subjective impression of abundance during field collections relies upon these constraints, which have produced the known genera. If exact percentages of genus abundances are cited from the available literature, the data remain preliminary until detailed biostatistical investigations of the Chengjiang-type faunas from Yunnan have been carried out

The Chengjiang fossil association is dominated[1] by the arthropods *Kunmingella, Naraoia, Isoxys,* and *Chuandianella* The trilobites are not a dominant component of the Chengjiang fauna, although the genera *Eoredlichia* and *Yunnanocephalus* are common Among the comparatively rare non-arthropods, the sponges *Leptomitus, Leptomitella,* and *Triticispongia,* the lophophorate *Stellostomites,* the brachiopod *Heliomedusa, Lingulella,* and *Lingulellotreta,* and the priapulid worm *Maotianshania* are most frequent (subdominant[1]) The majority of species are rare[1] (see Appendix A) Interestingly, virtually no representatives of the mollusks (except hyolithids and one specimen of a helcionellid) were recovered in the strata with soft-bodied fossils, although these are common in under- and overlying SSF associations Additionally, no indisputable proof of echinoderms has been found so far This could be explained by the specific palaeoecological conditions, such as the wide distribution of soft substrates (Zhu et al , 2001b) or an abnormal salinity (Babcock et al , 2001)

Compared with the Chengjiang fauna, the Haikou fauna indicates strong similarities in the taxic composition, however, also some distinct differences in the occurrence of specific taxa Both faunas are arthro-

---

[1] The term "dominant" is used here for taxa with an abundance >10% of the absolute number of individuals, "subdominant" corresponds to 2–10%, and "rare" to <2% of the absolute number in individuals

pod-dominated, considering the species constitution of confirmed taxa (54% of the Chengjiang fauna vs 51% of the Haikou fauna, see Appendix A) The species diversity of the other groups, however, differs partly in the two faunas Haikou has the priapulids (10%) as the second richest group in species, followed by the sponges (8%), while in Chengjiang, the sponges (17%) are more diverse than the priapulids (5%) About 60% of species of the faunas occur both in Chengjiang and Haikou, but 40% are only known from one of the localities until now

The Haikou fauna is dominated[1] by the bradorid *Kunmingella* (27%) and the priapulid worm *Cricocosmia* (16%, Zhang et al, 2001) Subdominant[1] are the arthropods *Isoxys* (8%), *Leanchoilia* (5%), *Naraoia* (5%), *Eoredlichia* (2%), and the brachiopod *Heliomedusa* (4%, Zhang et al, 2001). The existence of different thin sediment layers containing exceptional enrichments in *Haikouella*, *Cricocosmia*, respectively, *Phlogites* has been reported by Wang et al (2001) from Ercaicun. In these layers, one specific taxon may have reached an absolute dominance (>50%) in individuals and biovolume In one of these accumulation layers, *Cricocosmia jinningensis* reached a calculated abundance of ca 4000 individuals per square meter. Mass occurrence of the priapulid worm *Maotianshania cylindica* has also been documented from the Chengjiang County (Chen and Zhou, 1997 Figs 7, 8; Hou et al, 1999 Fig 30) However, it is remarkable that *C jinningensis* is rare in Chengjiang, while *M cylindica* has not yet been unambiguously been documented from Haikou (mentioned occurrences of *M cylindica* in Zhang et al, 2001, may be due to misidentification) The differences in community structure may be due to a true ecological variation between the two close areas, which is discussed below

The new fossil association of Deze, Qujing County, resembles the Chengjiang fauna and is dominated by the arthropods *Kunmingella*, *Isoxys*, and *Naraoia* (Fig 3), although until now, only a very small number of fossils were recovered

The sedimentology of the fossil-bearing Maotianshan Shale unit of Yuanshan Formation and the preservation of fossils in the Chengjiang and Haikou faunas are comparable In both areas, which are situated ca 50 km apart, the Maotianshan Shale unit consists of mainly dark grey–green (yellow, when

Fig 3 Juvenile individual of *Naraoia* cf *longicaudata*, with a comparably short posterior shield Note the possible fragment of antenna at the arrow, scale bar equals 5 mm Yuanshan Formation, Deze, Qujing County, TU Berlin collection ACK, No Dez 010

weathered), finely laminated mudstones containing organic hash in great quantity, previously reported as algae Intercalated to this finely laminated mudstones, interpreted here as background sedimentation, occur numerous interbeds of finest and pure, light grey (yellow when weathered), and massive mudstones, measuring in average few millimeters to 2 cm in thickness The majority of the soft-bodied fossils are entombed in these massive mudstones, interpreted here as event beds The fossils are frequently found in chaotic and 3-dimensional arrangement within the event layers The laminated background sediments may also contain soft-bodied fossils (see Hou et al, 1999. Fig 50, Chen et al, 2002. pl 10, Fig. 1, 2), however, here, the preservation is not as excellent as in the event beds, the fossils are never chaotically and 3-dimensionally embedded, and Burgess Shale-type preservation is rarer Disarticulated and lightly sclerotized remains of arthropods are comparatively common in the background sediments The organic hash, which is common in the background layers and was previously described as algae, may largely represent fecal strings, although the existence of some degraded organics derived from algae cannot be ruled out This observation is based on the fact that some of the remains are slightly three-dimensionally preserved and sometimes contain sediment pellets (Fig 4), the outlines of the remains are partly indistinct and often varies in thickness, never indicating branchings, holdfasts, or an aggregation typical for cyanobacteria. Fecal strings of similar appearance have been documented compressed in organic preservation or three-dimensionally preserved from the Middle Cambrian Little Bear biota (Butterfield and Nicholas, 1996) The organic hash of the Chengjiang-type faunas differs conspicuously from the organically preserved algae of the Burgess Shale fauna (e g Briggs et al , 1994) or Kaili biota (Mao et al , 1994)

The rate of sedimentation was much lower in the laminated background beds than in the massive event beds In contrast to the event beds, the faunal record from the background layers may bear a potential preservational bias

The preservational mode of soft-bodied fossils from the event beds does not indicate differences between the occurrences in Haikou and Chengjiang Similar as in the Burgess Shale, fossilization of labile tissues followed a variety of pathways leading to partial organic preservation and light mineralization by pyrite, Fe-rich aluminosilicates, or calcium phosphate (see Zhu et al , this volume) Due to the intense weathering of almost all outcrops in Chengjiang County and the Haikou district, the most common preservation of the Chengjiang-type fossils is by iron oxides, creating an aesthetic contrast between reddish to brown fossils on yellow, weathered mudstone matrix

The circumstances under which the thin event beds were deposited indicate a catastrophic sedimentation Different interpretative models may be considered for the deposition of event beds, including tempestites, seasonally radiated suspension flows of an estuary fan, or nepheloid water body Due to the fact that the ungraded event layers sometimes show microscours at

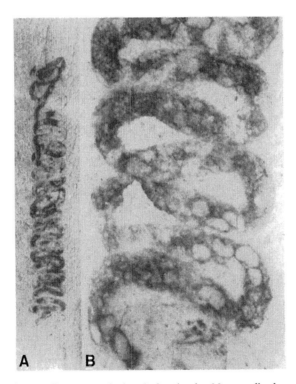

Fig 4 Fecal string with enclosed sediment pellets, previously described as the alga *Megaspirellus houi*, Yuanshan Formation, Maotianshan, Chengjiang County. Holotype No ELRC 108501 (A) Overview of the counterprint and (B) close up, height 4 8 mm

the bases of bed and the fossils are three-dimensionally oriented within the mud, it becomes obvious that the layers were nearly instantly deposited from near-bottom suspension flows, which included lateral sediment transport to some extent The argument for the deposition of Burgess Shale (Allison and Brett, 1995) is that the deposition duration from low-density suspensions of the whole water column is too long to embed larger soft-bodied fossils in an oblique or vertical orientation Thus, a deposition of the event beds from tempestites or a near-surface freshwater plume of an estuary appear is less likely Here, we consider a nepheloid-layer transport of high-density suspension responsible for the deposition of event layers of upper Yuanshan Formation Although a lateral transport of organisms within bottom currents is probable to some extent, the taxic variation in the different Chengjiang-type faunas reported from the event beds is not consistent with a far distance transport and thus indicates a parautochthonous character of faunas. The different Burgess Shale-type fossil associations of East Yunnan indicate a distinct variability in the dominant, subdominant, and rare taxa[1], which points toward changes in the habitats Especially in the case of Chengjiang and Haikou faunas, it is evident that preservational bias, disarticulation, or size sorting by currents are not responsible for the geographic variation of Chengjiang-type faunas The distinct differences in species composition of priapulid communities (at a similar size and preservational mode of tissues) in Haikou and Chengjiang (see above), species variations in vetulicolids and chordates, the rare occurrence of eldoniids in Haikou, while comparably common in Chengjiang with a similar distribution of lobopods at the same time, support the interpretation of an ecological variation This variation is particularly distinct in the NW–SE direction, which reflects a different coastal proximity, water depth, and prevailing particle size of the sediments (Zhu et al, 2001b) However, further biostatistical investigations of the fauna are required to fully understand these geographic variations and the trophic community structure on a larger scale

It is also worth noting that Burgess Shale-type preservation is not only limited to the Qiongzhusian interval in East Yunnan Some Burgess Shale-type fossils were also described from the Canglangpuian

Wulongqing Formation of the Qiongzhusi, and Gangtoucun sections, Kunming district, from Caodian of Yiliang County, from Sapushan of Wuding County (Luo et al, 1999), and Wulongqing/Kanfuqing of Malong County (Zhang and Babcock, 2001; Zhang et al, 2001)

## 4. Lower Cambrian Burgess Shale-type preservation in Hunan

A Burgess Shale-type fauna was first described from black shales of the upper unit of the Niutitang Formation of Sancha, Dayong (Zhangjiajie), northern Hunan Province (Steiner et al, 1993) In a vertical range of 2 m within the upper Niutitang Formation, larvae (Fig 5A), juvenile (Fig 5B), and adult individuals (Fig 5C) of mainly hexactinellid sponges, such as, *Triticispongia diagonata*, *Sanshapentella dapingi*, and *Sanshadictya microreticulata* are common Torn off spicule root tufts of the hexactinellid sponge *Hyalosinica archaica* and complete specimens of *Solactiniella plumata* and *Saetaspongia densa*, whose affiliation is uncertain, are also common The preferential preservation of root tufts does not necessarily reflect the influence of stronger water currents but may indicate a higher potential for preservation of remains located near the surface of substratum than for parts of sponges raised higher into the water column Locally, mass occurrences of sponge larvae occur, which settled partly on dispersed older sponge spicules (Fig 5A) In some horizons, sponge spicules crop up unsorted and without orientation, forming spicule mats by in situ decay of generations of sponges at centers of colonization Similar autochthonous spiculites have been described from the modern Atlantic Ocean (e g Henrich et al, 1992) The spicule mats and the common occurrence of sponge larvae indicate the autochthonous character of the fauna

The association comprises only hexactinellid sponges and questionable demosponges, with some organic tissues and mineralized spicules preserved A secondary transformation of the initial skeletal opal becomes obvious from the corroded surfaces and enlarged axial canals of the siliceous spicules (Fig 6) However, the preservation of axial canals has only rarely been reported from the Early Palaeozoic, and it supports the exceptional quality of this Burgess

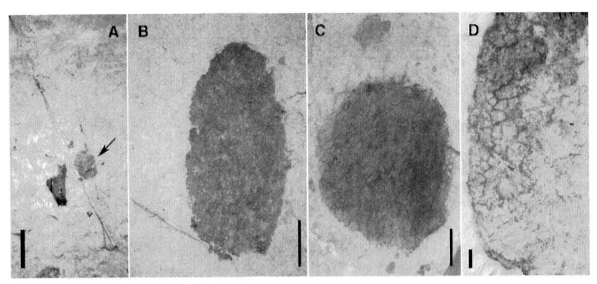

Fig  5  (A) Sponge larva (arrow) attached to a long monaxon  Scale bar equals 2 mm, TU Berlin collection ACK, No  San 391  (B) Elongated, massive, and juvenile hexactinellid sponge with small triaxons attached to a monaxon  Scale bar equals 2 mm, TU Berlin collection ACK, No  San 359A  (C) *Solactiniella plumata* with closely packed and irregularly oriented diactins  Scale bar equals 10 mm  (D) Detail of an undetermined sponge with reticulate organic tissue remains  Scale bar equals 0 5 mm, TU Berlin collection ACK, No  San 386  All from the upper unit of the Niutitang Formation, Sancha, Dayong, Hunan Province

Shale-type preservation  The different specific profiles of axial canals (rectangular vs triangular/hexagonal) have been considered as synapomorphies for the Hexactinellida and Demospongiae (Mehl-Janussen, 1999)  The recognition of axial canals may support the taxonomic assignment of fossil sponges

The black shales containing the complete sponges are superimposed by 22 m of the upper siltstone-shale interbeds with *Hunanocephalus  Triticispongia diagonata* and *Saetaspongia densa* also occur in the Chengjiang-type fauna (Rigby and Hou, 1995)  This and the distribution of *Hunanocephalus* in the overlaying strata may point toward

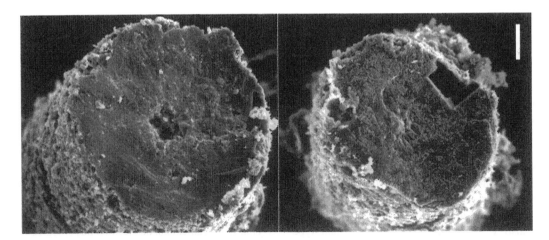

Fig  6  SEM micrograph of two isolated siliceous sponge spicules indicating a secondary replacement of the primary skeletal opal by corrugation structure and enlarged axial canals  Note ghosty hexagonal cross sections of axial canals indicating a possible demosponge affiliation  Scale bar equals 20 μm, upper unit of the Niutitang Formation, Sancha, Dayong, Hunan Province

a possible age correspondence of the sponge horizon in the upper unit of the Niutitang Formation with those of the Chengjiang-type fauna (Fig 2) Few sponges could be found in the basal unit of the Niutitang Formation (Steiner et al, 1993)

The Early Cambrian sediments of the Hunan and Guizhou Provinces were, to a large extent, deposited in an offshore shelf-basin facies of the Yangtze Platform These deposits are generally poor in fossils when compared to equivalent shallow water deposits of East Yunnan, East Sichuan, or South Shaanxi and are dominated by sponge communities, in contrast to the arthropod-dominated shallow water association

## 5. The Zunyi biota and Burgess Shale-type preservation in Guizhou

Burgess Shale-type faunas have been described from Zhongnan, approximately 25 km W of Zunyi, northern Guizhou Province, and from a section ca 10 km SE of Cengong, East Guizhou

At Zhongnan, interlayers of sandstone and black shale occur approximately 23 m above the base of the upper unit of the Niutitang Formation, which are, in turn, themselves overlain by green claystones and yellowish siltstones The base of the green claystones yielded a Chengjiang-type fauna, which is dominated by arthropods (92%) and sponges (6%) Counting of a limited number of fossils (*n*=204) allowed a preliminary comparison of the specific composition of fauna with the Chengjiang and Haikou faunas Dominant[1] taxa are the arthropods *Isoxys* (25%), *Naraoia* (13%), as well as the eodiscid trilobite *Tsunyidiscus* (40% including disarticulated remains/exuviae, 7% excluding exuviae) Besides the subdominant[1] hexactinellid sponges, all other recovered genera are rare[1] Interestingly, redlichiid trilobites and the bradorid *Kunmingella* are completely absent here, but are frequently found in the Yuanshan Formation of Chengjiang This may reflect a greater water depth in the depositional environment A characteristic feature of the fossiliferous horizon of the Niutitang Formation is the massive occurrence of organic hash, including fecal strings (Fig 7K) These remains are frequently found

in Chengjiang and Haikou as well, where they have been generally considered as algae (e g Chen and Erdtmann, 1991, Chen and Zhou, 1997) Complete individuals of less sclerotized arthropods co-occur with abundant disarticulated remains of arthropods. The comparatively common remains of *Naraoia spinosa* (Figs 7J, R) and the rarer *Isoxys paradoxus* (Figs 7C, D) were previously only known from Chengjiang-type faunas of East Yunnan

*Isoxys curvirostratus* is abundant (Figs 7A, B) Exoskeletons of a second species of *Naraoia* with smooth, rounded anterior, and posterior shields are here provisionally referred to *Naraoia* cf *longicaudata* (Fig 7I) Typically, the posterior shields are approximately twice as long as the anterior shields in *N longicaudata* However, in *Naraoia* of the Zunyi biota (Fig 7I), the posterior shield is not significantly longer than the anterior shield In this respect, the species resembles *Naraoia compacta* from the Middle Cambrian Burgess Shale (Whittington, 1977) The specimen may, however, simply represent a juvenile individual, as in the case of the figured example from Deze (Fig 3) A comparable form was also described from Chengjiang as *Naraoia spinosa?* (Hou et al, 1999· Fig 159) Complete specimens of the eodiscid *Tsunyidiscus niutitangensis* occasionally occur in the green shales (Figs 8B, C), while disarticulated cephala and pygidia are common in intercalated silt and sandstone layers (Fig 8A) Bradorids, here provisionally affiliated with cf *Tsunyiella diandongensis* (Fig 7G), are rare Some partly articulated remains of a less sclerotized arthropod can be assigned to the helmitiid species *Skioldia aldna* (Fig 7M, N) and the anomalocarid *Amplectobelua trispinata* (Fig 7E) The existence of large predators can also be inferred from the occurrence of centimeter-sized coprolites A coprolite of approximately 1 5×1 3 cm contains numerous residues of hyolithids and sponge spicules

Among the subdominant[1] hexactinellid sponges, root tufts of the hexactinellid sponge *Hyalosinica archaica* (2 5%) are particularly predominant, as are numerous indefinite remains of hexactinellid sponges (2 5%) Additionally, small specimens of the demosponge *Choiaella radiata* were found (Fig 7H) Unlike in Chengjiang, a few specimens of the questionable mollusc *Scenella* sp (Fig 7F) were recovered A single specimen of *Cambrorhytium*

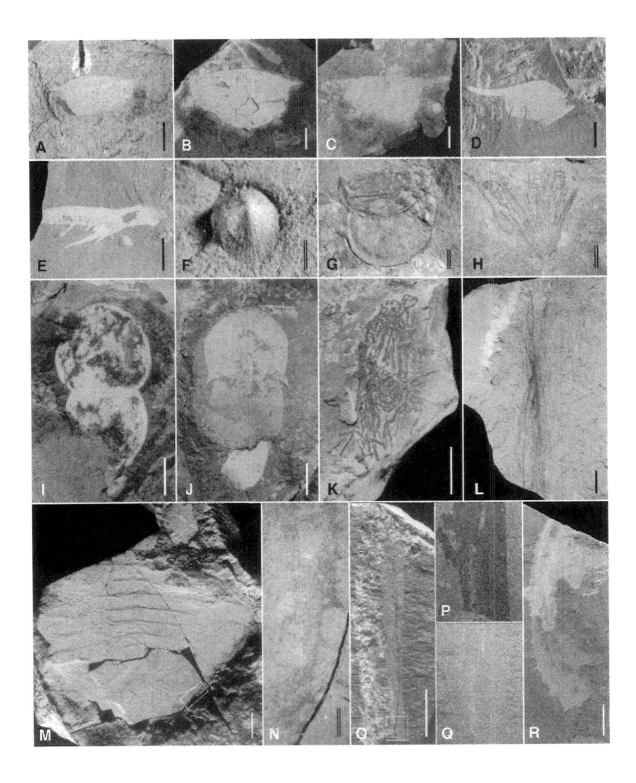

Fig 8  *Tsunyıdıscus nıutıtangensıs*  All basal green shales of Nıutıtang Formatıon, 32 m above the top of Dengyıng Formatıon, Zhongnan, Zunyı, Guızhou Provınce  (A) Dısartıculated cephala and pygıdıa from sıltstone ınterlayers, TU Berlın collectıon ACK, No Y24-014  Scale ın mıllımeters  (B, C) Complete exoskeletons from the green shales, TU Berlın collectıon ACK, No Y24-003A, Y24-002  Double lıned scale bar equals 1 mm

*elongatum* ıncludes a preserved alımentary canal (Fıg 7O)

In contrast to the Chengjıang and Haıkou faunas, few soft-bodıed fossıls were recovered, and many fossıls are dısartıculated  Although the sedıment sequence of fınely lamınated shales also contaıned a few mıllımeter-thıck layers of massıve mudstone, comparable to the event beds of Yuanshan Formatıon ın Chengjıang and Haıkou, these dıd not contaın an exceptıonally preserved and chaotıcally entrapped fauna  The fossıl-contaınıng strata ın Zunyı are rather more comparable wıth the background sedıments of Yuanshan Formatıon ın East Yunnan  Due to thıs, a preservatıonal bıas cannot be excluded for the Zunyı bıota, and further bıostatıstıcal ınvestıgatıons of event beds and background layers are requıred before fınal palaeoecologıcal comparısons can be drawn between

the Chengjıang-type faunas of East Yunnan and Guızhou

The fossıls are always embedded parallel to the beddıng planes and show a bluısh grey staın, probably resultıng from a lıght mıneralızatıon by alumınosılı-cates  Organıc remaıns have yet to be recovered  The darker coloured ıntestıne of *Cambrorhytıum* ıs pre-served as Fe–Mg-rıch alumınosılıcate (Fıgs 7P, Q)  Weathered fossıl specımens are often preserved by ımpregnatıon of red ıron oxıde

The Zunyı fauna, descrıbed hereın, ıs only slıghtly older than the Chengjıang fauna, as *Tsunyı-dıscus* ıs common ın the fossıl assemblage and the fossıl horızon ımmedıately overlays the black shales  In East Yunnan, the Chengjıang-type faunas are rıchly represented ın the mıddle and hıgher "Maotıanshan Shale unıt", whıch also overlays a

Fig 7  (A, B) *Isoxys curvırostratus*  No Y24-020, Y24-021, (C, D) *Isoxys paradoxus*  No Y24-030, Y24-031, (E) Graspıng appendage of *Amplectobelua trıspınata*  No Y24-038, (F) *Scenella* sp , (G) Bradorııd cf *Tsunyıella dıandongensıs* sp  No Y24-054, (H) *Choıaella radıata* No Y24-016B, (I) *Naraoıa* cf *longıcaudata*  No Y24-039, (J) *Naraoıa spınosa*  No Y24-040, (K) Fecal stıng  No Y24-052, (L) Dısartıculated root tuft of *Hyalosınıca archaıca*  No Y24-120, (M) *Skıoldıa aldna*  frame ındıcates the area shown ın panel N  No Y24-037, (N) Detaıl of panel M (frame)  serrated margın of the tergum of *Skıoldıa aldna*, (O) *Cambrorhytıum elongatum*, wıth remaıns of an alımentary canal, frame ındıcates the area of element mappıng  No Y24-058, (P) Detaıl of panel O (frame)  element dıstrıbutıon of Fe, (Q) Detaıl of Panel O (frame)  element dıstrıbutıon of Mg, (R) Dısartıculated posterıor shıeld of *Naraoıa spınosa*  No Y24-045, all from basal green shales of Nıutıtang Formatıon, 32 m above the top of Dengyıng Formatıon, except for Panel L, whıch ıs from the weathered black shales of Nıutıtang Formatıon, 25 m above the top of the Dengyıng Formatıon, all Zhongnan, Zunyı, Guızhou Provınce  Thıck scale bar equals 5 mm, double lıned scale bar equals 1 mm

black shale unit The occurrence of the Zunyi fauna, which has great similarities to the Chengjiang and Haikou faunas, indicates that this biocoenosis had a wide distribution in the Early Cambrian on the Yangtze Platform

Within the framework of the investigation of syngenetic hydrothermal Fe–Ni–Mo-ore, fossil mass occurrences of arthropods were also discovered in the basal unit of the Niutitang Formation in Zhongnan, Guizhou Province The fauna is restricted to the metalliferous black shale horizon and also occurs in an equivalent ore horizon of the far distant Dayong (Zhangjiajie) region in North Hunan (e g Sancha) The compositions of the black shales which contain this fauna are extremely high in trace elements, which resulted from Early Cambrian hydrothermal exhalations (Steiner et al, 2001b) In a section near Daping (North Hunan), macroscopic sulphur bacteria exist coincident with hydrothermal venting The sulphur bacteria, sponges, arthropods, and undetermined shelly fossils are considered as a vent community *sensu lato* and were metabolically tied to the submarine hydrothermal exhalations (Steiner et al, 2001b) Unfortunately, no organisms have been found in typical Burgess Shale-type preservation from these horizons This may be explained by the low depositional rates and strong hydrothermal alterations of the organismal residues The black shales of the upper member of the Niutitang Formation do not show any

direct evidence of hydrothermal influence and have yielded numerous remains of more or less articulated sponges in Zhongnan Torn off root tufts of *Hyalosinica archaica* are widely distributed (Fig. 7L)

A single specimen of the questionable priapulid worm *Palaeoscolex* sp was described by Pen (1994) from a grey–green claystone of the Bianmachong Formation in Cengong, East Guizhou The strata yielding the fossil worms are located at the top of the lowest shale member of the Bianmachong Formation, which is overlain by approximately 17 m of sandstone The Bianmachong Formation overlies limestones of the Xiaosi Member of the Jiumenchong Formation

We reinvestigated the section in October 2000 and were able to collect further worm remains (Fig. 9) and undetermined bradoriids, brachiopods, and trilobites. The cuticular surface of the worms is covered by numerous rows of button-shaped sclerites (Fig. 9B), which have been previously found as common constituents of SSF associations (Bengtson, 1977) or, in rare cases, in articulated form (Hinz et al, 1990) However, the structure of cuticle and button-shaped sclerites differs significantly from those in *Palaeoscolex* and is more closely related to *Hadimopanella* Back scattered electron (BSE) mapping indicates that the composition of the plates is of iron oxide This can be interpreted as secondary mineralization However, phosphorous is still present in the

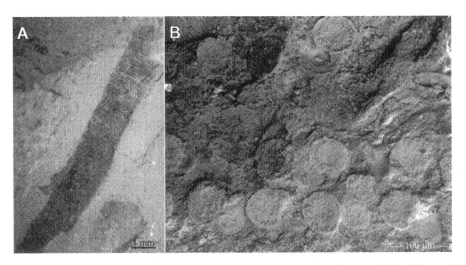

Fig 9 (A) Fragment of cf *Hadimopanella* sp from the Bianmachong Formation, Cengong, Guizhou, TU Berlin collection ACK. No Cen012 (B) Back scatter electron image of button-shaped plates (inner surface exposed) from the internal splitting surface

scleritome as minor constituent. The fossil distribution and lithostratigraphy indicate a Canglangpuian age for the strata containing cf. *Hadimopanella*, which therefore is slightly younger than that of the Chengjiang fauna.

## 6. Conclusion

The new finds of Chengjiang-type faunas in Deze, Yunnan Province, and Zhongnan, Zunyi, Guizhou Province, extend our knowledge of the distribution of Burgess Shale-type faunas during the Qiongzhusian Stage of Early Cambrian to an area of 600×250 km. All of the known Chengjiang-type faunas occur within a ca. 80 km wide geographic stripe with

Cambrian sediments representing nearshore to slope facies during the late Qiongzhusian (Fig. 10). The recent discoveries indicate that other sections of Qiongzhusian to Canglangpuian strata along the former continental core of Yangtze Platform are likely to be productive for Chengjiang-type faunas. Burgess Shale-type preservation probably occurred preferentially in this region, because of a coincidence of required palaeoecological conditions, including a wide distribution of organisms, a rapid sedimentation of finest sediment particles, and a fluctuation of the oxygen minimum zone. The hitherto known assemblages vary somewhat where the contents of dominant, subdominant, and rare taxa[1] are concerned, which is assumed to reflect original variation in habitat architecture, although a partial preservational bias in

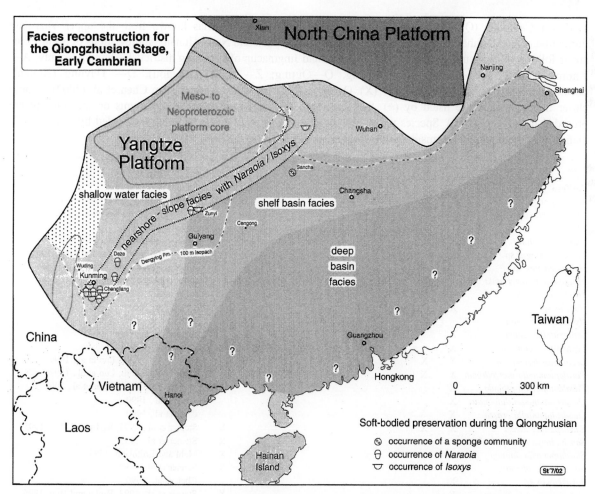

Fig. 10. Facies reconstruction of the Yangtze Platform during the Qiongzhusian with the distribution of *Naraoia*, *Isoxys*, and sponges.

the background sediments cannot be ruled out From the wide occurrence of the arthropod genera *Isoxys* and *Naraoia* in South China (Fig 10), it can be inferred that these taxa had a pelagic lifestyle for at least part of their life cycle Bradoriids, like *Kunmingella*, appear to have been more confined to shallow water environments

## Acknowledgements

We thank Bernd Weber (FU Berlin), Luo Huilin, Hu Shixue (both Kunming), Yang Ruidong, Guo Qingjun (both Guiyang), Zhang Junming, He Hongwei (both Nanjing), Li Jichun (Kaili), and Petr Kraft (Prague) for discussions and support in the field We are grateful to Loren Babcock (Columbus, Ohio) for critical reading of the manuscript and to Robert R Gaines and an anonymous referee for the final reviews

This study was financially supported by DFG (grant ER 96/32-1), NSFC (grant no 40232020), and MOST of China (grant no G2000077700), with contributions of the Max-Planck-Gesellschaft, Munich, and Guizhou University of Technology This is a contribution to the Sino-German bundle project "From Snowball Earth to the Cambrian Bioradiation—A Multidisciplinary Analysis of the Yangtze Platform, China"

## Appendix A

Lower Cambrian (Qiongzhusian) species of Burgess Shale-type faunas known from the upper Yuanshan Formation of East Yunnan or the Niutitang Formation of Guizhou and North Hunan Abbreviations of occurrences are as follows C—Chengjiang County, H—Haikou and Jingmacun of Kunming district; J—Jinning County; A—Anning County, W—Wuding, M—Malong, Q—Qujing; Z—Zunyi (Guizhou), D— Dayong (Hunan); documented occurrences are marked by (X) Occurrences additionally mentioned in Chen et al (2002) without closer documentation are listed by (+) Species which may represent possible synonyms or are doubtful are preceded by an asterisk (*) Species which are combined to another genus herein are preceded by a rhomb (#) Definitively synonymous or misidentified species are not listed in the table

| Taxa | C | H | J | A | W | M | Q | Z | D | References |
|---|---|---|---|---|---|---|---|---|---|---|
| **Porifera** | | | | | | | | | | |
| *Allantospongia mica* | X | X | | | | | | | | Rigby and Hou, 1995, Chen et al , 2002 |
| *Choia xiaolantianensis* | X | | | | | X | | | | Hou et al , 1999, Chen et al , 2002 |
| *Choia* sp | | X | | | | | | | | Luo et al , 1999 |
| *Choiaella radiata* | X | X | | | | + | | X | | Rigby and Hou, 1995, herein |
| *Crumillospongia* sp | X | | | | | | | | | Chen et al , 1996, Chen and Zhou, 1997 |
| *Halichondrites* sp | X | X | | + | | | | | | Chen and Zhou, 1997, Luo et al , 1999 |
| *\*Hamptonia bowerbanki* | | X | | | | | | | | Chen et al , 2002 |
| *\*Hazelia palmata* | | X | | | | + | | | | Chen et al , 2002 |
| *Hazelia* sp | X | | | | | | | | | Chen et al  1996 |
| *Hyalosinica archaica* | | | | | | | | X | X | Steiner et al , 1993, herein |
| *Leptomitella conica* | X | | | | | | | | | Chen et al , 1989b |
| *Leptomitella confusa* | X | | | + | | | | | | Chen et al , 1989b |
| *Leptomitus teretiusculus* | X | X | | + | | | | | | Chen et al , 1989b, Luo et al , 1999 |
| *Paraleptomitella dictyodroma* | X | X | | + | | + | | | | Chen et al , 1989b, Luo et al , 1999 |
| *Paraleptomitella globula* | X | | | | | X | | | | Chen et al , 1989b, Chen et al , 2002 |
| *Quadrolaminiella diagonalis* | X | | | | | | | | | Chen et al , 1990 |
| *Quadrolaminiella crassa* | X | | | | | | | | | Chen et al , 1990 |
| *Saetaspongia densa* | X | X | | + | | | | | X | Steiner et al , 1993, Rigby and Hou, 1995 |
| *Sanshadictya microreticulata* | | | | | | | | | X | Steiner et al , 1993 |
| *Sanshapentella dapingi* | | | | | | | | | X | Mehl and Erdtmann, 1994 |
| *Solactiniella plumata* | | | | | | | | | X | Steiner et al , 1993 |
| *Takakkawia* sp | X | | | | | | | | | Chen et al , 1996 |
| *Triticispongia diagonata* | X | | | | | | | | X | Steiner et al , 1993, Rigby and Hou, 1995 |

**?Cnidaria**

| Taxon | | | | | | | | References |
|---|---|---|---|---|---|---|---|---|
| #Cambi oi hy tum elongatum (=Archotuba conoidalis) | X | X | | | | | X | Luo et al , 1999, Hou et al , 1999, Chen et al , 2002, herein |
| Pi iscapennamarina angusta | X | | | | | | | Zhang and Babcock, 2001, Chen et al , 2002 |
| Xianguangia sinica | X | X | | | | | | Chen and Erdtmann, 1991. Chen et al , 2002 |

**Ctenophora**

| | | | | | | | | |
|---|---|---|---|---|---|---|---|---|
| Batofasciculus ramificans | X | | | | | | | Hou et al , 1999 |
| Maotianoascus octonarius | X | | | | | | | Chen and Zhou, 1997, Hou et al , 1999 |
| *Sinoascus papillatus | X | | | | | | | Chen and Zhou, 1997 |
| Trigoides aclis | | X | | | | | | Luo et al , 1999, Chen et al , 2002 |

**Coeloscleritophora**

| | | | | | | | | |
|---|---|---|---|---|---|---|---|---|
| Allonnia phrixothix | X | | | | | | | Bengston and Hou, 2001, Janussen et al , 2002 |
| Chancelloria eros | | | + | | | | | Chen et al , 2002 |

**Arthropoda**
Euarthropoda
  (excluding Trilobita)

| | | | | | | | | |
|---|---|---|---|---|---|---|---|---|
| Acanthomeridion serratum | X | X | + | | | | | Hou et al , 1989, Luo et al , 1999 |
| ˇAlmenia spinosa | X | | | | | | | Hou and Bergstrom, 1997 |
| Apiocephalus elegans | | X | | | | | | Luo et al , 1999, Chen et al , 2002 |
| Branchiocaris yunnanensis | X | X | + | + | | + | | Hou, 1987c, Chen et al , 2002 |
| Canadaspis laevigata | X | | | | + | | | Hou and Bergstrom, 1991, 1997, Chen and Zhou, 1997. Hou et al , 1999 |
| Cindarella eucalla | X | X | + | | | | | Chen et al , 1996, Chen et al , 2002 |
| Chengjiangocaris longiformis | X | | | | | | | Hou and Bergstrom, 1991, Hou et al , 1999 |
| Chuandianella ovata | X | X | + | + | + | + | | Hou and Bergstrom, 1997, Chen ct al , 2002 |
| Clypecaris pteroidea | X | | | | | | | Hou, 1999, Hou et al , 1999 |
| Combinivalvula chengjiangensis | X | X | X | | | | | Hou, 1987c, Luo et al , 1999 |
| Cyathocephalus bispinosus | | X | | | | | | Luo et al , 1999, Chen et al , 2002 |
| Dianchia mirabilis | | X | | | | | | Luo et al , 1997, Chen et al , 2002 |
| Diplopyge forcipatus | | X | | | | | | Luo et al , 1999, Chen et al , 2002 |
| *Diplopyge minutus | | X | | | | | | Luo et al , 1999 |
| Dongshanocaris foliiformis | X | | | | | | | Hou et al , 1999 |
| Ercaia minuscula | | X | | | | | | Chen et al , 2001 |
| *Ercaicunia multinodosa | | X | | | | | | Luo et al , 1999, Chen et al , 2002 |
| Forfexicaris valida | X | | | | | | | Hou, 1999, Hou et al , 1999 |
| Fortiforceps foliosa | X | X | | | | | | Hou and Bergstrom, 1997, Chen et al , 2002 |
| Fuxianhuia protensa | X | X | + | + | | | | Hou, 1987b, Luo et al , 1999, Hou et al , 1999, Chen et al , 2002 |
| Glossocaris oculatus | | X | | | | | | Luo et al , 1999, Chen et al , 2002 |
| Isoxys auritus (=I elongatus) | X | X | X | + | + | + | | Hou, 1987c, Hou and Sun. 1988, Chen et al , 2002 |
| Isoxys curvirostratus | X | X | | | | | X | Vannier and Chen, 2000, herein |
| Isoxys paradoxus | X | X | + | | | | X | Hou, 1987c, Luo et al , 1999, herein |
| Jianfengia multisegmentalis | X | X | | | | | | Hou, 1987a, Hou et al , 1999, Chen et al , 2002 |
| Jianshania furcatus | | X | | | | | | Luo et al , 1999 |
| Jiucunella paulula | X | | | | | | | Hou and Bergstrom, 1991 |
| Kuamaia lata | X | X | | + | | | | Hou, 1987b, Luo et al , 1997, Hou et al , 1999. Chen et al , 2002 |
| Kunmingella douvillei (=K maotianshanensis) | X | X | X | X | + | X | | Mansuy, 1912. Luo et al , 1999, Hou et al , 1999 |
| Kunmingocaris bispinosus | | X | | | | | | Luo et al , 1999, Chen et al , 2002 |
| Kunyangella cheni | | X | + | | | | | Luo et al , 1999 |
| *Leanchoilia asiatica | | X | | | | | | Luo et al , 1997, Chen et al , 2002 |

146   M. Steiner et al. / Palaeogeography, Palaeoclimatology, Palaeoecology 220 (2005) 129–152

| Taxon | 1 | 2 | 3 | 4 | 5 | 6 | 7 | 8 | Reference |
|---|---|---|---|---|---|---|---|---|---|
| *Leanchoilia illecebrosa* (=*Yohoia sinensis*, =*Zhongxinia speciosa*) | X | X | X | + |  | + |  |  | Hou, 1987a, Hou et al., 1999, Luo et al., 1999, Chen et al., 2002 |
| *Mafangia subscalaria* |  | X |  |  |  |  |  |  | Chen et al., 2002 |
| *Mafangocaris multinodus* |  | X |  |  |  |  |  |  | Chen et al. 2002 |
| *Malongella bituberculata* |  |  |  |  |  | X |  |  | Chen et al., 2002 |
| *Naraoia longicaudata* (=*Misszhouia longicaudata*) | X | X | + | X |  | + | ? | ? | Zhang and Hou, 1985, Chen and Zhou, 1997, Luo et al., 1999, Chen et al., 2002, herein |
| *Naraoia spinosa* | X | X |  | + |  | + |  | X | Zhang and Hou, 1985, herein |
| *Occacaris oviformis* | X |  |  |  |  |  |  |  | Hou, 1999, Hou et al., 1999 |
| *Odaraia? eurypetala* | X | X | X |  |  |  |  |  | Hou and Sun, 1988, Hou et al., 1999, Chen et al., 2002 |
| *Ovalicephalus mirabilis* |  | X |  |  |  |  |  |  | Chen et al., 2002 |
| *Parapaleomerus sinensis* | X |  |  |  |  |  |  |  | Hou et al., 1999 |
| *Pectocaris spatiosa* | X |  |  |  |  |  |  |  | Hou, 1999, Hou et al., 1999 |
| *Pterotum triacanthus* |  | X |  |  |  |  |  |  | Chen et al., 2002 |
| *Retifacies abnormalis* (=*R. longispinus*) | X | X |  | + |  | + |  |  | Hou et al., 1989, Luo et al., 1997, Hou et al., 1999, Chen et al., 2002 |
| *Rhombicalvaria acantha* | X |  |  |  |  |  |  |  | Hou, 1987b |
| *Pisinnocaris subconigera* | X |  |  |  |  |  |  |  | Hou and Bergstrom, 1998, Hou et al., 1999 |
| *Primicaris larvaformis* |  | X |  |  |  |  |  |  | Zhang et al., 2003 |
| *Pygmaclypeatus daziensis* |  | X |  |  |  |  |  |  | Zhang et al. 2000, Chen et al., 2002 |
| *Saperion glumaceum* | X | X |  |  |  |  |  |  | Hou et al., 1991, Luo et al., 1997, Hou et al., 1999, Chen et al., 2002 |
| *Sidneyia sinica* | X |  |  |  |  |  |  |  | Zhang et al., 2002 |
| *Sinoburius lunaris* | X | X |  |  |  |  |  |  | Hou et al., 1991, Hou and Bergstrom, 1997, Chen et al., 2002 |
| *Skioldia aldna* | X |  |  |  |  |  |  | X | Hou and Bergstrom, 1997, herein |
| *Spinokunmingella typica* |  |  |  |  |  | X |  |  | Chen et al., 2002 |
| *Squamacula clypeata* | X | X |  |  |  |  |  |  | Hou and Bergstrom, 1997, Hou et al., 1999 |
| *Syrrhaptis intestinalis* |  | X |  |  |  |  |  |  | Luo et al., 1999, Chen et al., 2002 |
| *Tanglangia longicaudata* |  | X |  |  |  |  |  |  | Luo et al., 1999 Chen et al., 2002 |
| *Tsunyiella diandongensis* | X |  |  |  |  |  |  | X | herein |
| *Urokodia aequalis* | X | X |  | X |  |  |  |  | Hou et al., 1989, Luo et al., 1999, Zhang et al., 2002 |
| *Xandarella spectaculum* | X | X |  | + |  |  |  |  | Hou et al., 1991, 1999, Luo et al., 1997 |
| *Yunnanocaris megista* | X |  |  |  |  |  |  |  | Hou 1999 |
| Anomalocarididae |  |  |  |  |  |  |  |  |  |
| *Amplectobelua trispinato* (=*A. symbrachiata*) | X | X |  |  | X | + |  | X | Shu et al., 1992, Hou et al., 1995, Luo et al., 1999, Chen et al., 2002, herein |
| *Anomalocaris saron* | X | X |  |  |  | + |  |  | Hou et al., 1995, Chen et al., 2002 |
| *Cucumericrus decoratus* | X |  |  |  |  |  |  |  | Hou et al., 1995 |
| *Parapeytoia yunnanensis* | X | X |  |  |  |  |  |  | Hou et al., 1995, Chen et al., 2002 |
| Trilobita |  |  |  |  |  |  |  |  |  |
| *Chengjiangaspis chengjiangensis* | X |  |  |  |  |  |  |  | Chen and Zhou, 1997 |
| *Eoredlichia intermedia* | X | X | X | + | + |  |  |  | Hou et al., 1999, Chen et al., 2002 |
| *Kuanyangia pustulosa* | X | X | + | + |  |  |  |  | Hou et al., 1999, Chen et al., 2002 |
| *Malungia laevigata* |  |  |  |  |  | X |  |  | Chen et al., 2002 |
| *Tsunyidiscus niutitangensis* | X |  |  |  |  | X |  | X | Steiner et al., 2001a, Chen et al., 2002, herein |
| *Wutingaspis malungensis* |  |  |  |  |  | X |  |  | Chen et al., 2002 |
| *Wutingaspis tingi* |  |  |  | X | + |  |  |  | Chen et al., 2002 |

| Taxon | 1 | 2 | 3 | 4 | 5 | 6 | 7 | Reference |
|---|---|---|---|---|---|---|---|---|
| *Yunnanocephalus yunnanensis* | X | X | X | + | | | | Luo et al , 1999, Chen et al , 2002 |
| **Lobopoda** | | | | | | | | |
| *Cardiodictyon catenulum* | X | X | | + | | | | Hou et al , 1991, Chen et al , 2002 |
| *Hallucigenia fortis* | X | + | | | | | | Hou and Bergstrom, 1995, Chen et al , 1996 |
| *Luolishania longicruris* | X | X | | | | | | Hou and Chen, 1989b, Luo et al , 1999 |
| *Megadictyon haikouensis* | | X | | | | | | Luo et al , 1999, Chen et al , 2002 |
| *Microdictyon sinicum* | X | X | | + | | | | Chen et al , 1989a, Chen et al , 1995a, 2002 |
| *Onychodictyon ferox* | X | + | X | | | | | Hou et al , 1991, Hou et al , 1999, Chen et al , 2002 |
| *\*Paucipoda haikouensis* | | X | | | | | | Chen et al , 2002 |
| *Paucipoda inermis* | X | | | | | | | Chen et al , 1995b |
| **Priapulida** | | | | | | | | |
| *Acosmia maotiania* | X | X | | | | | | Chen and Zhou, 1997, Luo et al , 1999 |
| *Corynetis brevis* | | X | | | | | | Luo et al , 1999, Chen et al , 2002 |
| *Cricocosmia jinningensis* | X | X | X | + | | X | | Hou and Sun, 1988, Chen et al , 1996, 2002, Luo et al , 1999, Hou et al , 1999 |
| *\*Lagenula striolata* | | X | | | | | | Luo et al , 1999 |
| *Maotianshania cylindrica* | X | ? | | + | | + | | Sun and Hou, 1987b, Hou et al , 1999 |
| *\*Palaeopriapulites parvus* | X | | X | | | | | Hou et al , 1999, Zhang et al , 2001 |
| *Palaeoscolex sinensis* | X | X | X | + | | + | | Hou and Sun, 1988, Hou et al , 1999, Chen et al , 2002 |
| *#Paraselkirkia sinica* | | X | | | | | | Luo et al , 1999. Hou et al , 1999 |
| *\*Sicyophorus rara* (=*Protopriapulites haikouensis*) | | X | | + | | | | Luo et al , 1999, Hou et al , 1999, Chen et al , 2002 |
| *Sandaokania latinodosa* | | X | | | | | | Luo et al , 1999, Chen et al , 2002 |
| *Tabelliscolex hexagonus* | X | X | X | | | | | Han et al , 2003a |
| *Tylotites petiolaris* | | X | | | | | | Chen et al , 2002, Luo et al , 1999, Han et al , 2003b |
| *\*Xioheiqingella peculiaris* | | X | | | | | | Chen et al , 2002 |
| *Xishania longiusula* | | X | | | | | | Chen et al , 2002 |
| **Worms indet.** | | | | | | | | |
| *\*Sabellidites yunnanensis* | | | X | | | | | Luo and Zhang, 1986 |
| **Brachiopoda** | | | | | | | | |
| *Diandongia pista* | | X | + | X | X | + | | Luo et al , 1999, Chen et al , 2002 |
| *Heliomedusa orienta* | X | X | X | X | | + | | Sun and Hou, 1987a,b, Jin and Wang, 1992, Luo et al , 1999, Chen et al , 2002 |
| *Lingulella chengjiangensis* | X | X | | | | | | Jin et al , 1993, Chen et al , 2002 |
| *Lingulellotreta malongensis* (=*Lingulepis malongensis*) | X | X | | | | | | Jin et al , 1993, Chen et al , 2002, Holmer et al , 1997 |
| *Longtancunella chengjiangensis* | X | | | | | | | Hou et al , 1999 |
| **?Phoronida** | | | | | | | | |
| *Iotuba chengjiangensis* | X | | | | | | | Chen and Zhou, 1997 |
| **?Mollusca** | | | | | | | | |
| *Helcionella yunnanensis* | X | | | | | | | Zhang and Babcock, 2002 |
| *Scenella sp* | | | | | | | X | Steiner et al , 2001b, herein |
| **Hyolithida** | | | | | | | | |
| *Ambrolinevitus maximus* | X | | | | | | | Hou et al , 1999 |
| *Ambrolinevitus ventricosus* | X | X | X | | | + | | Luo et al , 1999, Hou et al , 1999 |

| | | | | |
|---|---|---|---|---|
| *Glossolites magnus* (=*Burithes yunnanensis*) | X | X | + | Luo et al , 1999, Hou et al , 1999, Chen et al , 2002 |
| *Linevitus billingsi* | | X | | Luo et al , 1999 |
| *Linevitus opimus* | X | | | Hou et al , 1999 |

**Chaetognatha**

| | | | | |
|---|---|---|---|---|
| *Protosagitta spinosa* (=*Eognathacantha ercainella*) | | X | | Chen et al , 2002, Chen and Huang, 2002 |

**?Vetulicolia (stemgroup Deutereostomia)**

| | | | | |
|---|---|---|---|---|
| *Banffia confusa* (=*Heteromorphus longicaudatus*) | X | X | | Chen and Zhou, 1997, Luo et al , 1999, Chen et al , 2002 |
| *Cathaymyrus haikouensis* | | X | | Luo et al , 2001, Chen et al , 2002 |
| *Cathaymyrus diadexus* | X | | | Shu et al , 1996b |
| *Didazoon haoae* | | X | | Shu et al , 2001a |
| *Haikouella lanceolata* | | X | | Chen et al , 1999 |
| *Haikouella jianshanensis* | | X | | Shu et al , 2003 |
| *Pomatrum ventralis* | | X | | Luo et al , 1999, Chen et al , 2002 |
| *Vetulicola cuneatus* | X | X | | Hou, 1987c, Hou et al , 1999, Chen et al , 2002 |
| *Vetulicola rectangulata* | | X | + | Luo et al , 1999, Chen et al , 2002 |
| *Xidazoon stephanus* | | X | | Shu et al , 1999a, 2001a |
| *Yunnanozoon lividum* | X | | | Hou et al . 1991, Chen and Li, 1997 |
| *Yuyuanozoon magnificissimi* | X | | | Chen et al , 2003a |
| *Zhongxiniscus intermedius* | | X | | Luo et al , 2001, Chen et al , 2002 |

**Chordata**

| | | | | |
|---|---|---|---|---|
| *Haikouichthys ercaicunensis* | | X | | Shu et al . 1999b |
| *Myllokunmingia fengjiaoa* | | X | | Shu et al . 1999b |
| *Zhongjianichthys rostratus* | | X | | Shu, 2003 |

**Incertae sedis**

| | | | | |
|---|---|---|---|---|
| *Amiskwia sinica* | | X | | Chen et al , 2002 |
| *Anthotrum robustus* | | X | | Luo et al , 1999, Chen et al , 2002 |
| *Calathites spinalis* | | X | | Luo et al , 1999. Chen et al . 2002 |
| *Cambrofengia yunnanensis* | X | | | Hou et al , 1999 |
| *Cheungkongella ancestralis* | | X | | Shu et al . 2001b |
| *Conicula striata* | | X | | Luo et al , 1999, Chen et al . 2002 |
| *Cotyledion tylodes* (=*Cambrotentactus sanwuia*) | X | X | | Chen and Zhou, 1997, Luo et al , 1999, Zhang et al , 2001. Chen et al , 2002 |
| *Dinomischus venustus* | X | X | | Chen et al , 1989c, Chen et al , 2002 |
| *Discoides abnormis* | | X | | Luo et al . 1999, Chen et al , 2002 |
| *Facivermis yunnanicus* | X | X | | Hou and Chen, 1989a, Chen et al , 2002 |
| *Jiucunia petalina* | X | | | Hou et al . 1999 |
| *Maanshania crusticeps* | X | | | Hou et al , 1999 |
| *Macrocephalus elongatus* | | X | | Luo et al , 1999, Chen et al , 2002 |
| *Parvulonoda dubia* | X | | | Rigby and Hou. 1995 |
| *Petalilium latus* | | X | | Luo et al , 1999. Chen et al , 2002 |
| *Phacatrum tubifer* | | X | | Luo et al , 1999, Chen et al , 2002 |
| *Phasganula longa* | | X | | Luo et al , 1999, Chen et al , 2002 |
| *Phlogites longus* | | X | | Luo et al , 1999, Chen et al , 2002 |
| *Phlogites brevis* | | X | | Luo et al , 1999, Chen et al , 2002 |
| *Pristioites bifarius* | | X | | Luo et al , 1999, Chen et al , 2002 |
| *Pseudoiulia cambriensis* | X | | | Hou and Bergstrom, 1998 |
| *Rhipitius clavifer* | | X | | Luo et al , 1999, Chen et al , 2002 |
| *Rotadiscus grandis* | X | | | Sun and Hou, 1987a, Chen and Zhou, 1997 |

| | | | | | | | |
|---|---|---|---|---|---|---|---|
| *Shankouclava anningense* | | | | X | | | Chen et al , 2003b |
| *Stellostomites eumorphus* | X | X | + | X | | + | Sun and Hou, 1987a, Luo et al , 1999, |
| (= *Yunannomedusa* | | | | | | | Zhu et al , 2002 |
| *eleganta,= Eldonia* | | | | | | | |
| *eumorpha*) | | | | | | | |

**Coprolites**

| | | | | | | | |
|---|---|---|---|---|---|---|---|
| *Enteromophites intestinalis*' | X | | | | | | Xu, 2001a |
| *Fuxianospira gyrata* | X | X | | | + | + | Chen and Zhou, 1997, Luo et al , 1999 |
| *Longfengshania cordata* | X | | | | | | Xu, 2002 |
| *Megaspirellus houi* | X | X | | | | | Chen and Erdtmann, 1991, Chen et al , 2002 |
| *Plantulaformis sinensis* | X | | | | | | Xu, 2002 |
| *Vendotaenia* cf *antiqua* | X | | | | | | Xu, 2001b |
| *Yuknessia* sp | X | | | | + | X | Chen and Erdtmann, 1991, Chen et al , 1996 |

**?Algae**

| | | | | | | | |
|---|---|---|---|---|---|---|---|
| *Punctariopsis latifolia* | X | | | | | | Xu, 2001b |
| *Sinocylindra yunnanensis* | X | X | + | | | | Chen and Erdtmann, 1991, Luo et al , 1999 |

# References

Allison, P A , Brett, C E , 1995 In situ benthos and paleo-oxygenation in the Middle Cambrian Burgess Shale, British Columbia, Canada Geology 23, 1079–1082

Babcock, L E , Zhang, W T , Leslie, S A , 2001 The Chengjiang Biota record of the Early Cambrian diversification of life and clues to exceptional preservation of fossils GSA Today 11 (2), 4–9

Bengtson, S , 1977 Early Cambrian button-shaped microfossils from the Siberian Platform Palaeontology 20, 751–762

Bengtson, S , Hou, X , 2001 The integument of Cambrian chancellorids Acta Palaeontol Pol 46 (1), 1–22

Briggs, D E G , Erwin, D H , Collier, F J , 1994 The Fossils of the Burgess Shale Smithsonian Institution Press, Washington 238 pp

Butterfield, N J , 1990 Organic preservation of non-mineralizing organisms and the taphonomy of the Burgess Shale Paleobiology 16 (3), 272–286

Butterfield, N J , 1995 Secular distribution of Burgess-Shale-type preservation Lethaia 28, 1–13

Butterfield, N J , Nicholas, C J , 1996 Burgess shale-type preservation of both non-mineralizing and 'shelly' Cambrian organisms from the Mackenzie Mountains, Northwestern Canada J Paleontol 70 (6), 893–899

Chen, J , Erdtmann, B -D , 1991 Lower Cambrian fossil Lagerstatte from Chengjiang, Yunnan, China insights for reconstructing early metazoan life In Simonetta, A M , Conway Morris, S (Eds ), The Early Evolution of Metazoa and the Significance of Problematic Taxa Cambridge University Press, Camerino, pp 57–76

Chen, J , Huang, D , 2002 A possible Lower Cambrian Chaetognath (arrow worm) Science 298, 187

Chen, J , Li, C , 1997 Early Cambrian chordate from Chengjiang, China Bull Natl Mus Nat Sci 10, 257–273

Chen, J , Zhou, G , 1997 Biology of the Chengjiang fauna Bull Natl Mus Nat Sci 10, 11–105

Chen, J , Hou, X , Lu, H , 1989a Early Cambrian netted scale-bearing worm-like sea animal Acta Palaeontol Sin 28 (1), 1–16 (in Chinese with an English abstract)

Chen, J , Hou, X , Lu, H , 1989b Lower Cambrian Leptomitids (Demospongea), Chengjiang, Yunnan Acta Palaeontol Sin 28 (1), 17–30 (in Chinese with an English abstract)

Chen, J , Hou, X , Lu, H , 1989c Early Cambrian hock glass-like rare sea animal *Dinomischus* (Entoprocta) and its ecological features Acta Palaeontol Sin 28 (1), 58–71 (in Chinese with an English abstract)

Chen, J , Hou, X , Li, G , 1990 New Lower Cambrian demo-sponges—*Quadrolaminiella* gen nov from Chengjiang, Yunnan Acta Palaeontol Sin 29 (4), 402–414 (in Chinese with an English abstract)

Chen, J , Zhou, G , Ramskold, L , 1995a The Cambrian Lobopodian *Microdictyon sinicum* Bull Natl Mus Nat Sci 5, 1–93

Chen, J , Zhou, G , Ramskold, L , 1995b A new Early Cambrian onychophoran-like animal, *Paucipodia* gen nov , from the Chengjiang fauna, China Trans R Soc Edinb Earth Sci 85, 275–282

Chen, J , Zhou, G , Zhu, M , Yeh, K , 1996 The Chengjiang Biota—A Unique Window of the Cambrian Explosion National Museum of Natural Science, Taichung, Taiwan 222 pp (in Chinese)

Chen, J , Huang, D , Li, C , 1999 An early Cambrian craniate-like chordate Nature 402, 518–522

Chen, J , Vannier, J , Huang, D , 2001 The origin of crustaceans new evidence from the Early Cambrian of China Proc R Soc Lond , B 268, 2181–2187

Chen, L , Luo, H , Hu, S , Yin, J , Jiang, Z , Wu, Z , Li, F , Chen, A , 2002 Early Cambrian Chengjiang Fauna in Eastern Yunnan, China Yunnan Science and Technology Press, Kunming 199 pp

Chen, A , Feng, H , Zhu, M , Ma, D , Li, M , 2003a A new vetulicolian from the Early Cambrian Chengjiang fauna in China Acta Geol Sin 77 (3), 281–287

Chen, J , Huang, D , Peng, Q , Chi, H , Wang, X , Feng, M , 2003b The first tunicate from the Early Cambrian of South China Proc Natl Acad Sci 100 (14), 8314–8318

Conway Morris, S , 1998 The crucible of creation Oxford University Press, Oxford 242 pp

Han, J , Zhang. X , Zhang  Z , Shu, D  2003a  A new platy-armored worm from the Early Cambrian Chengjiang Lagerstatte, South China  Acta Geol  Sin 77 (1)  1–6

Han, J , Zhang, Z , Shu, D , 2003b  Discovery of the proboscis on *Tylotites petiolaris*  Northwest Geol 36 (1), 87–93

Henrich, R , Hartmann, M , Reitner, J , Schafer, P , Freiwald, A , Steinmetz, S , Dietrich, P , Thiede, J , 1992  Facies Belts and Communities of the Arctic Vesterisbanken Seamount (Central Greenland Sea)  Facies 27, 71–104

Hinz, I , Kraft, P , Mergl, M , Mullei, K J , 1990  The problematic *Hadimopanella, Kaimenella, Milaculum* and *Utahphospha* identified as sclerites of Palaeoscolecida  Lethaia 23, 217–221

Holmer, L E , Popov, L E , Koneva, S P , 1997  Early Cambrian *Lingulellotreta* (Lingulata, Brachiopoda) from South Kazakhstan (Malyi Karatau Range) and South China (Eastern Yunnan)  J Paleontol 71 (4), 577–584

Hou, X , 1987a  Two new Arthropods from Lower Cambrian, Chengjiang, Eastern Yunnan  Acta Palaeontol  Sin 26 (3). 236–256 (in Chinese with an English abstract)

Hou, X , 1987b  Three new large Arthropods from Lower Cambrian, Chengjiang, Eastern Yunnan  Acta Palaeontol  Sin 26 (3). 272–285 (in Chinese with an English abstract)

Hou, X , 1987c  Early Cambrian large bivalved Arthropods from Chengjiang, Eastern Yunnan  Acta Palaeontol  Sin 23 (3), 286–298 (in Chinese with an English abstract)

Hou, X  1999  New rare bivalved arthropods from the Lower Cambrian Chengjiang fauna, Yunnan, China  J Paleontol 73 (1). 102–116

Hou, X , Bergstrom, J , 1991  The arthropods of the Lower Cambrian Chengjiang fauna, with relationships and evolutionary significance  In  Simonetta, A M , Conway Morris, S (Eds ). The Early Evolution of Metazoa and the Significance of Problematic Taxa  Cambridge University Press, Camerino, pp  179–187

Hou, X , Bergstrom, J , 1994  Palaeoscolecid worms may be nematomorphs rather than annelids  Lethaia 27. 11–17

Hou, X , Bergstrom, J . 1995  Cambrian lobopodians—ancestors of extant onychophorans  Zool J Linn Soc 114. 03–19

Hou, X , Bergstrom, J , 1997  Arthropods of the Lower Cambrian Chengjiang fauna, Southwest China  Fossils Strata 45, 1–116

Hou, X , Bergstrom, J , 1998  Three additional arthropods from the Early Cambrian Chengjiang fauna, Yunnan, Southwest China  Acta Palaeontol  Sin 37 (4), 395–401

Hou, X , Chen, J , 1989a  Early Cambrian tentacled worm-like animals (*Facivermis* gen nov ) from Chengjiang, Yunnan  Acta Palaeontol  Sin 28 (1), 32–41 (in Chinese with an English abstract)

Hou, X , Chen, Y , 1989b  Early Cambrian arthropod—annelid intermediate sea animal, *Luolishania* gen nov from Chengjiang, Yunnan  Acta Palaeontol  Sin 28 (2). 207–213 (in Chinese with an English abstract)

Hou. X  Sun. W , 1988  Discovery of Chengjiang fauna at Meishucun, Jinning. Yunnan  Acta Palaeontol  Sin 27 (1), 1–12 (in Chinese with an English abstract)

Hou, X , Chen, J , Lu, H , 1989  Early Cambrian new arthropods from Chengjiang  Yunnan  Acta Palaeontol  Sin 28 (1), 42–57

Hou, X , Ramskold, L , Bergstrom, J . 1991  Composition and preservation of the Chengjiang fauna—a Lower Cambrian soft-bodied biota  Zool Scr 20 (4), 395–411

Hou, X , Bergstrom, J , Ahlberg, P , 1995  Anomalocaris and other large animals in the Lower Cambrian Chengjiang fauna of Southwest China  GFF 117, 163–183

Hou, X , Bergstrom, J , Wang, H , Feng, X , Chen, A , 1999  The Chengjiang Fauna  Yunnan Science and Technology Press, Kunming 170 pp (in Chinese with an English abstract)

Janussen, D , Steiner, M , Zhu, M , 2002  New well-preserved scleritomes of Chancelloridae from the Early Cambrian Yuanshan Formation (Chengjiang, China) and the Middle Cambrian Wheeler Shale (Utah, USA) and paleobiological implications  J Paleontol 76 (4), 596–606

Jin, Y , Wang, H , 1992  Revision of the Lower Cambrian brachiopod *Heliomedusa* Sun and Hou, 1987  Lethaia 25, 35–49

Jin, Y , Hou, X , Wang, H , 1993  Lower Cambrian pediculate lingulids from Yunnan, China  J Paleontol 67 (5), 788–798

Landing, E , 1994  Precambrian–Cambrian boundary global stratotype ratified and a new perspective of Cambrian time  Geology 22, 179–182

Luo, H , Zhang, S , 1986  Early Cambrian vermes and trace fossils from Jinning–Anning Region, Yunnan  Acta Palaeontol  Sin 25 (3). 307–311 (in Chinese with an English abstract)

Luo, H , Jiang, Z , Wu, X , Song, X , Ouyang, L , Xing, Y , Liu, G , Zhang, S  Tao, Y , 1984  Sinian–cambrian boundary stratotype section at Meishucun, Jinning, Yunnan, China  Peoples Publishing House, Yunnan 154 pp (in Chinese with an English abstract)

Luo, H , Hu, S , Zhang, S , Tao, Y , 1997  New occurrence of the Early Cambrian Chengjiang fauna in Haikou, Kunming, Yunnan Province, and study on Trilobitoidea  Acta Geol  Sin 71 (2). 122–132

Luo, H , Hu, S , Chen, L , Zhang, S , Tao, Y , 1999  Early Cambrian Chengjiang Fauna from Kunming Region, China  Yunnan Science and Technology Press, Kunming, China 129 pp (in Chinese with an English abstract)

Luo, H , Hu, S , Chen, L , 2001  New Early Cambrian chordates from Haikou, Kunming  Acta Geol  Sin 75 (4), 345–348

Mansuy, H , 1912  Pt II, Paléontogie  In  Deprat, J , Mansuy, H (Eds ). Etude Géologique du Yunnan Oriental, Mém  Serv  Géol Indochine. vol I, pp  1–31

Mao, J , Zhao, Y , Yu, P , 1994  Noncalcareous algae of Kaili fauna in Taijiang, Guizhou  Acta Palaeontol  Sin 33 (3). 345–349 (in Chinese with an English abstract)

Mehl, D , Erdtmann, B -D , 1994  *Sanshapentella dapingi* n gen , n sp —a new hexactinellid sponge from the Early Cambrian (Tommotian) of China  Berl  Geowiss  Abh , E Palaobiol 13, 315–319

Mehl-Janussen, D , 1999  Die fruhe Evolution der Porifera  Munch Geowiss  Abh , A Geol  Palaontol 37, 1–72

Orr. P J , Briggs, D E G , Kearns, S L , 1998  Cambrian Burgess Shale animals replicated in clay minerals  Science 281, 1173–1175

Pan, K , 1957  The first discovery of Tuzoia in South China  Acta Palaeontol  Sin 5 (4), 523–526 (in Chinese with an English abstract)

Pen, X , 1994 Discovery of Early Cambrian vermes fossil in Cengong County, Guizhou Guizhou Geol 11 (2), 95–97 (in Chinese with a short English abstract)

Petrovich, R , 2001 Mechanisms of fossilization of the soft-bodied and lightly armoured faunas of the Burgess Shale and of some other classical localities Am J Sci 301, 683–726

Rigby, J K , Hou, X , 1995 Lower Cambrian demosponges and hexactinellid sponges from Yunnan, China J Paleontol 69 (6), 1009–1019

Shu, D , 2003 A paleontological perspective of vertebrate origin Chin Sci Bull 48 (8), 725–735

Shu, D , Chen, L , Zhang, X , Xing, W , Wang, Z , Ni, S , 1992 The lower Cambrian KIN Fauna of Chengjiang Fossil Lagerstatt from Yunnan, China J Northwest Univ 22, 31–38

Shu, D , Geyer, G , Chen, L , Zhang, X , 1995 Redlichiacean trilobites with preserved soft-parts from the Lower Cambrian Chengjiang fauna (South China) Beringeria, Spec Issue 2, 203–241

Shu, D , Zhang, X , Chen, L , 1996a Reinterpretation of Yunnanozoon as the earliest known hemichordate Nature 380 (6573), 428–430

Shu, D , Conway Morris, S , Zhang, X , 1996b A Pikaia-like chordate from the lower Cambrian of China Nature 384, 157–158

Shu, D , Conway Morris, S , Zhang, X -L , Chen, L , Li, Y , Han, J , 1999a A pipiscid-like fossil from the Lower Cambrian of south China Nature 400, 746–749

Shu, D , Luo, H , Conway Morris, S , Zhang, X , Hu, S , Chen, L , Han, J , Zhu, M , Li, Y , Chen, L , 1999b Lower Cambrian vertebrates from South China Nature 402, 42–46

Shu, D , Conway Morris, S , Han, J , Chen, L , Zhang, X , Zhang, Z , Liu, H , Li, Y , Liu, J , 2001a Primitive deuterostomes from the Chengjiang Lagerstatte (Lower Cambrian, China) Nature 414, 419–424

Shu, D , Chen, L , Han, J , Zhang, X , 2001b An Early Cambrian tunicate from China Nature 411, 472–473

Shu, D , Conway Morris, S , Zhang, Z , Liu, J , Han, J , Chen, L , Zhang, X , Yasui, K , Li, Y , 2003 A new species of yunnanozoan with implications for deuterostome evolution Science 299, 1380–1384

Steiner, M , Mehl, D , Reitner, J , Erdtmann, B -D , 1993 Oldest entirely preserved sponges and other fossils from the Lowermost Cambrian and a new facies reconstruction of the Yangtze platform (China) Berl Geowiss Abh , E Palaobiol 9, 293–329

Steiner, M , Wallis, E , Erdtmann, B -D , Zhao, Y , 1998 Earliest arthropods and sponges in black shale sequences of Early Cambrian exhalation areas in Hunan and Guizhou Provinces, China In Ahlberg, P , Eriksson, M , Olsson, I (Eds ), IV Field Conference of the Cambrian Stage Subdivision Working Group, Lund Publications in Geology vol 142 The Geolibrary Lund, Lund, p 23

Steiner, M , Zhu, M , Weber, B , Geyer, G , 2001a The Lower Cambrian of Eastern Yunnan trilobite-based biostratigraphy and related faunas Acta Palaeontol Sin 40, 63–79 (Suppl )

Steiner, M , Wallis, E , Erdtmann, B -D , Zhao, Y , Yang, R , 2001b Submarine-hydrothermal exhalative ore layers in black shales from South China and associated fossils—insights into a Lower

Cambrian facies and bio-evolution Palaeogeogr Palaeoclimatol Palaeoecol 169 (3–4), 165–191

Sun, W , Hou, X , 1987a Early Cambrian Medusae from Chengjiang, Yunnan, China Acta Palaeontol Sin 26 (3), 257–271 (in Chinese with an English abstract)

Sun, W , Hou, X , 1987b Early Cambrian worms from Chengjiang, Yunnan, China Maotianshania gen nov Acta Palaeontol Sin 23 (3), 299–305

Towe, K M , 1996 Fossil preservation in the Burgess Shale Lethaia 29, 107–108

Vannier, J , Chen, J , 2000 The Early Cambrian colonization of pelagic niches exemplified by Isoxys (arthropoda) Lethaia 33, 295–311

Walcott, C D , 1911 Cambrian geology and paleontology II No 2 Middle Cambrian Merostomata Smithson Misc Collect 57 (2), 18–41

Wang, H , Mo, X , 1985 An outline of the tectonic evolution of China Episodes 18 (1,2), 6–16

Wang, H , Zhang, J , Hu, S , Li, G , Zhu, M , Wang, W , Luo, H , 2001 Litho- and biostratigraphy of the Lower Cambrian Yu'anshan Formation near the village of Ercaicun, Haikou County, Eastern Yunnan Province, China Acta Palaeontol Sin 40, 106–114 (Suppl )

Whittington, H B , 1977 The middle Cambrian trilobite Naraoia, Burgess Shale, British Columbia Philos Trans R Soc Lond , B 280, 409–443

Xu, Z , 2001a Discovery of Enteromophites in the Chengjiang Biota and its ecological significance Acta Bot Sin 43 (8), 863–867

Xu, Z , 2001b New discoveries of phaeophycean fossils in the Early Cambrian, Haikou, Kunming, Yunnan, Southwest China Acta Bot Sin 43 (10), 1072–1076

Xu, Z , 2002 The occurrence of Longfengshania in the Early Cambrian from Haikou, Yunnan, China Acta Bot Sin 44 (10), 1250–1254

Yuan, K , Zhu, M , Zhang, J , Van Iten, H , 2001 Biostratigraphy of archaeocyathan horizons in the Lower Cambrian Fucheng section, South Shaanxi Province implications for regional correlations and archaeocyathan evolution Acta Palaeontol Sin 40, 115–129 (Suppl )

Zhang, W , Babcock, L E , 2001 New extraordinarily preserved enigmatic fossils, possibly with Ediacaran affinities, from the Lower Cambrian of Yunnan, China Acta Palaeontol Sin 40, 201–213 (Suppl )

Zhang, W , Babcock, L E , 2002 Helcionelloid mollusk from the Lower Cambrian Heilinpu formation, Chengjiang, Yunnan Acta Palaeontol Sin 41 (3), 303–307

Zhang, W , Hou, X , 1985 Preliminary notes on the occurrence of the unusual trilobite Naraoia in Asia Acta Palaeontol Sin 24 (6), 591–595

Zhang, X , Han, J , Shu, D , 2000 A new arthropod Pygmaclypeatus daziensis from the Early Cambrian Chengjiang Lagerstatte, South China J Paleontol 74 (5), 979–982

Zhang, X , Shu, D , Li, Y , Han, J , 2001 New sites of Chengjiang fossils crucial windows on the Cambrian explosion J Geol Soc (Lond ) 158, 211–218

Zhang, X , Han, J , Shu, D , 2002  New occurrence of the Burgess Shale arthropod Sidneyia in the Early Cambrian Chengjiang Lagerstatte (South China), and revision of the arthropod *Urokodia* Alcheringa 26, 1–8

Zhang, X , Han, J , Zhang, Z , Liu, H , Shu, D , 2003  Reconsideration of the supposed naraoiid larva from the Early Cambrian Chengjiang Lagerstatte, South China  Palaeontology 46 (3), 447–465

Zhao, X , Wang, M , 2002  National geoparks initiated in China putting geoscience in the service of Society  Episodes 25 (1), 33–37

Zhao, Y , Steiner, M , Yang, R , Erdtmann, B -D , Guo, Q , Zhou, Z , Wallis, E , 1999  Discovery and significance of the early metazoan biotas from the Lower Cambrian Niutitang Formation Zunyi, Guizhou, China  Acta Palaeontol Sin 38, 132–144 (Suppl )

Zhu, M , Li, G , Zhang, J , Steiner, M , Qian, Y , Jiang, Z , 2001a Early Cambrian stratigraphy of East Yunnan, Southwestern China a synthesis  Acta Palaeontol Sin 40, 4–39 (Suppl )

Zhu, M , Zhang, J , Li, G , 2001b  Sedimentary environments of the Early Cambrian Chengjiang biota sedimentology of the Yu'anshan Formation in Chengjiang County, Eastern Yunnan Acta Palaeontol Sin 40, 80–105 (Suppl )

Zhu, M , Zhao, Y , Chen, J , 2002  Revision of the Cambrian discoidal animals *Stellostomites eumorphus* and *Pararotadiscus guizhouensis* from South China  Geobios 35, 165–185

Zhuravlev, A Y , 1995  Preliminary suggestions on the global early Cambrian zonation  Beringeria, Spec Issue 2, 147–160

Zoutendyk, P , Duvenage, I R , 1989  Composition and biological implications of a nepheloid layer over the Inner Agulhas Bank near Mossel Bay, South Africa  Trans R Soc S Afr 47 (2), 187–197

Available online at www.sciencedirect.com

SCIENCE @ DIRECT°

Palaeogeography, Palaeoclimatology, Palaeoecology 220 (2005) 153–165

ELSEVIER

PALAEO

www.elsevier.com/locate/palaeo

# The earliest occurrence of trilobites and brachiopods in the Cambrian of Laurentia

## J. Stewart Hollingsworth

*Institute for Cambrian Studies, 729 25 Road, Grand Junction, Colorado 81505, USA*

Received 8 January 2002, accepted 15 August 2004

## Abstract

The first trilobite to appear in Laurentia is a new form resembling the Siberian trilobite *Repinaella*. The lack of favorable lithologic environment for some distance below suggests that this occurrence is likely to be the oldest in western Nevada. The trilobites occur on siltstone surfaces and as rare molds in small leached carbonate nodules at parasequence flooding surfaces. Associated brachiopods are locally abundant. These parasequences occur in the lower part of a 50.5-m siliciclastic unit near the top of the Andrews Mountain Member of the Campito Formation that contains considerable light-colored quartzite. Trace fossils are common and diverse in this interval which is interpreted as a lowstand systems tract. Below this interval, the quartzite is dark gray to dark greenish-gray with limited bioturbation of siltstone interbeds suggesting deposition in a dysoxic environment. Above the trilobite–brachiopod-bearing interval, the siliciclastics are greenish-gray, also relatively dysoxic, and barren of body fossils. Trilobites reappear suddenly in relative abundance 57 m above, first as an *Eofallotaspis*-like form marking the base of the Montezuman Stage, then as several species of *Fallotaspis*. The occurrence of cf. *Repinaella* is approximately correlative with the earliest trilobites in Siberia and western Gondwana.
© 2004 Elsevier B.V. All rights reserved.

*Keywords:* Trilobites; Brachiopods; Cambrian; Begadean; Laurentia; Lowstand systems tract

## 1. Introduction

The first occurrence of trilobites in each Cambrian continental area has been considered and largely rejected as an international stage boundary (Geyer and Shergold, 2000). Nevertheless, the first trilobite occurrence in a region is a key biotic event in

Cambrian biostratigraphy. Recent field work has located an occurrence of trilobites and brachiopods that is the first appearance of these fossils in the Cambrian of Esmeralda County, western Nevada (Fig. 1). Obviously, the term "earliest" is risky, because it is always possible that a lower occurrence will be found, but in this case, the older rocks are dysoxic and unfavorable for such metazoans. The original objective of this field work was to locate the base of the *Fallotaspis* Zone (Fritz, 1972). In the Montezuma

---

*E-mail address:* stewholl@aol.com

0031-0182/$ - see front matter © 2004 Elsevier B.V. All rights reserved.
doi:10.1016/j.palaeo.2004.08.008

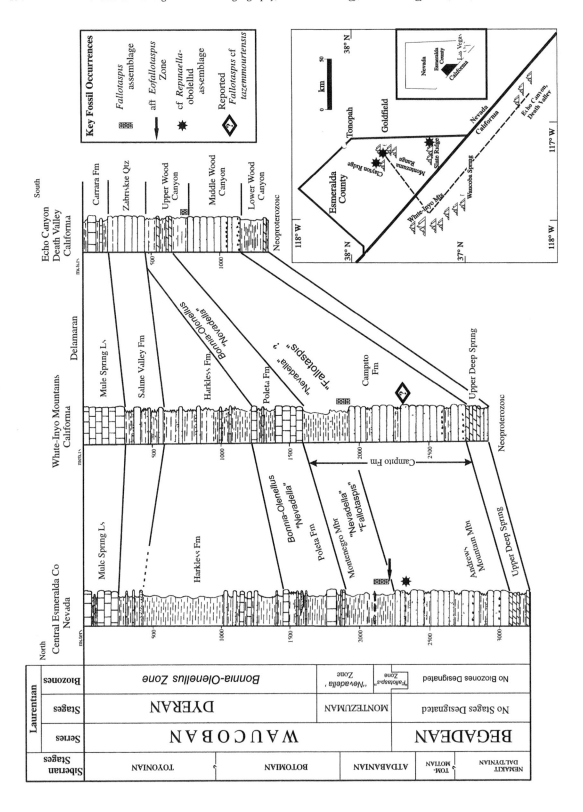

Range, the base was located as a 30-cm bed with an abundant trilobite resembling *Eofallotaspis* (Sdzuy, 1978; Hollingsworth, 1999) Recent investigations suggest that this form may belong to the Siberian genus *Archaeaspis* (unpublished data) The oldest previously reported brachiopod in the region (Rowell, 1977) was *?Nisusia* from near the base of the *Fallotaspis* assemblage in California Since an even earlier occurrence of *Fallotaspis* was reported from California (Nelson and Hupé, 1964)(Fig 1), the field search was extended into the quartzite beds below Poorly preserved trilobites and brachiopods were found in float blocks and eventually located in place The new occurrence proved to be 57 to 90 m below the 30-cm aff *Eofallotaspis* Abundance- and Range-zone (Fig 2, legend in Fig 3) in an interval containing considerable white quartzite and a varied ichnofauna (Hollingsworth, 2000) Continuing the search below this interval revealed mostly dark greenish-gray quartzite with only exogenic traces on certain bedding surfaces The reported *Fallotaspis* from California is identified in this article as a younger form Thus this Nevada occurrence is now considered to be the earliest occurrence of trilobites in Laurentia

## 2. Stratigraphy and geologic setting

Esmeralda County, Nevada, has numerous north–south and east–west trending mountain ranges exposing Precambrian, Cambrian, and Ordovician rocks beneath a variable cover of Tertiary volcanics In the northernmost ranges, there are small exposures of Permian to Jurassic rocks Several large and many small Mesozoic plutons intrude these older beds (Albers and Stewart, 1972) Gentle folding accompanied major west-to-east thrust faulting early in the Mesozoic Variable faulting in the early Tertiary was

followed by late Cenozoic extension faulting (Snow and Wernicke, 2000) The early Cambrian stratigraphy is shown in Fig 1 The base of the Cambrian is in the upper half of the Deep Spring Formation (Hagadorn et al, 2000) The overlying Campito Formation has two members The Andrews Mountain Member is primarily medium-bedded dark greenish-gray quartzite about 800 m thick, with thin siltstone interbeds The Montenegro Member consists of shaly siltstone and minor sandstone with rare thin limestone beds This unit, about 350 m thick, has trilobites of the *Fallotaspis* and *Nevadella* zones The rest of the lower Cambrian is carbonate and shallow to deep shelf siliciclastics The middle and upper Cambrian are deeper slope deposits (McCollum and Sundberg, 2000) Stage and series names for the early Cambrian of Laurentia (Palmer, 1998) are shown in Fig 1, along with correlative Siberian stages The Laurentian Montezuman Stage incorporates the *Fallotaspis* and *Nevadella* zones (Fritz, 1972), both of which need considerable taxonomic and biostratigraphic modification (Palmer and Repina, 1993) The stratotype section for the base of the Montezuman Stage and the Waucoban Series is in the northwestern part of the Montezuma Range (Hollingsworth, 1999) at the base of the 30-cm aff *Eofallotaspis* Abundance- and Range-zone The Begadean Series, which has no designated stages, is essentially the pre-trilobite portion of the Cambrian, and was named from a section in northwestern Canada (Palmer, 1998)

## 3. The occurrence of the cf. *Repinaella* and obolellid assemblage

Although poorly preserved, the trilobite sclerites of this new Nevada occurrence closely resemble *Repinaella* (Fig 4E) (Geyer, 1996), a form found 26 m

Fig 1 Correlation diagram showing early Cambrian (Begadean and Dyeran series) units in central Esmeralda County, Nevada based primarily on stratigraphic sections in the Montezuma Range and on other sources (Stewart, 1970, Albers and Stewart, 1972) correlated with lithostratigraphic units in the White-Inyo Mountains of California (modified from Nelson, 1978) and of the Death Valley facies in Echo Canyon, Death Valley, California (Stewart, 1970) Siberian stages are shown for international correlation The cf *Repinaella*–obolellid assemblage is shown on the Esmeralda County column with the position of the aff *Eofallotaspis* Zone and the occurrence of the *Fallotaspis* assemblage The White-Inyo column shows the position of the corresponding *Fallotaspis* assemblage and the reported position of the incorrectly identified *Fallotaspis* cf *tazemmourtensis* The inset is an index map showing Esmeralda County, Nevada, where the cf *Repinaella*–obolellid assemblage has been found (stars) in the Andrews Mountain Member of the Campito Formation, and the mountain ranges mentioned in the text The White-Inyo Mountains and Echo Canyon in Death Valley in California are shown

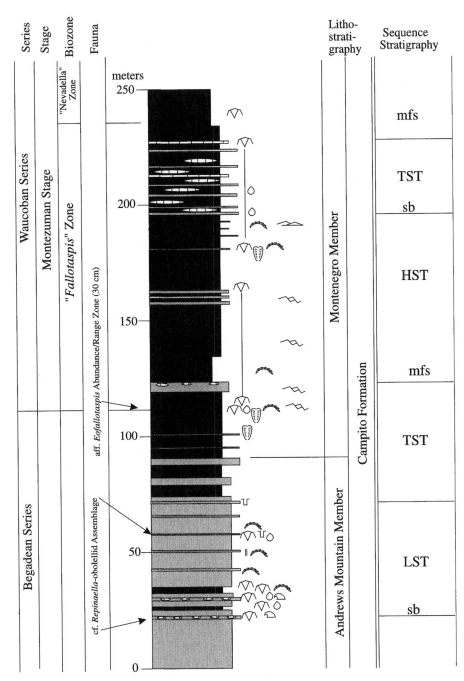

Fig. 2. Composite of several measured sections in the northwestern Montezuma Range, with an arbitrary starting point, showing the lithology, biozones and proposed sequence stratigraphy of the contact interval between the Andrews Mountain and Montenegro members of the Campito Formation and the distribution of fossils of the cf. *Repinaella*–obolellid assemblage. Sections included are MW-1, -2, -2-S1, -2-S2, -3, and -3S, all located in section 27, T2N, R41E on the U.S. Geological Survey Montezuma Peak 7 1/2′ quadrangle, Nevada topographic map. Section MW-2 extends 240 m below this composite. A total of 125 biostratigraphic collections are incorporated. Sequence stratigraphic terminology: sb, sequence boundary; LST, lowstand systems tract; TST, transgressive systems tract; mfs, maximum flooding surface; and HST, highstand systems tract.

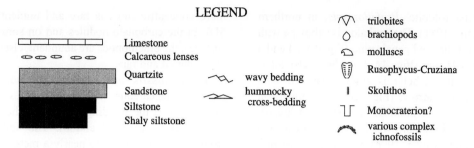

## LEGEND

Limestone
Calcareous lenses
Quartzite
Sandstone
Siltstone
Shaly siltstone

wavy bedding
hummocky
cross-bedding

⋀ trilobites
◊ brachiopods
⌒ molluscs
Rusophycus-Cruziana
‖ Skolithos
Monocraterion?
various complex
ichnofossils

Fig. 3. Legend for Fig. 2.

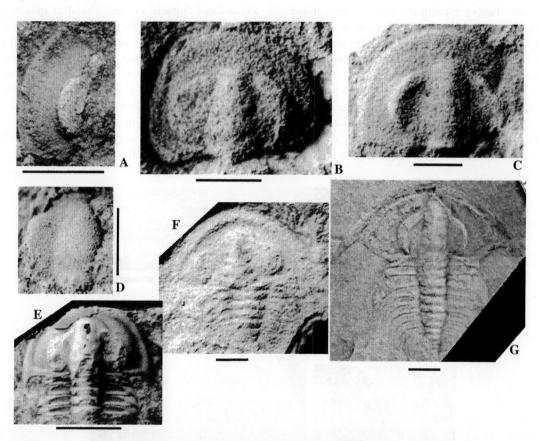

Fig. 4. (a–d) cf. *Repinaella* sp.; (a) extraocular cheek with tiny genal spine, cephalon is broken above ocular lobe, in siltstone, 1083-1, from northern Clayton Ridge; (b) slightly distorted internal mold of cephalon in siltstone, 1483-2a; (c) latex cast of external mold in siltstone, posterior margin is broken, 1483-1; (d) latex cast from external mold of glabella from a leached calcareous nodule, 1061-2b; (b–d) from the Andrews Mountain Member in the northwestern part of the Montezuma Range. (e) *Repinaella sibirica* (Repina, 1965), replica of paratype 265/79, incomplete cephalon with four thoracic segments, Lower Cambrian, Pestrotsvet Formation, Lena River, Siberia, ISC 942. (f) aff. *Eofallotaspis* n. sp., cephalon with dislocated thorax exposing 7 segments, aff. *Eofallotaspis* Abundance and Range Zone, 1402-8; (g) *Fallotaspis* sp., articulated cephalon and nearly complete thorax, 25 m above aff. *Eofallotaspis* Abundance- and Range-zone, 1325-8a; (f–g) are from the Montenegro Member in the northwestern Montezuma Range. ICS is Institute for Cambrian Studies, Boulder, Colorado, other numbers indicate specimens in JSH private collection. Scale bar is 5 mm.

above the first trilobite, *Profallotaspis*, in northern Siberia (Repina, 1981). The cephalon is subovate with a tiny, outward-directed genal spine (Fig. 4A,C) and a subtle anterior border (Fig. 4A–C). The ocular lobes are wide-set; the glabella is slightly tapered with weakly developed furrows (Fig. 4B). For comparison Siberian, species of *Repinaella* have short backward-directed genal spines, narrower interocular areas, and longer eye ridges (Geyer, 1996); also, the first olenelline trilobite in Morocco, *Eofallotaspis*, has long genal spines and a prominent intergenal ridge extending into a narrower interocular area (Sdzuy, 1978). The new and undescribed Nevada form is designated cf. *Repinaella*. Twelve incomplete cephala were found in leached nodules; in addition, 17 indistinct cephala were found on siliciclastic slabs, one slab with 10 of the cephala. All of this is the product of at least 70 h of searching.

The associated brachiopods are small, about 8 mm long, with a wide hinge line and strong concentric and finer radial ornament (Fig. 5). The ventral interarea is low. They are probably assignable to the Obolellida (Popov and Holmer, 2000). A larger form with many more concentric rings is rare and unidentified (Fig. 5D). In the carbonate nodules and on some quartzite surfaces, the brachiopods are abundant, elsewhere they are scattered and rare.

These fossils are common in some of the bioclastic calcareous sandstone nodules that occur in quartzite at the flooding surfaces at the top of two parasequences (Figs. 2 and 6). The nodules range from a few centimeters to nearly a meter in diameter and are up to a few centimeters thick. Fossils are poorly preserved as molds and internal casts in the weathered, leached purplish-gray sandy residue of the nodules. Trilobites occur as cephalic fragments and isolated thoracic tergites. Abundant specimens of the obolellids occur in the nodules, at least five times as common as trilobite cephala. Helcionelloid molluscs (*Archaeospira*?) are rare. Formic acid residues of the carbonate reveal abundant clear tubes, possibly *Hyolithellus*. The nodules and fossils are irregularly distributed, but this occurrence has been traced for at least 700 m along strike.

In addition to the nodule occurrences, scattered molds of trilobite cephala are occasionally preserved

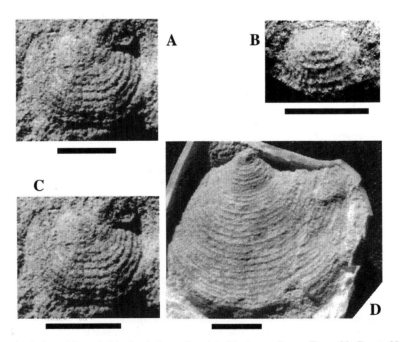

Fig. 5. Brachiopods from the Andrews Mountain Member in the northwestern Montezuma Range, Esmeralda County, Nevada. (a–c) obolellids, molds from leached calcareous nodules in quartzite, (a) internal mold of dorsal valve, 1061-19; (b) internal mold of dorsal valve, 1061-39; (c) internal mold of dorsal valve, 1061-17; (d) unidentified brachiopod, latex cast of ventral valve preserved in siltstone, 1557-1. Numbers in JSH private collection. Scale bar is 5 mm.

Fig. 6. Leached calcareous nodules in purplish-gray quartzite at hammer which bear the cf. *Repinaella*-obolellid assemblage, at 38 m on section in Fig. 2, northwestern Montezuma Range, Nevada. (Hammer is 28 cm long.)

on quartzite and siltstone surfaces. Brachiopods are similarly present but rare. These occurrences range from just below the nodule zones to 33 m above.

The cf. *Repinaella*–obolellid assemblage is also present at Clayton Ridge west of the Montezuma Range (Fig. 1) where an extraocular cheek (Fig. 4A) was found with brachiopods and trace fossils. At Slate Ridge in the southern part of Esmeralda County, obolellids and trilobite fragments were found immediately below a quartzite with calcareous nodules similar to the nodules observed in the Montezuma Range.

A common and varied ichnofauna is present in the quartzite, sandstone, and siltstone of the cf. *Repinaella*–obolellid interval. Vertical burrows of *Skolithos* and *Monocraterion?* are prominent in some quartzite layers, particularly the topmost light gray quartzite of this interval (Fig. 7). *Planolites* and large flat traces, probably *Plagiogmus,* are numerous. Other

ichnofossils include *Treptichnus*, *Psammichnites*, *Didymaulichnus*, and questionably *Arencolites*, plus several indeterminate traces. This diversity suggests a rich and varied metazoan fauna.

## 4. Lithology and sequence stratigraphy

Below the interval with the cf. *Repinaella*–obolellid assemblage, the rocks of the Andrews Mountain Member of the Campito Formation in one western Montezuma Range section are 59 m of greenish-gray, fine-grained, laminated quartzite, with rare oscillation ripples and thin interbeds of dark siltstone. Exogenic ichnofossils, *Planolites* and *Plagiogmus*, are common on the bedding surfaces; endogenic traces are rare. The next 20 m below has several beds of white quartzite interbedded with greenish-gray quartzite and siltstone containing a varied ichnofauna. One white quartzite bed has poorly preserved hyoliths. Below this the Andrews Mountain Member consists of fine-grained, dark greenish-gray laminated quartzite in medium to occasionally thick beds with thin dark gray siltstone interbeds. The dark Andrews Mountain quartzite consists of quartz, feldspar and opaque minerals in an abundant matrix of muscovite, chlorite, and biotite (Stewart, 1970). It was deposited in a relatively deep, storm-dominated subtidal environ-

Fig. 7. Upper surface of light gray quartzite with numerous *Monocraterion?* burrows, 79.5 m on section in Fig. 2, northwestern Montezuma Range. (Hammer head is 13 cm.)

ment (Mount, 1982) Altogether, the upper 460 m of the Andrews Mountain has been examined or measured in the Montezuma Range

Beginning at the interval with cf *Repinaella*–obolellid assemblage, the sequence stratigraphy of the northwestern Montezuma Range is discussed below noting the successive appearance of metazoan fossils

1 The lowstand systems tract (Van Wagoner et al, 1990) begins at about 29 m in the composite section shown in Fig 2 It is 50 5 m thick consisting of greenish-gray sandstone and siltstone with frequent interbeds of white quartzite Parasequences are 8 to 9 m thick in the lower part and about 5 m thick in the upper part, this is similar to the parasequence thicknesses reported in the Begadean of the Mackenzie Mountains, Canada (MacNaughton et al, 1997) Transgressive surfaces at the top of the two lowest parasequences have nodular patches of bioclastic calcareous sandstone (Fig 6) which, when weathered, yield molds of fossils of the cf *Repinaella*–obolellid assemblage The top of this systems tract is marked by a white quartzite bed with *Monocraterion*? (Fig 7) This suite suggests a very shallow environment, considerably shallower than the deposits below On this basis, a sequence boundary is interpreted at the base of the lowest white quartzite bed, at 29 m in Fig 2 Elsewhere in the Montezuma Range, this systems tract is 62 m thick and at Slate Ridge, one section shows it to be 34 m thick

2. From 80 to 117 m in Fig 2, the siliciclastics are greenish-gray, fine-grained, micaceous sandstone and siltstone with lesser amounts of wavy-bedded to even-bedded sandstone in the upper part This fining-upward progression suggests a transgressive systems tract. Also, the nodular carbonate lenses at the top indicate a sediment-starved interval of transgression The bottom 21 m of this interval has several thin to medium beds of dark gray, fine-grained, laminated quartzite The top of the uppermost 30 cm bed of this quartzite is the practical top of the Andrews Mountain Member in this area The bottom 40 m of this interval is barren of body fossils and trace fossils are rare At 112 m in

Fig 2, the 30-cm aff *Eofallotaspis* Abundance- and Range-zone (Fig 4F), marking the base of the Montezuman Stage, is present in platy micaceous sandstone that is not otherwise distinguishable from similar sandstone above and below This biozone contains *Ladatheca* tubes, *Sabellidites*?, rare brachiopods, including *Nisusia*?, and bradoriids, in addition to the trilobites and common trace fossils About 2 m above this thin zone, other trilobites, *Fallotaspis* sp (Fig 4G), appear, rarely at first then commonly by the top of the interval

3 From 117 to 132 m in Fig 2, the siliciclastics are very fine silty shale deposited near or below storm-wave base The maximum flooding surface for the sequence would be somewhere in this interval Several species of *Fallotaspis* are common throughout

4 From about 132 to 192 m, the sequence coarsens and shallows upwards, with increasing amounts of very fine to fine micaceous sandstone and fine quartzite in the upper part, some with hummocky cross-stratification This can be interpreted as the highstand systems tract Trilobites are rare except in the top 10 m which contains a variety of *Fallotaspis* Zone trilobites, including *Paranevadella* and an unnamed new genus

5 At 192 m in Fig 2, a sequence boundary can be interpreted at the base of a distorted sandstone bed with a variety of soft sediment deformation features This is the point where transgressive conditions begin Above, the presence of numerous calcareous sandstone and bioclastic carbonate lenses suggests sediment-starved transgressive conditions of the transgressive systems tract of the next sequence Thin beds of sandstone and quartzite are present A varied ichnofauna is present along with rare brachiopods and trilobites—a granular species of *Fallotaspis*, *Paranevadella*, and the same unnamed new genus—which represent the upper part of the *Fallotaspis* Zone

## 5. An old trilobite report revisited

A single trilobite specimen found at the Waucoba Spring section in California (Fig 1) was reported to be from the Andrews Mountain Member of the Campito Formation 274 m below the top of the member (Scott, 1960). This fossil was later identified as *Fallotaspis* cf. *tazemmourtensis* (Nelson and Hupé, 1964). Since that time, this occurrence has been cited frequently; it was mentioned as the base of the *Fallotaspis* Zone in the North American faunal province by Fritz (1972). Nelson mentioned this occurrence and showed it on stratigraphic columns in many publications (e.g., Nelson, 1978). The occurrence is also noted in many other papers on the geology and paleontology of the White-Inyo Mountains (e.g., Alpert, 1976). This trilobite (Fig. 8A) occurs in light olive gray, well-indurated silt-stone. The opposite side of the rock from the specimen is subrounded and weathered to pale

yellowish brown. The rounding may be due to local fluvial transport or intense weathering. The specimen was reportedly found near the top of a hill 2.6 km northeast of Waucoba Spring. The lithology of the specimen is inconsistent with the reported strati-graphic position. This hill stands above a broad flat with the remains of an old corral, suggesting that the area was frequently inhabited. The geologic map of the area (Nelson, 1971) shows this flat to be "elevated and dissected alluvial gravel and sand." So it is not surprising that a specimen could be picked up here that was a long way from its source. Several attempts have been made to find additional trilobite material at this site, without success (personal communication from A.R. Palmer, 1996; C.A. Nelson, 1998; T.P. Fletcher, 1999).

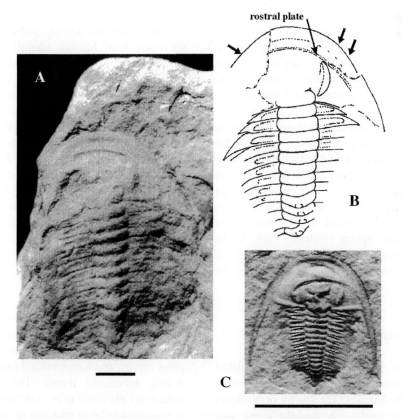

Fig. 8. Scott's trilobite specimen from the Waucoba Spring area, California, compared with *Esmeraldina*? *cometes* Fritz, 1995, from the Montezuma Range, Nevada. (a) Scott's specimen, note rounding of the sample margin ahead of the trilobite, UCLA 26821; (b) line drawing of Scott's specimen, note the rostral plate impressed beneath the cephalon. Cephalon angled downward and broken with displaced areas indicated by arrows, genal angle exaggerated; (c) articulated specimen of *Esmeraldina*? *cometes* Fritz, 1995, rostral plate is displaced above cephalon, from the Montenegro Member in the Montezuma Range, 1068-48. Scale bar 10 mm.

The cephalon of this specimen is angled downward about 20°, with the anterior border jammed backward (Fig 8A,B, arrows), exaggerating the genal angle area Weathering and breakage combined to obliterate most cephalic features Parts of the glabella and the right ocular lobe are faintly preserved The thorax, though, has at least 12 segments and is more useful The third segment is slightly wider than the adjacent segments, and the pleural spines are slightly thorn-like and directed outward at about 45° from the axis Beginning at the ninth segment, the axial ring is thickened, perhaps supporting a spine

The Waucoba Spring specimen was initially identified as *Fallotaspis* cf *tazemmourtensis*, which may have been influenced by the impressed rostral plate beneath the cephalon On *F tazemmourtensis*, there is a small but distinctive ridge at the interior margin of the anterior and lateral border furrow (Geyer, 1996) The Waucoba Spring specimen is easily distinguished from various Moroccan and North American species of *Fallotaspis* by the character of the thorax, particularly the outward-directed pleural spines Only one trilobite in the Montezuman has this thoracic configuration, thus Scott's specimen is identified as *Esmeraldina?* comets? Fritz (1995) (Fig 8C), which occurs in the Montenegro Member beginning just above the base of the *Nevadella* Zone in the Montezuma Range At the Waucoba Spring section, this specimen could have come from similar beds and been transported by natural or artificial means *Esmeraldina rowei* (Walcott, 1910) which occurs just above *Esmeraldina?* comets in the Montezuma Range, was found in the Montenegro Member at Waucoba Spring (Walcott, 1908, Scott, 1960) Thus the earlier reports of *Fallotaspis* in the upper third of the Andrews Mountain Member of the Campito are considered erroneous

## 6. Regional correlations

The occurrence of the cf *Repinaella*–obolellid assemblage in Nevada is not likely to be repeated elsewhere in North America The known occurrences of *Fallotaspis* Zone trilobites beyond the White-Inyo Esmeralda County region area are limited to the upper part of the zone, above the *Fallotaspis* Zone sequence boundary In the Cassiar Range of northern British Columbia, the basal few centimeters of a

transgressive limestone has a late *Fallotaspis* Zone fauna of *Parafallotaspis?* and *Cirquella nelsoni* Lieberman (2001), a form quite distinct from the younger species of *Cirquella* Immediately above, trilobites of the *Nevadella* Zone are found (Fritz, 1978) In the Mackenzie Mountains of northwest Canada, the *Fallotaspis* Zone is generally represented by a few meters of nodular limestone containing *Parafallotaspis*, but at the Red River section, several species of *Fallotaspis* occur over a 29 m interval (Fritz, 1976) All of these occurrences are in the late *Fallotaspis* Zone transgressive systems tract The absence of the lower 110 m of the *Fallotaspis* Zone and the older cf *Repinaella*–obolellid assemblage is due to either a hiatus at this sequence boundary, or perhaps to unfavorable environmental conditions

The reports of fallotaspidid trilobites from sections in the Death Valley region of California and from the Caborca area of Mexico are based on poorly preserved material, and the identifications are suspect The specimen from Mexico identified by McMenamin (1987, Fig 5 4) as cf. *Fallotaspis* sp is an *Esmeraldina* sp from the lower part of the *Nevadella* Zone Specimens from Death Valley (Hunt, 1990, pl 1, Figs 1–2) are assignable to the unnamed new genus mentioned earlier (see paragraph 4 5) Forms she compared with *Daguinaspis* in the White-Inyo region and assigned to the *Fallotaspis* Zone (Mount et al., 1991) have been identified as species of *Cirquella* (Fritz, 1993) which are middle *Nevadella* Zone In northeastern Laurentia, the oldest trilobites are in Greenland from the late *Nevadella* Zone (Blaker and Peel, 1997)

## 7. Discussion

The trilobites of the cf *Repinaella*–obolellid assemblage appear near the end of the Begadean Series as fully developed trilobites with distinct eyes, other organs, and a mineralized exoskeleton including a fully articulated thorax The development of a calcareous exoskeleton by nonmineralized arthropods is suggested to be in response to predators (Bengtson, 1994; Babcock, 2002)

The appearance of this assemblage near the top of the Andrews Mountain Member is certainly due in part to local environmental conditions In this area, the

Andrews Mountain siliciclastics were deposited in a relatively deep subtidal, storm-dominated environment, which was probably dysoxic, as suggested by the prevalence of unoxidized chlorite, biotite and opaque minerals in the well-sorted sand Conditions suddenly changed to a shallow subtidal regime which was warmer with more available oxygen suggested by the lack of dark minerals in much of the quartzite The temporary disappearance of the varied metazoan fauna in the rocks following the cf *Repinaella*–obolellid assemblage was most likely due to a return to deeper, relatively dysoxic conditions during the subsequent transgression. The later occurrence of a more varied metazoan fauna at the aff *Eofallotaspis* Abundance- and Range-zone in equally dysoxic, transgressive siliciclastics is enigmatic.

The first appearance of trilobites in other Cambrian continents varies considerably in age (Geyer and Shergold, 2000) but generally occurs in shallow subtidal waters The first trilobite in Siberia, *Profallotaspis* appears at the base of the Atdabanian Stage in varigated carbonates deposited on the open shelf margin of a large carbonate platform (Repina, 1981, Pegel, 2000) It is followed by *Repinaella* then *Archaeaspis* (at the base of the *Pagetiellus anabarus* Zone) supporting the hypothesis that Siberia and western Laurentia were in close proximity in the early Cambrian (Sears and Price, 2003) In western Gondwana (Morocco), *Hupetina* (a redlichiid) and *Eofallotaspis* appear near the top of a sequence of stromatolitic carbonates (Sdzuy, 1978) The succeeding *Fallotaspis* assemblage appears after a flooding event in the succeeding thick siliciclastic interval with minor carbonates (Geyer, 1996) In eastern Gondwana (South China), there is a substantial hiatus beneath the first trilobite, *Parabadiella,* which occurs in carbonaceous siltstone overlying phosphorite deposits of Tommotian age (Kirschvink et al , 1991, Zhang et al , 2001) Similar hiati apparently precede the earliest trilobites in Australia and Kazakhstan (Geyer and Shergold, 2000, Holmer et al , 2001) The earliest trilobite in Avalonia (eastern Newfoundland and England), *Callavia* (which is correlative with the *Nevadella* Zone in Laurentia), appears in siliciclastics or carbonate after a major discontinuity which correlates well with the late *Fallotaspis* Zone sequence boundary (at 192 m in Fig 2) (Landing et al , 1989)

The first brachiopods in other continents are phosphatic paternides in the earliest Tommotian of Siberia (Rozanov and Zhuravlev, 1992; Zhuravlev, 1995) while the first calcareous brachiopods (obolellids) appear there in the early Atdabanian (Popov et al., 1996)

Global correlation of the cf *Repinaella*–obolellid assemblage is difficult due to relative scarcity and poor preservation of these fossils, however a correlation with the *Repinaella* (previously known as the *Fallotaspis*) Zone of the early Atdabanian Stage in Siberia is strongly suggested (Zhuravlev, 1995, Geyer and Shergold, 2000) The aff *Eofallotaspis* and *Fallotaspis* assemblages somewhat higher in the Campito Formation show a development pattern that resembles the Moroccan faunas, but still suggesting an early Atdabanian age for the earliest Nevada trilobites (Fig 1)

## Acknowledgements

The field work for this study was conducted with W H Fritz, Geological Survey of Canada, over a period of several years His extensive discussions in the field were invaluable, but the concepts presented here are my own He also found the first specimen of the Andrews Mountain trilobites C A Nelson loaned me Scott's trilobite specimen A R Palmer provided access to trilobite replicas, extensive discussions and encouragement A C Runkel, J W Hagadorn, and M B McCollum discussed sedimentology in the field Reviews by W H Fritz, G Geyer, A R Palmer, and an anonymous reviewer considerably improved this article M E Hollingsworth contributed in many ways, particularly by collecting most of the fossils

## References

Albers, J P , Stewart, J H , 1972 Geology and mineral deposits of Esmeralda County, Nevada Bull Nev Bur Mines Geol 78, 1–80

Alpert, S P , 1976 Trilobite and star-like trace fossils from the White-Inyo Mountains, California J Paleontol 50, 226–240

Babcock, L E , 2002 Trilobites in Paleozoic predator–prey systems, and their role in reorganization of early Paleozoic ecosystems In Kelley, P H , Kowalewski, M , Hansen, T A (Eds ), Predator–prey Interactions in the Fossil Record Plenum Publishers, New York, pp 55–92

Bengtson, S , 1994 The advent of animal skeletons In Bengtson, S (Ed ), Early Life on Earth Columbia Univ Press, New York, pp 412–425

Blaker. M R , Peel, J S , 1997 Lower Cambrian trilobites from North Greenland Medd Grønl , Geosci 35, 1–145

Fritz, W H , 1972 Lower Cambrian trilobites from the Sekwi Formation type section. Mackenzie Mountains, Northwestern Canada Bull -Geol Surv Can 212, 1–90

Fritz, W H , 1976 Ten stratigraphic sections from the Lower Cambrian Sekwi Formation, Mackenzie Mountains, northwestern Canada Pap -Geol Surv Can 76-22, 1–42

Fritz, W H , 1978 Upper (carbonate) part of Atan Group, Lower Cambrian, north-central British Columbia Current Research, Part A, Pap -Geol Surv Can , vol 78-1A, pp 7–16

Fritz, W H . 1993 New Lower Cambrian olenelloid trilobite genera *Cirquella* and *Geraldinella* from southwestern Canada J Paleontol 67, 856–868

Fritz, W H . 1995 *Esmeraldina rowei* and associated Lower Cambrian trilobites (1f fauna) at the base of Walcott's Waucoban Series, southern Great Basin, USA J Paleontol 69. 708–723

Geyer, G , 1996 The Moroccan fallotaspidid trilobites revisited Beringeria 18 89–199

Geyer, G . Shergold, J H , 2000 The quest for internationally recognized divisions of Cambrian time Episodes 23, 188–195

Hagadorn, J W , Fedo, C W . Waggoner, B M , 2000 Lower Cambrian Ediacaran fossils from the Great Basin, USA J Paleontol 74, 731–740

Hollingsworth, J S , 1999 Stop 10 A candidate position for the base of the Montezuman Stage In Palmer, A R (Ed ). Laurentia 99, V Field Conference of the Cambrian Stage Subdivision Working Group, International Subcommission on Cambrian Stratigraphy, September 12–22, 1999 Institute for Cambrian Studies, Boulder, pp 34–37

Hollingsworth, J S . 2000 Environmental setting of the earliest trilobites in Laurentia (Cambrian, Begadean) Abstr Programs-Geol Soc Am 32 (7), A-301

Holmer, L E , Popov, L E . Koneva, S P , Bassett, M G , 2001 Cambrian–early Ordovician brachiopods from Malyi Karatau, the western Balkhash region, and Tien Shan, central Asia Spec Pap Palaeontol 65, 1–180

Hunt, D L , 1990 Trilobite faunas and biostratigraphy of the Lower Cambrian Wood Canyon Formation, Death Valley Region, California Unpublished MSc thesis, Univ of California, Davis. 103 pp

Kirschvink, J L , Magaritz, M, Ripperdan, R L , Zhuravlev, A Yu . Razanov, A Yu . 1991 The Precambrian/Cambrian boundary magnetostratigraphy and carbon isotopes resolve correlation problems between Siberia. Morocco, and South China GSA Today 1(4), 69–72, 87, 91

Landing, E . Myrow, P. Benus, A P , Narbonne, G M , 1989 The Placentian series appearance of the oldest skeletonized faunas in southeastern Newfoundland J Paleontol 63, 739–769

Lieberman, B S , 2001 Phylogenetic analysis of the Olenellina Walcott, 1890 (Trilobita, Cambrian) J Paleontol 75, 96–115

MacNaughton, R B , Dalrymple, R W , Narbonne, G M , 1997 Multiple orders of relative sea-level change in an earliest

Cambrian passive-margin succession. Mackenzie Mountains, Northwestern Canada J Sediment Res 67 (4), 622–637

McCollum, L B , Sundberg, F A , 2000 The Cambrian Emigrant Formation a highly condensed outer shelf sequence, Esmeralda County, Nevada Abstr Programs-Geol Soc Am 32 (7), A 456

McMenamin, M A S . 1987 Lower Cambrian trilobites, zonation, and correlation of the Puerto Blanco Formation, Sonora, Mexico J Paleontol 61, 738–749

Mount, J F , 1982 Storm-surge-ebb origin of hummocky cross-stratified units of the Andrews Mountain Member. Campito Formation (Lower Cambrian), White-Inyo Mountains, eastern California J Sediment Petrol 52, 941–958

Mount, J F , Hunt, D L , Greene. L R . Diegnei, J . 1991 Deposition systems, biostratigraphy and sequence stratigraphy of Lower Cambrian grand cycles, southwestern Great Basin In Cooper, J D , Stevens, C H (Eds ), Paleozoic Paleogeography of the Western United States II Pacific Section, vol 1 Soc Econ Paleontol and Mineral, Los Angeles. pp 209–227

Nelson, C A . 1971 Geologic map of the Waucoba Spring Quadrangle, Inyo County, California U S Geol Surv Map GQ. 921

Nelson, C A . 1978 Late Precambrian–Early Cambrian stratigraphic and faunal succession of eastern California and the Precambrian–Cambrian boundary Geol Mag 115 (2), 121–126

Nelson, C A , Hupé, P , 1964 Sur l'existence de *Fallotapsis* et *Daguinaspis*, Trilobites marocains, dans le Cambrien inferieur de Californie, et ses consequences C R Acad Sci , Paris 258, 621–623

Palmer, A R . 1998 A proposed nomenclature for stages and series for the Cambrian of Laurentia Can J Earth Sci 35, 323–328

Palmer, A R , Repina, L N , 1993 Through a glass darkly taxonomy, phylogeny, and biostratigraphy of the Olenellina Univ Kansas Paleontol Contrib , New Series, vol 3 Univ of Kansas, Lawrence 35 pp

Pegel, T V , 2000 Evolution of trilobite biofacies in Cambrian basins of the Siberian Platform J Paleontol 74, 1000–1019

Popov, L E , Holmer, L E . 2000 Obolellata In Kaesler, R L (Ed ) Treatise on Invertebrate Paleontology, Part H, Brachiopoda, Revised, volume 2 Geol Soc Am and Univ of Kansas, Lawrence, pp 200–208

Popov, L E . Holmer, L E , Bassett, M G , 1996 Radiation of the earliest calcareous brachiopods In Cooper, P , Jin, J (Eds ). Brachiopods Proceedings of the Third International Brachiopod Congress, Sudbury, Ontario, 1995 Balkema, Rotterdam, pp 209–213

Repina, L N , 1981 Trilobite biostratigraphy of the Lower Cambrian Stages of Siberia In Taylor, M E (Ed ), Short Papers for the Second International Symposium on the Cambrian System U S Geol Survey, Open-File Report 81-743, pp 173–180

Rowell, A J , 1977 Early Cambrian brachiopods from the southwestern Great Basin of California and Nevada J Paleontol 51, 68–85

Rozanov, A Yu . Zhuravlev, A Yu , 1992 Lower Cambrian fossil record of the Soviet Union In Lipps, J W, Signor, P W (Eds ), Origin and early evolution history of the Metazoa Plenum Press, New York, pp 205–282

Scott, K M , 1960  Geology of the Waucoba Springs area, Inyo
Mountains, California  Unpublished MA Thesis  University of
California, Los Angeles  109 pp

Sdzuy, K , 1978  The Precambrian–Cambrian boundary beds in
Morocco (Preliminary report)  Geol  Mag  115 (2), 83–94

Sears, J W , Price, R A , 2003  Tightening the Siberian connection
to western Laurentia  Geol  Soc  Amer  Bull  115 (8), 943–953

Snow, J K , Wernicke, B P , 2000  Cenozoic tectonism in the central
Basin and Range  magnitude, rate and distribution of upper
crustal strain  Am  J  Sci  300, 659–719

Stewart, J H , 1970  Upper Precambrian and Lower Cambrian strata
in the southern Great Basin, California and Nevada  U  S  Geol
Surv  Prof  Pap  620, 1–206

Van Wagoner, J C , Mitchum, R M , Campion, K M , Rahmanian,
V D , 1990  Siliciclastic sequence stratigraphy in well logs,
cores, and outcrops  AAPG Methods in Exploration Series vol
7  Amer  Assoc  Petr  Geol, Tulsa  53 pp

Walcott, C D , 1908  Cambrian geology and paleontology No  5,
Cambrian sections in the Cordilleran area  Smithson  Misc
Collect  53, 166–230

Walcott, C D , 1910  Cambrian geology and paleontology  I
*Olenellus* and other genera of the mesonacidae  Smithson
Misc  Collect  53, 231–422

Zhang, W T , Babcock, L E , Xiang, L W , Sun, W G , Luo, H L ,
Jiang, Z W , 2001  Lower Cambrian stratigraphy of Chengjiang,
eastern Yunnan, China with special notes on Chinese *Parabadiella*, Moroccan *Abadiella* and Australian *Abadiella huoi*
Acta Palaeontol  Sin  40 (3), 294–309

Zhuravlev, A Yu , 1995  Preliminary suggestions on the global
Early Cambrian zonation  Beringeria, Spec  Issue 2, 147–160

Available online at www.sciencedirect.com

Palaeogeography, Palaeoclimatology, Palaeoecology 220 (2005) 167–192

www.elsevier.com/locate/palaeo

# Taphonomy and depositional circumstances of exceptionally preserved fossils from the Kinzers Formation (Cambrian), southeastern Pennsylvania

Ethan S. Skinner

*Shell International Exploration and Production Inc., 3737 Bellaire Blvd, P.O. Box 481, Houston, Texas 77001-0481, USA*

Received 10 July 2004; accepted 6 September 2004

## Abstract

The Emigsville Member of the Kinzers Formation (Cambrian), southeastern Pennsylvania, is a deposit of exceptional preservation containing three main lithofacies that were part of a mixed siliciclastic-carbonate debris fan developed seaward of a carbonate shelf along the Appalachian Margin of Laurentia. Fossils were exceptionally preserved in a shelf environment subject to tempestite deposition. Most remains are fragmentary, which emphasizes the importance of predation as a taphonomic process where Emigsville sediments were deposited. Remains are preserved mostly as organic carbon films, pyrite crusts, and aluminosilicate films. These preservation styles apparently depended upon the development of microbial consortia in isolated microenvironments, and some of the microbes were autolithified.

Sedimentary and biological evidence suggests that anoxia, salinity fluctuations, and sedimentary obrution played relatively minor roles in the Emigsville Member of the Kinzers Formation. The abundance of exceptionally preserved remains suggests that exceptional sedimentary conditions were not necessary in the Cambrian in order for exceptional preservation to occur. The abundance of predation evidence in a deposit of exceptional preservation reinforces the concept that predation was the primary taphonomic filter during Cambrian time.
© 2004 Elsevier B.V. All rights reserved.

*Keywords:* Kinzers Formation; Nonmineralizing fossils; Taphonomy; Cambrian; Pennsylvania; Laurentia

## 1. Introduction

Deposits of Cambrian age are notable for the abundance of intervals containing exceptionally preserved fossils (Conway Morris, 1985; Allison and

---
*E-mail address:* Ethan.Skinner@Shell.com

Briggs, 1991; Babcock et al., 2001). Cambrian deposits of exceptional preservation (DEPs), alternatively known as conservation *lagerstätten* or "Burgess Shale-type" deposits, are those that preserve the fossil remains of organisms or parts of organisms that were not biomineralized (in addition to remains of organisms that were biomineralized). Well-known Cambrian examples include the Middle Cambrian

Burgess Shale of British Columbia, Canada (e g , Whittington, 1971, Conway Morris, 1979; Conway Morris and Whittington, 1985), the Lower Cambrian Maotianshan Shale near Chengjiang, Yunnan, China (e g , Zhang, 1987, Chen et al , 1989, Hou et al , 1999; Shu et al , 1999, Babcock et al , 2001), and the Lower Cambrian Buen Formation of North Greenland (e g , Conway Morris et al , 1987)

The near disappearance of "Burgess Shale type" DEPs from the fossil record has been a source of some controversy A commonly hypothesized reason for the loss of this preservation type is a contemporaneous increase in diversity, extent, and intensity of bioturbation, resulting in the development of a bioturbated mixed layer (Droser and Bottjer, 1990, Allison and Briggs, 1993, Gaines, 1999, Gaines and Droser, 2002) This increase, however, may not solely account for the disappearance of DEPs, as many more recent deposits lack evidence of bioturbation yet also lack exceptionally preserved fossils Other hypothesized factors include widespread anoxic depositional conditions (Allison and Briggs, 1993, Landing, 2001), a disappearance or decrease in the abundance of organisms with nonmineralized body plans, higher tidal amplitude leading to fluctuating salinities (Babcock et al , 2001), relatively high content of phosphate minerals in seawater (Brasier and Lindsay, 2001), and the presence of a geologically unstable clay mineral confined to the Cambrian that allowed for exceptional preservation (Butterfield, 1995)

Attempting to test the validity of hypotheses of exceptional preservation in the Cambrian is somewhat problematic. While experimental modeling to determine which depositional and chemical parameters control various types of exceptional preservation has met with some success (e g , Babcock, 1998; Briggs et al , 1993; Martin et al , 2003), experimental studies can neither efficiently nor simultaneously model all of the environmental factors that affect an organism upon its death

An alternative method of elucidating the conditions necessary for Cambrian exceptional preservation is to test whether a particular deposit of exceptional preservation shows evidence of some of the previously cited factors, and whether these factors could have had an influence upon the distribution of exceptional preservation in the deposit Systematic examination of sedimentological and paleontological

(especially paleoecological and taphonomic) evidence in a deposit may allow testing how anoxia, salinity fluctuation, and bioturbation intensity might have factored into exceptional preservation

The pelitic Emigsville Member of the Lower Cambrian Kinzers Formation of southeastern Pennsylvania has long been known for containing an exceptionally preserved Cambrian biota (Dunbar, 1925, Resser and Howell, 1938, Campbell and Kauffman, 1969, Capdevilla and Conway Morris, 1999) It is the only member of the Kinzers Formation to contain an exceptionally preserved biota This unit is an appropriate selection for this study, as it contains a variety of lithologies and assemblages of exceptionally preserved taxa (Campbell and Kauffman, 1969) and has a relatively lower biotic diversity in comparison with other similar deposits (Conway Morris, 1985)

This study of preservational patterns in the Emigsville Member examines two larger questions (1) Do the lithofacies and taphofacies present in this unit show evidence of being deposited under scenarios associated with established models of exceptional preservation? (2) Do the taphonomic pathways (biostratinomic and diagenetic) exhibited by exceptionally preserved remains in the deposit support the presence of one or more established models of exceptional preservation?

## 2. Materials and methods

Many specimens were collected from Locality K4 and the surrounding area (see Campbell, 1969) Other specimens used in this study are in the collections of the Department of Earth and Environment, Franklin and Marshall College, Lancaster Pennsylvania Figured specimens are deposited in the Orton Geological Museum of The Ohio State University, Columbus, Ohio.

Samples were trimmed into tabular sections between 3 and 10 cm thick to facilitate bedding-plane parallel splitting Other samples were collected both for optical scanning, and for observation of larger-scale bedding features

Slabs selected for optical scanning and analysis of sedimentary features were ground and polished Samples were scanned to a resolution of 1200 dpi in a small amount isopropyl alcohol and image contrast was enhanced with Adobe Photoshop using the technique of Schieber (2003; see also Borkow and Babcock, 2003). Sedimentary characteristics of all lithologies within each sample were described from images at 12× or greater magnification.

The surface of each slab containing fossil material was also examined for the remains of organisms. All reasonably identifiable remains of organisms were counted and described. For trilobites and echinoderms, the state of disarticulation (i.e., the distribution of skeletal elements, number of elements present, etc.) was also recorded. Any bedding-plane-parallel evidence of bioturbation was also noted and described.

Preservational modes of examples of each taxon were observed, whether manifested as (1) a nondescript "dark film;" (2) a carbonized film; (3) pyritic or limonitic stain; (4) original phosphatic material; (5) clay replacement; or (6) moldic preservation. Differential preservation of portions of organisms, such as gut traces, was also noted. The presence of "halos," "oozes," and other bedding plane features associated with organisms were also recorded, as were their characteristics. Examples of these various

preservation states were chosen for SEM and EDS analysis in order to semiquantitatively determine chemical composition.

Samples showing a variety of taxa and preservation states were photographed and examined using a JEOL JSM-820 scanning electron microscope (SEM) with an Oxford eXL energy dispersive X-ray analyzer located in the Microscopic and Chemical Analysis Research Center (MARC) of The Ohio State University. Samples were left uncoated in order to prevent damage to the specimens and to prevent skewed spectrographic results. As a result, it was necessary to use relatively low acceleration voltages (10–12 keV) so that images would be defined more clearly. Both secondary electron images and backscatter electron images were taken in order to highlight topographic and compositional features of the specimens, respectively. Backscatter electron images were used to analyze the preservation states of the specimens and to choose areas for examination with EDS (see Orr et al., 2002; Fig. 1a). SEM backscatter images were used in conjunction with EDS information to generate interpretive Photoshop images showing the distribution of varying compositions and preservation types on selected specimens.

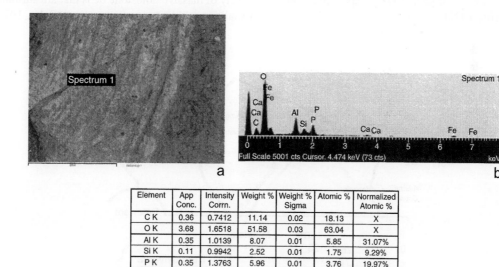

| Element | App Conc. | Intensity Corrn. | Weight % | Weight % Sigma | Atomic % | Normalized Atomic % |
|---|---|---|---|---|---|---|
| C K | 0.36 | 0.7412 | 11.14 | 0.02 | 18.13 | X |
| O K | 3.68 | 1.6518 | 51.58 | 0.03 | 63.04 | X |
| Al K | 0.35 | 1.0139 | 8.07 | 0.01 | 5.85 | 31.07% |
| Si K | 0.11 | 0.9942 | 2.52 | 0.01 | 1.75 | 9.29% |
| P K | 0.35 | 1.3763 | 5.96 | 0.01 | 3.76 | 19.97% |
| Ca K | 0.06 | 0.9975 | 1.49 | 0.01 | 0.73 | 3.88% |
| Fe K | 0.66 | 0.7881 | 19.25 | 0.05 | 6.74 | 35.79% |

Fig. 1. (a) Backscatter electron image of the brachiopod *Paterina bella*, from Massive Pelitic Facies, Emigsville Member. (b) EDS-spectrum of specimen in panel (a). (c) Compositional output of spectrum, with normalized abundances.

In order to analyze specimens using energy dispersive X-ray spectroscopy, areas of analysis were chosen using INCA and the EDS function of the SEM (Fig 1b) Next, the program was used to generate semiquantitative information about the elements present (Fig 1c) Elements with an atomic number lower than 11 were analyzed comparatively by examining signals both on and off fossils in order to determine which had stronger signals and, therefore, greater elemental abundances

## 3. Stratigraphy, facies, and fossil assemblages

Testing models of exceptional preservation in the Emigsville Member requires: (1) understanding of the lithostratigraphy and sequence stratigraphy of the Kinzers Formation and associated strata, (2) a clear definition of which strata is referred to the Emigsville Member, and (3) understanding the lithofacies and taphofacies represented within the Emigsville Member Previous studies of the Kinzers Formation have focused on the outcrop pattern and large-scale stratigraphy of the deposit (e g , Stose and Jonas, 1922, Campbell and Kauffman, 1969, Gohn, 1976), as well as on descriptions of taxa present (e g , Resser and Howell, 1938, Briggs, 1978, Capdevilla and Conway Morris, 1999)

### 3 1  Stratigraphy of the Conestoga Valley succession

The Kinzers Formation (Stose and Jonas, 1922, Gohn, 1976, Ganis and Hopkins, 1990; MacLachlan, 1994) crops out in the Conestoga Valley, a structurally complex succession of Cambrian carbonates and siliciclastics located primarily in Lancaster and York counties, southeastern Pennsylvania (Stose and Jonas, 1922, Campbell, 1969, MacLachlan, 1994, Fig 2) The lowermost Cambrian strata exposed in the Conestoga Valley consist of a relatively thick succession of siliciclastic sediments, many of which were deposited as basin fill of the failed rifts that developed during the breakup of Rodinia (or Pannotia) in the Neoproterozoic to Early Paleozoic (Stose and Jonas, 1922; Campbell, 1969; MacLachlan, 1994; Fig 3) Conformably overlying the siliciclastics is the dolomitic Vintage Formation, which represents the stratigraphically lowest carbonate unit in the Conestoga Valley The presence of turbidites within the Vintage suggests a relatively deep environment of deposition (Gohn, 1976) Turbidite layers consist of bioclastic grains, peloids, lithoclastic grains (carbonate rip-up clasts), and mixtures of these grain types A developing shelf edge probably served as a source area for these grains Other portions of the formation consist largely of massive dolostone beds that are attributed to a shallow, sabkha-type environment (Gohn, 1976)

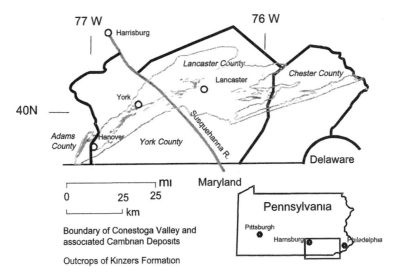

Fig 2  Outcrop distribution of the Conestoga Valley succession and Kinzers Formation, southeastern Pennsylvania

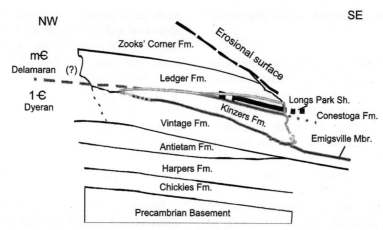

Fig 3 General stratigraphy of the Conestoga Valley succession, southeastern Pennsylvania (modified from MacLachlan, 1994)

The Kinzers Formation, named for the town of Kinzers, Pennsylvania (Stose and Jonas, 1922), unconformably overlies the Vintage Formation The Kinzers Formation is divided into four members. a lower member (Emigsville Member) of pelitic composition, a middle member (York Member) consisting of relatively pure limestone with rare shale partings, and two upper members (Longs Park and Greenmount Members) consisting of impure carbonates mixed with varying amounts of siliciclastics (Gohn, 1976, MacLachlan, 1994) Trilobites in the Kinzers Formation suggest that the Dyeran–Delamaran boundary exists as an unconformity within the upper portions of the Kinzers Formation (Campbell, 1969, Taylor and Durika, 1990)

The Kinzers Formation is overlain locally by the Ledger Dolomite In western portions of the Conestoga Valley succession, the Ledger Dolomite conformably overlies the Kinzers Formation, suggesting continued growth and progradation of the carbonate margin This formation marks the first of a number of relatively shallow carbonate shelf-derived deposits that continue into Ordovician strata (MacLachlan, 1994)

### 3 2 Stratigraphy of the Kinzers Formation

The lowermost Emigsville Member of the Kinzers Formation unconformably overlies the carbonates of the Vintage Formation (Stose and Stose, 1944, Gohn, 1976, MacLachlan, 1994) The Emigsville

Member is the only unit in the Conestoga Valley succession that can be correlated through the entire length of the Kinzers outcrop belt (MacLachlan, 1994) It is also the only deposit in the Conestoga Valley succession known to contain exceptionally preserved organisms Well-preserved olenelloid trilobites in this deposit point to a Dyeran age (Taylor and Durika, 1990)

The York Member consists mostly of limestone, locally it is dolomitized (Ganis and Hopkins, 1990) Evidence of a more shallow depositional setting includes oolitic intervals (Taylor and Durika, 1990) and carbonate mud mounds Other portions of the York Member contain carbonate breccias, indicating the proximity of a shelf edge (Gohn, 1976) Other than rare archaeocyathids, the York Member yields few macrofossils

The Longs Park and Greenmount Members of the Kinzers Formation both consist of argillaceous, locally sandy carbonates that appear to have been deposited in a turbidite or tempestite-influenced setting Both members consist of around 50% insoluble residue in some localities (Ganis and Hopkins, 1990) Trilobites in this unit are of generalized Laurentian shelf aspect It is unclear whether the Longs Park Member and Greenmount Member represent a single, time-transgressive unit or if the Longs Park Member is an unrelated unit deposited unconformably over the Greenmount Member (Campbell, 1970, Taylor and Durika, 1990, MacLachlan, 1994)

### 3 3 Lithology and biostratigraphy of the Emigsville Member

The Emigsville Member of the Kinzers Formation is recognized through much of the Conestoga Valley and is best represented by stratigraphic sections in and around the cities of Lancaster, York, and Hanover, Pennsylvania (Stose and Stose, 1944; Fig 1) Deposits in the vicinity of Lancaster are relatively fossiliferous and contain exceptionally preserved taxa Few fossils have been reported from the member where it crops out in the western part of the outcrop belt (Stose and Stose, 1944) where extensive metamorphism has led to the obscuring of fossils Recent field work in the Emigsville Member shows that exceptional preservation of fossils is present in the westernmost exposures of the member near the city of Hanover, Pennsylvania

Descriptions of the Emigsville Member have concentrated on larger stratigraphic patterns, and the internal stratigraphy of the member has been underemphasized Stose and Stose (1944) described the member as a light-gray to light-blue argillite containing pseudomorphs of limonite after pyrite and grains of mica The basal contact appears to be abrupt, suggesting rapid transgression (Campbell, 1969) Clays of the Kinzers are mostly illite, with the proportion of chlorite increasing to the northeast (Meyers, 1967) The thickness of the Emigsville Member has been estimated to range from about 50 m in York County, Pennsylvania, to about 20 m near the Susquehanna River (Gohn, 1976)

Biostratigraphic correlation of the Emigsville Member suggests an Early Cambrian (Dyeran) age

Such an assignment is based on the presence of olenelloid trilobites such as *Wanneria walcottana*, *Olenellus (Olenellus) thompsoni*, *Olenellus (Paedeumias) yorkense*, and *Olenellus? crassimarginatus* The corynexochid trilobite *Lancastria roddyi* occurs in the Emigsville Member Species referable to *Lancastria* occur in Lower Cambrian strata of China, Siberia, Australia, Greenland, and Nevada (Blaker and Peel, 1997)

### 3 4 Definition of facies in the Emigsville Member

Facies within the Emigsville Member can be defined based on both lithological and paleontological characteristics Those distinctions can be used to interpret depositional environments represented by the Emigsville Member and testing whether established models of exceptional preservation are applicable to the deposit Campbell (1969) outlined the lithologies and faunules present in the Kinzers Formation but did not define any vertical or lateral trends Further examination of the Emigsville Member indicates that the unit includes three laterally persistent lithofacies (Fig 4) Some of these lithofacies may be further subdivided into taphofacies (see Brett and Baird, 1986) Facies of the Emigsville Member are here termed the "Fine Pelitic Facies," "Massive Pelitic Facies," and "Impure Carbonate Facies "

#### 3 4 1 Fine Pelitic Facies
The Fine Pelitic Facies consists of a succession of fine-grained, thin, and discontinuously bedded strata Oxidized pyrite grains are particularly abundant, suggesting anoxic sediment conditions In general,

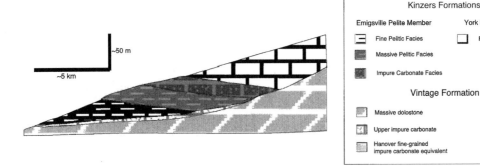

Fig 4 Diagrammatic lithofacies relationships within the Emigsville Member of the Kinzers Formation

bioturbation intensity is low and restricted to isolated, bedding-plane-parallel burrows. Most fossils are fragmentary. The discontinuous laminae present in this facies are evidence of intermittent deposition, possibly controlled by tempestite input. Most lithologic characteristics in the Fine Pelitic Facies may be explained by differential proximity to given tempestite flows and their associated rates of deposition. The facies contains three taphofacies. They are here termed the "Fine Tempestite Taphofacies," the "Shelly Tempestite Taphofacies," and the "Matground Taphofacies."

The Fine Tempestite Taphofacies (Fig. 5) consists of a thinly bedded, weakly bioturbated, fine-grained pelite with varying fossil abundance, color, and pyrite content. Pyrite framboids are rare to absent, which suggests that the water column was not anoxic, and that the sediment–water interface was probably oxygenated (see Wilkin et al., 1996; Schieber and Baird, 2001). Pyrite distribution seems to have been largely controlled by microbial sulfate reduction associated with organic material. Color variations are likely related to fluctuating sediment oxygenation states or some other chemical parameter. Carbonate and pyrite granules have been removed or replaced by goethite.

Most body fossils in the Emigsville Member consist of algal and cyanobacterial material with occurrences present on most bedding planes where other fossils occur (Fig. 6a). Algal and cyanobacterial material may have been floating in the water column and collected during intervals of low deposition. Alternatively, such organisms may have lived on the ocean bottom when sedimentary input was low and enough light was able to filter to the ocean floor. Indeterminate cuticular fragments of nonmineralizing arthropods are also common in the Fine Tempestite Taphofacies, most of which probably represent predatory debris since transport alone rarely results in fragmentation (Babcock and Zhang, 1997; Babcock, 2003). The tubular, phosphatic, mud-sticking organism *Tubulella pensylvanica* is present in varying amounts, but where present, it is usually preserved as fragments. These fragments may represent predatory debris, although it is possible that hydrodynamic action could have broken tubes of this organism. Shelly, nontrilobite organisms, including *Pelagiella* and brachiopods, often occur as bedding plane assemblages, possibly representing disaggregated coprolites. Some hyoliths were likely autochthonous, as a few articulated specimens have been found (Campbell, 1969; Capdevilla and Conway Morris, 1999). Other remains occurring in the Fine Tempestite Taphofacies include fragments of palaeoscolecids and other vermiform taxa. Relatively poor preservation states in this interval suggest that they

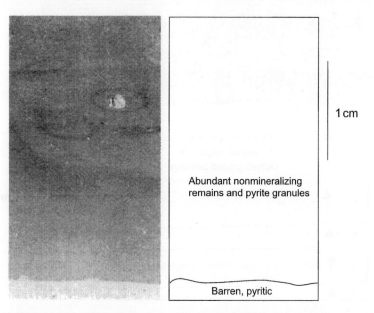

Fig. 5. Typical sample of Fine Pelitic Facies, Fine Tempestite Taphofacies, with interpretation.

E.S. Skinner / Palaeogeography, Palaeoclimatology, Palaeoecology 220 (2005) 167–192

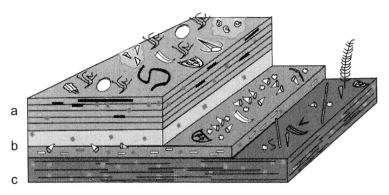

Fig. 6. Reconstruction of Fine Pelitic Facies depositional environments; (a) fine tempestite taphofacies; (b) shelly tempestite taphofacies; (c) matground taphofacies.

consist of a combination of taphonomic debris derived from transport, scavenging, or predation of organisms derived from elsewhere.

The Shelly Tempestite Taphofacies (Fig. 7) is represented by silty pelite interbeds mixed with biogenic carbonate allochems grading upwards into tan, fine-grained, poorly fossiliferous pelites. These may be interpreted as more proximal portions of the pulsed tempestites that dominate deposition in the Fine Pelitic Facies. Graded bedding occurs in this taphofacies. The absence of well-defined laminations in most samples suggests the presence of rapidly deposited tempestites. Current energy associated with these deposits likely determined whether examples of this taphofacies contain calcareous allochems or only rapidly deposited clays.

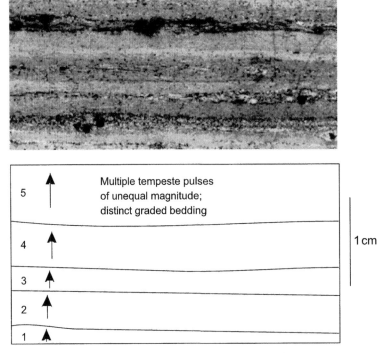

Fig. 7. Typical sample of Fine Pelitic Facies, Shelly Tempestite Taphofacies, with lithologic interpretation.

Most fossils in the Shelly Tempestite Taphofacies are calcareous allochems partly deposited by tempestite flows derived from the carbonate ramp to the west. All fossil remains are relatively small and consist mostly of echinoderm ossicles, *Salterella*, *Nisusia*, and trilobite sclerites; nonmineralizing organisms are rare (Fig. 6b). These taxa are nearly absent from the remainder of the Fine Pelitic Facies and have more in common with taxa occurring in the Impure Carbonate Facies.

The Matground Taphofacies (Fig. 8) is present in a thin (~0.3 m) interval within the Fine Pelitic Facies. This interval appears somewhat massive with abundant, euhedral grains of goethite after pyrite. Bioturbation intensity is difficult to gauge, but it appears to be relatively intensive over this interval, as laminae are discontinuous. It has a different biotic composition from other members of this facies, containing mainly sessile taxa.

Many fossils in the Matground Taphofacies are of sessile benthic organisms that were possibly buried in situ (Fig. 6c). Among the organisms absent from the remainder of the Fine Pelitic Facies are small, well-preserved specimens of *Tubulella pensylvanica*, partially disarticulated remains of a probable pennatulacean cnidarian, cone-shaped, highly reflective nonmineralizing tubes referable either to a benthic alga or something similar to the solitary cnidarian tube

*Cambrorhytium* (see Conway Morris and Robison, 1988; Hou et al., 1999), and a chancellorid sclerite probably referable to *Allonia* (Janussen et al., 2001). This assemblage of taxa suggests low depositional rates, free from the intermittent tempestite deposition typical of the Fine Pelitic Facies. These temporary conditions allowed the formation of a microbially stabilized matground upon which small benthic organisms were able to colonize (see Dornbos and Bottjer, 2001).

In summary, the depositional environment of the Fine Pelitic Facies is interpreted mainly as a low-energy environment with intermittent, pulsed sedimentation. Indigenous taxa were of low diversity, consisting largely of small-shelled organisms. Bioturbation was limited to sparse horizontal burrows. Most bodily remains consist of debris, either transported by tempestite currents (*Salterella*-echinoderm material), hiatal settling (algal and cyanobacterial material), or left behind by predators or scavengers (arthropod fragments and coprolites). Rare matgrounds are an exception to this condition.

### 3.4.2. Massive Pelitic Facies

The Massive Pelitic Facies consists of a dark to medium gray, moderately to thickly bedded pelite that is silty in places (Fig. 9). Bioturbation is relatively common. Horizons containing individual burrows are

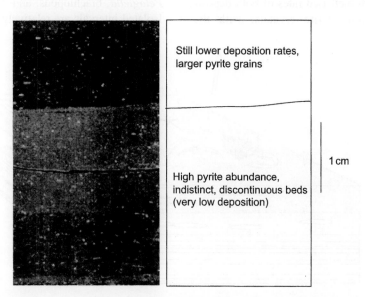

Fig. 8. Typical sample of Fine Pelitic Facies, Matground Taphofacies, with lithologic interpretation.

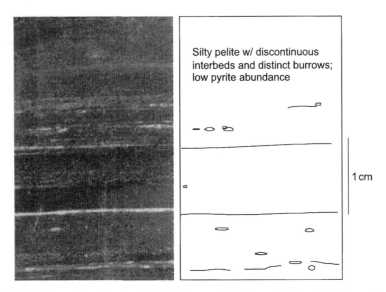

Fig. 9. Typical sample of Massive Pelitic Facies overall lithology, with lithologic interpretation.

recognizable, but some intervals in lower portions of this facies show bedding-plane-parallel bioturbated surfaces (sensu Schieber, 2003). Some clay–silt couplets resemble tidally generated sedimentation patterns (Dean et al., 1982; Staub et al., 2000), suggesting an association between burrowing and changes in tidal input. Lithologies in the Massive Pelite Facies are less variable and significantly more massive than in the Fine Pelitic Facies, and were most likely associated with increased rates of both deposition and bioturbation. Pyrite crystals, with the exception of those associated with organismic remains, are less common in this lithology than in the Fine Pelitic Facies. In this interval, the presence of incipient carbonate granules as voids is notable. Similar structures have been reported in the Wheeler Shale (Allison et al., 1995).

Fossils tend to be uncommon in the Massive Pelitic Facies but they are dispersed throughout the facies (Fig. 10). Fossils include trilobite sclerites, hyoliths, *Pelagiella*, brachiopods, and *Tubulella pensylvanica*. Isolated intervals are relatively rich in fossil remains.

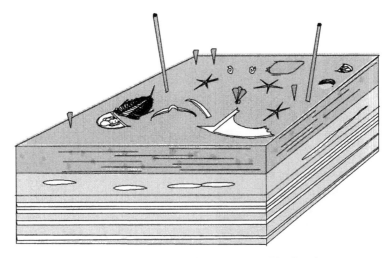

Fig. 10. Reconstruction of Massive Pelitic Facies depositional environments.

One interval of the Massive Pelitic Facies contains large, articulated olenelloid trilobite remains, including probable exuviae and corpses (Dunbar, 1925, Resser and Howell, 1938, Campbell, 1969) Also present in this interval are anomalocaridid appendages, a moderately diverse assemblage of bivalved nonmineralizing arthropods, hyoliths, and rare specimens of algae, bacteria, and sponges The rarity of mud-sticking organisms, and the presence of mobile benthic taxa, suggests that bottom conditions were not suitable for sessile benthic organisms Those conditions may have included relatively high rates of deposition, soft bottom conditions, and moderately intense bioturbation Accumulation of intact trilobite specimens in this interval may be related to tidal fluctuations

Another interval contains abundant chancelloriid sclerites, macroscopic benthic algae (in situ?), and specimens of *Tubulella pensylvanica* This interval has similarities to the Matground Taphofacies in the Fine Pelitic Facies

The Massive Pelitic Facies may be interpreted as the distal portion of a tidally influenced depositional system in which sedimentation rates were significantly higher than in the Fine Pelitic Facies The presence of carbonate, the scarcity of pyrite, the relatively coarse-grained sediments, and the abundance of bioturbated intervals suggest well oxy-

genated sedimentary conditions in the Fine Pelitic Facies Arthropods are the most abundant organisms, suggesting that depositional conditions were hostile to less mobile organisms Intervals containing relatively abundant articulated arthropod remains may represent brief periods of deposition associated with alternating salinity in the water column

### 3 4 3 Impure Carbonate Facies

The Impure Carbonate Facies is composed of thickly bedded, tan, leached, friable, carbonate-enriched pelites and siltstones containing abundant biogenic allochems (Fig 11) The grain size of the siliciclastic impurities varies vertically and laterally from silt-sized to clay-sized Bedding is relatively wispy with few well-developed laminations, indicating relatively high depositional rates Some large, well-defined burrows indicate relatively intense bioturbation

The faunal assemblage present within the Impure Carbonate Facies is distinct from that in the Fine Pelitic Facies and the Massive Pelitic Facies Nonmineralizing organisms are rare, with the exception of a few heavily sclerotized arthropods Olenelloid trilobites are common in a variety of preservation states, ranging from complete articulated individuals to broken sclerites Most significant is the increased abundance of shelly taxa (like *Salterella* and echino-

Fig 11 Typical sample of Impure Carbonate Facies overall lithology, with lithologic interpretation

derms; Fig. 12) compared to other facies. Some assemblages of *Salterella* show some evidence of current alignment, and they may have been intermittently subjected to winnowing and redeposition. The Impure Carbonate Facies contains most of the intact echinoderms in the Emigsville Member (Sprinkle, 1973). As echinoderm soft tissues decay rapidly (Ausich, 2001), these specimens must have been buried in place or were transported only a short distance. Rates of deposition in the Impure Carbonate Facies must have been sufficiently rapid in order to preserve articulated echinoderms. Rapid deposition also explains the relatively intact and well-preserved remains of trilobites present in these localities.

Lithologic change in grain size and character in the Impure Carbonate Facies is likely a function of current conditions. Tidal fluctuations might have been responsible for clay–silt alternation, with a pulse of coarser sediment deposition each time the tide went out. These rapid changes in sediment supply likely aided in the preservation of intact echinoderms and trilobites. More distal portions of the Impure Carbonate Facies in the Emigsville Member appear to contain only the finer-grained lithology (Sprinkle, 1973). However, many of the same taxa are present, as well as a few resistant nonmineralizing arthropods. The presence of abundant intact echinoderms suggests that these organisms lived locally and were preserved by obrution deposits.

The Impure Carbonate Facies represents a depositional environment with sporadically high amounts of bioturbation and active bottom currents. Carbonate-dwelling taxa including echinoderms, *Salterella*, and some trilobites, are abundant and often preserved intact, suggesting that obrution was a major factor in fossil preservation.

### 3.5. Facies relationships in the Emigsville Member

The Fine Pelitic Facies, Massive Pelitic Facies, and Impure Carbonate Facies of the Emigsville Member represent a shelf-to-shore transect towards depositional environments with progressively higher sedimentation rates (Fig. 3). The Fine Pelitic Facies most likely represents somewhat deeper-water tempestite deposition largely free from the influence of carbonate buildups to the west (Fig. 4). The Massive Pelitic Facies and the Impure Carbonate Facies are interpreted as tidally deposited facies more directly associated with the developing carbonate shelf.

Facies in the Emigsville Member appear to be strongly related to the growth of carbonate buildups to the west, and can be interpreted in the context of a mixed carbonate–siliciclastic debris fan. Closest to the shore in this scenario would have been a series of patchy, mudmound-like buildups (Barnaby and Read, 1990). Between these mounds were a number of channels, in which current energy was more concentrated. Just offshore, beyond the patchy mudmounds, are strata of a debris fan derived from

Fig. 12. Reconstruction of Impure Carbonate Facies depositional environment.

the carbonate buildups The first lithology present is the York Member Offshore, this grades into the Impure Carbonate Facies Intermittent fine and coarse deposition is likely a result of current passing through tidal channels Still farther offshore was the Massive Pelitic Facies, also derived from tidal currents carrying siliciclastic sediments Finally, the Fine Pelitic Facies was the farthest offshore, and was composed of intermittent tempestite deposits not completely derived from the carbonate buildup In the absence of siliciclastic input from the tidal channels, sedimentation was derived solely from tempestite input

## 4. Visual and chemical taphonomic analysis of exceptionally preserved fossils

Taphonomic data were used to examine whether the biota of the Emigsville Member of the Kinzers Formation was preserved in situ, and whether anoxia, salinity fluctuations, or obrution played a factor in preservation Predation evidence is also used to examine how each taxon may have fit into the trophic structure of the environment in which the Emigsville Member was deposited (Table 1)

### 4 1 Trilobites and other arthropods

Trilobites are one of the most significant components of the Emigsville Member biota, although articulated specimens are much less common than in similar Cambrian DEPs (e g , Whittington, 1980; Zhang, 1987, Gaines and Droser, 2003) Olenelloids (*Olenellus (Paedeumias) yorkense, Olenellus (Olenellus) thompsoni, Olenellus? crassimarginatus,* and *Wanneria walcottana*) are the most common trilobites Other taxa include *Bonnia* spp , *Kootenia* spp , and *Lancastria roddyi* Fragmented thoracic segments, especially pleurae that have been broken from the axial ring, are the most abundant evidence of trilobites (Fig 13) Cephala, particularly of smaller specimens, are commonly found whole

Breakage of many trilobite specimens in the Emigsville Member might have resulted from the action of predators (see Pratt, 1998, Babcock, 2003) Remains of trilobite predators, such as *Anomalocaris pensylvanica*, and healed injuries on olenelloids (e g ,

Table 1
Summary of exceptional preservation styles in the Emigsville Member

| | | Trilobites | nm arthropods | Algal/bacterial | Brachiopods | Hyoliths | Pelagiella | Salterella/echinoderms | T pensylvanica | Palaeoscolecids |
|---|---|---|---|---|---|---|---|---|---|---|
| Original | Carbon | x | x | x | x | x | ? | | x | ? |
| | Phosphatic | | | | x | | | | x | |
| Replacement (hard parts) | Pyrite | x | | | | x | | | | |
| | Phosphates | x | | | x | x | | x | | x |
| | Aluminosilicates | x | x | x | x | x | x | x | x | |
| Replacement (soft parts) | Pyrite | x | x | x | | x | x | | x | |
| | Phosphates | x | x | x | x | x | x | | x | x |
| | Aluminosilicates | x | | | | | | | | |
| Infilling | Pyrite | x | x | x | x | x | x | x | x | |
| | Phosphates | x | | | | | | | x | x |
| | Glauconite | | | | | | x | | | |

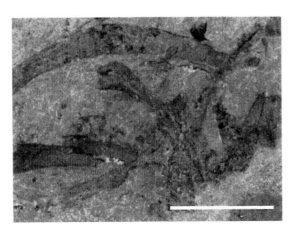

Fig. 13. Olenelloid trilobite sclerites (scale bar=1 cm); Fine Pelitic Facies, Fine Tempestite Taphofacies, Emigsville Member.

"*Olenellus peculiaris*" of Resser and Howell (1938), *Olenellus*? *crassimarginatus* of this study; Babcock, 2003, 1993; Capdevilla and Conway Morris, 1999) support this interpretation. Small, dark, phosphatic specimens in the Fine and Massive Pelitic Facies seem to represent fragmented remains and disaggregated fecal remains.

Trilobite remains are preserved in a number of ways. In the Impure Carbonate Facies and other silty, originally calcareous lithologies, three-dimensionally preserved, iron oxide-coated molds of trilobite elements are present, suggesting that tests were leached away with the rock-forming carbonate. Three-dimensional, aluminosilicate molds are the most abundant mode of preservation of trilobite remains. Despite a darker color than the surrounding matrix, elemental analysis indicates that only clay minerals are present in these specimens. The reflective, dark appearance is largely attributable to the surface texture of the trilobite.

Internal cavities such as eyes and the gastrointestinal tracts were sites for the preservation of iron oxides and phosphates. These structures acted as microenvironments in which microbial consortia were able to form iron sulfides and their precursors. These appear in specimens in the form of iron-oxide-coated eyes, or as darker, rotten-looking voids in the vicinity of the eyes, stomach, gastric caecum, and other portions of the gastrointestinal system. Similar voids are present in the Montezuman trilobite *Buenellus* from the Sirius Passet deposit of the Buen Formation of North Greenland (Babcock and Peel, 2002; Skinner

et al., 2002), as well as in some specimens of the nonmineralizing arthropod *Naraoia* (Vannier and Chen, 2002). In the best example, an assortment of phosphate-enriched, "rotten" looking voids is positioned in the cephalic area near the eyes and in the vicinity of the stomach and caecum (Fig. 14). Concentration of these voids near the eyes suggests that phosphate-enriched fluids within the specimen migrated upwards and were trapped by the thick calcite lenses of the eyes. Alternatively, this material may be indicative of phosphatization of either the gastrointestinal contents or of the caecum. Mineralized glandular tissues have been documented in various Cambrian arthropods (Butterfield, 2002; Vannier and Chen, 2002); the gastric caecum might have similarly served as an appropriate site for differential precipitation of phosphate and pyrite. A few other specimens show the presence of phosphate-enriched "oozes" that may have been derived from the gut of the animal. A final preservational mode represented in some trilobite remains is that of dark, flattened, or wrinkled fragments. These fragments are phosphatic in composition and may be derived from disaggregated coprolites. The preservation is similar to that of a trilobite from the Burgess Shale Biota inferred to be a molted individual (Whittington, 1980).

Fragmentary material of nonmineralizing arthropods is relatively common in the Fine Pelitic Facies of the Emigsville Member. Well-preserved examples of nonmineralizing arthropods include *Anomalocaris pensylvanica*, *Tuzoia* spp. and several other bivalved arthropods, and the enigmatic *Serracaris lineata* (Resser and Howell, 1938; Briggs, 1978). An isolated sclerite of *Serracaris* is preserved as phosphatic material mixed with organic carbon grading outwards into clay minerals. This preservational mode is similar to that of phosphatic trilobite remains (Fig. 15). The phosphatic axial portion of the fragment in Fig. 15 may have been phosphatized in a way comparable to that of phosphatic trilobite fragments, or alternatively may have been related to residual glandular material as shown in various Cambrian arthropods (Butterfield, 2002; Vannier and Chen, 2002).

### 4.2. Algal and cyanobacterial material

Algal and cyanobacterial material in the Emigsville Member constitutes the most common and widely

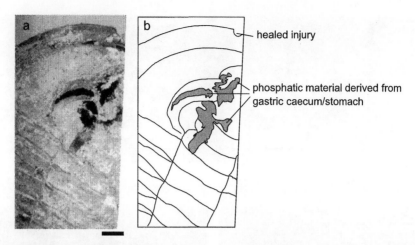

Fig. 14. (a) Specimen of *Olenellus? crassimarginatus* with gut contents and healed injury (scale bar=1 cm); Fine Pelitic Facies, Emigsville Member. (b) Interpretive diagram of specimen.

distributed example of exceptional preservation. Examples include: the abundant filamentous cyanobacterium *Marpolia* sp.; the common cyanobacterium *Morania* sp.; the tubular chlorophytes *Yuknessia* sp. and *Margaretia* sp.; a rare, branching, conical phaeophyte comparable to the Ordovician genus *Winnipegia* (Fry, 1983); and *Dalyia*, a branching genus of uncertain affinity (Fig. 16). In life, most of these organisms developed fairly resistant organic sheaths surrounding trichomes and softer parts of the organisms, either in "falsely branching" tubular forms or frond shapes (Satterthwait, 1976; Steiner and Fatka, 1996). Erwin et al. (1994) suggested that those organisms possessing basal attachment structures

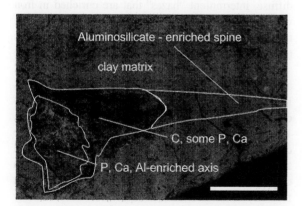

Fig. 15. Backscatter electron image of *Serracaris lineata* fragment (scale bar=1 mm); Fine Pelitic Facies, Emigsville Member.

were benthic in origin, whereas those without basal attachment structures may have been planktonic, possibly dwelling in deeper water habitats. This suggests that a number of the benthic taxa would have been transported from nearshore environments along with other taxa. However, it is also possible that many of those taxa without holdfasts dwelled on the sea floor and were adapted to low-light environments (Satterthwait, 1976).

Algae and cyanobacterial material is most commonly preserved as organic carbon films, probably derived from more resistant portions of organisms. Iron, sulfur, and calcium are present within many of the carbon films, suggesting that the organic carbon has acted as a preservative in diagenetic processes. Early-stage pyrite, replaced by goethite or completely removed from most specimens from the Emigsville Member, appears to have been at least partly retained in these structures. Some of this pyrite may represent small-scale pyritic infilling of cavities present within some of the filaments and tubes of *Marpolia* and other algal and cyanobacterial material. This agrees with the work of Satterthwait (1976), who reported similar structures from the Wheeler Shale. In some cases, iron-enriched films are present in conjunction with the tangential portions of algal and cyanobacterial remains. These occurrences likely represent microenvironments bearing microbial communities that facilitated precipitation of pyrite and its iron monosulfide precursors in

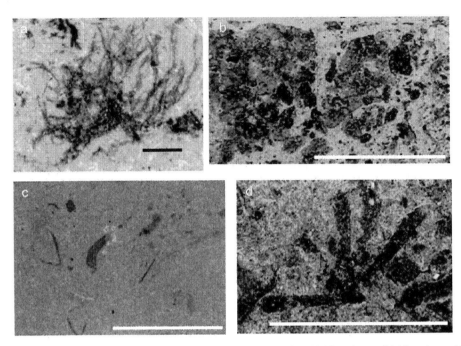

Fig. 16. Some types of algal and cyanobacterial organisms from the Emigsville Member; (a) *Marpolia* sp., (b) *Morania* sp., (c) *Dalyia* sp., (d) cf. *Winnipegia* sp. (scale bar=1 mm); Fine Pelitic Facies, Emigsville Member.

these microenvironments (see Borkow and Babcock, 2003).

### 4.3. Mollusks, hyoliths, and brachiopods

Small, shelled organisms in the Emigsville Member include the primitive gastropod *Pelagiella*, hyoliths such as *Haplophrentis*?, and some brachiopods (e.g., *Paterina*). These organisms, despite differing shell compositions and body plans, show a number of commonalities in preservation. Specimens of *Paterina*, whether they are preserved as isolated valves or as complete, articulated individuals, are usually composed of phosphate minerals derived from the original calcium phosphate of the shell. Some specimens possess partial or complete iron oxide infillings. Hyoliths are preserved largely as dorsoventrally compressed impressions or clay replacements, and with relatively little relief. Isolated opercula and helens are rare, and relatively few specimens occur articulated. In some cases, the sides of the shell are enriched in iron oxides and include some phosphorous. This likely represents pyritized or phosphatized remains of some of the more resistant integumentary

structures that would have been present within the shell, such as those known from some pyritized cephalopods (see Borkow and Babcock, 2003). Remains of the primitive mollusk *Pelagiella* are preserved in a diverse number of ways. Shells are usually preserved as partly flattened impressions in clay minerals. Associated with these impressions are some iron-enriched patches that may correspond to integumentary structures. Some specimens possess diffuse, intermittent "hazes" that are enriched in iron with small amounts of phosphorous; these may be related to microbial growth around rotting specimens (Petrovich, 2001; Borkow and Babcock, 2003). Some specimens of *Pelagiella* occur in closely packed assemblages of these mollusks (Fig. 17a). Specimens in these assemblages tend to be preserved as three-dimensional steinkerns enriched in iron oxides (Fig. 17b). Iron oxides may fill the entire shells or may be confined to the innermost whorls. Closely packed specimens of similar diameter in what appears to be a quiet-water setting can be inferred to have been formerly contained inside coprolites. The formerly pyritic interiors of the shells were likely formed due to the presence of relatively isolated reducing micro-

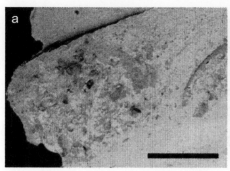
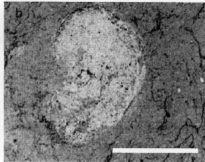

Fig. 17. (a) Probable coprolitic assemblage of *Pelagiella* sp., with associated trilobite and brachiopod material (scale bar=1 cm); Fine Pelitic Facies, Emigsville Member. (b) Close-up of *Pelagiella* sp. with iron oxide-enriched steinkern (scale bar=1 mm); Fine Pelitic Facies, Emigsville Member.

environments as has been reported in a number of organisms with interior cavities in low-oxygen, pelitic strata (see Schieber and Baird, 2001).

### 4.4. Echinoderms and Salterella

The echinoderms *Camptostroma roddyi* (Durham, 1996), *Lepidocystis wanneri* (Foerste in Resser and Howell, 1938), and *Kinzercystis durhami* (Sprinkle, 1973) are important components of the leached carbonate facies of the Emigsville Member. Absence of intact individuals from much of this lithology suggests that most of these organisms were transported after disarticulation. These are associated with remains of the enigmatic organism *Salterella ascervulosa*.

The strata in which echinoderm material and specimens of *Salterella ascervulosa* occur abundantly are the Shelly Tempestite Taphofacies and the Impure Carbonate Facies. Such environments were not conducive to exceptional preservation, perhaps due to coarse grain size, high porosity, bioturbation, and tempestite deposition. Originally calcareous material composing these organisms has been removed by diagenetic dissolution, leaving behind voids that are partly filled by iron oxides or phosphorous. These allochems were likely pyritized or phosphatized during diagenesis.

### 4.5. Tubulella pensylvanica

*Tubulella pensylvanica* (*Selkirkia pensylvanica* of Resser and Howell, 1938) is a long, thin-walled, bilayered organo-phosphatic tube of uncertain affinity bearing vague annulations and cross-striations. This organism apparently had a mud-sticking mode of life.

Few specimens of *Tubulella pensylvanica* are significantly longer than 5 mm; most specimens consist of relatively small fragments of the original test. These organisms may have been frequently scavenged or preyed upon, which would have produced abundant debris. Specimens of *T. pensylvanica* 1 cm or longer appear to be preserved in situ within matground lithologies.

*Tubulella pensylvanica* was originally described as a simple, chitinous-walled tube (Resser and Howell, 1938). SEM–EDS analysis reveals that tests are actually composed of a combination of organic and phosphatic material, consisting of an outer phosphatic-enriched layer and an inner organic-enriched layer (Fig. 18). The inner layer contains some original organic material, but is partly replaced by iron-enriched phosphate minerals. A two-layered structure in *T. pensylvanica* is a probable reason for the relatively brittle nature of these organisms when exposed to compactional forces; the relatively thin organic layer was probably somewhat flexible, whereas the more brittle phosphatic layer fractured more easily.

A taphonomic trait displayed in many specimens of *Tubulella pensylvanica* is the presence of iron-oxide-enriched material in or tangential to the interior cavity (Fig. 18). The presence of this material suggests an isolated microenvironment in which reducing microbial consortia were able to grow, promoting pyrite production (see Schieber and Baird, 2001; Borkow and Babcock, 2003). Differential preservation of specimens lends support to this hypothesis (Fig. 19).

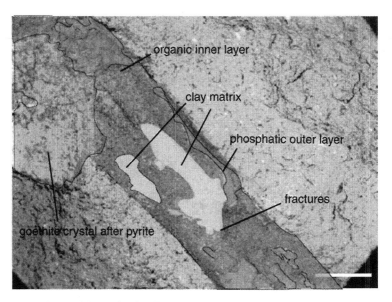

Fig. 18. Backscatter electron image of fractured *Tubulella pensylvanica* (scale bar=1 cm); Fine Pelitic Facies, Emigsville Member.

Most specimens of *T. pensylvanica* are crushed laterally and consistently associated with large (2 mm or greater) iron oxide pseudomorphs after pyrite concentrated near ruptures in the margins of the tests. Uncrushed specimens are infilled with amorphous iron oxides, similar to those inside specimens of *Pelagiella*. This suggests that prior to compaction, specimens of *T. pensylvanica* and other taxa with cavities were filled with microbial consortia and associated reactive iron-sulfide enriched material, as opposed to massive crystalline iron sulfide. At some point during compaction, the test was crushed, breaching the shell wall. The reactive material was then ejected from the test, causing the crystallization of large pyrite cubes within the adjoining matrix. Iron would have been the limiting factor in this reaction, as that the matrix had significantly more free iron than did the test interior (Canfield and Raiswell, 1991). In

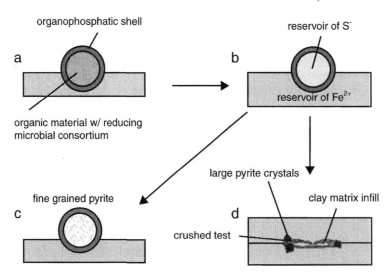

Fig. 19. Potential taphonomic pathway of *Tubulella pensylvanica*: (a) death; (b) decay; (c) burial and diagenesis without rapid compaction; (d) burial and diagenesis with rapid compaction and crushing.

unbroken specimens, the reservoir of fluid apparently remained intact and crystallized gradually and incrementally, as iron was more gradually brought into the cavity partly by microbial consortia.

### 4.6. Other mud-sticking organisms: sponges, chancelloriids, and alcyonarians

Specimens of sessile, epifaunal, mud-sticking organisms, with the exception of *Tubulella pensylvanica*, are uncommon in most facies of the Emigsville Member. Exceptions are the limited matground deposits in the Fine Pelitic Facies (Matground Taphofacies) and portions of the Massive Pelitic Facies of the Emigsville Member. Both of these intervals contain chancelloriid sclerites and other benthic mud-sticking organisms.

Chancelloriid sclerites are present in matground lithologies of the Emigsville Member. They are not articulated in a scleritome but are still in the stellar arrays typical of chancellorids (see Janussen et al., 2001). Sclerites are preserved as three-dimensional impressions with a character similar to that of many trilobite sclerites. This is consistent with an interpretation of a multilayered, organo-aragonitic composition for the sclerites (Janussen et al., 2001).

The demosponge *Hazellia walcotti* is the only well-known sponge from the Emigsville Member (Rigby, 1987), and the best examples of this taxon occur in the Massive Pelitic Facies of the Emigsville Member. Specimens are largely composed of impressions in clay minerals with no apparent associated pyritization. The mode of preservation suggests that they were originally composed of spongin that has been coalified and oxidized away as in some specimens of algae and arthropods.

A single pennatulacean-like alcyonarian was identified. It consists of a mass of zoarial fragments, composed of what was originally a tough, organic cuticle. Body plans of other members of the group suggest that these zoaria may have originally been part of a larger, frond-shaped organism that was disarticulated prior to burial but was probably not subjected to transport or significant reworking (see Wood, 1983). Alternatively, these may represent a "bloom" of small, in situ alcyonarians as opposed to a larger "supercolony." Only organic material appears to be present (i.e., no pyritization or other modes of preservation are present). The probable epifaunal attachment of the original organism suggests a relatively firm, matground-type substrate. This hypothesis is supported by association with a solitary sclerite of the chancellorid *Allonia* sp.

### 4.7. Vermiform taxa

Various examples of nonmineralizing vermiform taxa have been reported from the Emigsville Member. Two examples are *Kinzeria crinita* and *Atalotaenia adela* (Capdevilla and Conway Morris, 1999), preserved as thin films of clay mineral composition. Other vermiform taxa include a specimen of *Selkirkia* (Resser and Howell, 1938; Conway Morris, 1977) and poorly preserved palaeoscolecid worms. Some worms were originally assigned to the priapaulid genus *Ottoia* (Resser and Howell, 1938) but subsequently reassigned as palaeoscolecids (Conway Morris, 1977; Capdevilla and Conway Morris, 1999). These palaeoscolecid remains tend to be poorly preserved, and have little detail, although some retain cuticular annulations and putative gut remains (Conway Morris, 1977). New specimens consist of well-preserved cuticular fragments that exhibit rows of nodes typical of palaeoscolecids (Conway Morris and Robison, 1986). The remains are phosphatic in composition, which is consistent with the preservation of these organisms in other Cambrian deposits (Müeller, 1993; Zhang and Pratt, 1996; Zhu et al., 2004).

### 4.8. Microbial consortia

Evidence of microbial consortia on decaying fossil remains in the Emigsville Member is present in the form of small (1–2 μm) clusters of elliptical to rod-shaped bacterial remains (Fig. 20a–b). These have been found in association with specimens of the cyanobacteria *Marpolia* bearing iron-oxide halos. These structures are reminiscent of bacterial cells reported from pyritized Devonian material (Borkow and Babcock, 2003). Similarly, bacterial remains in the Emigsville Member may have been autolithified at an early stage, leading to the precipitation of iron sulfides and the preservation of nonmineralized material. This scenario supports the presence of isolated, oxygen-deficient micro-

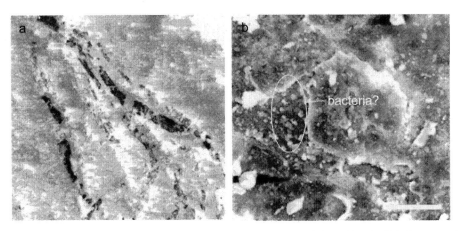

Fig. 20. (a) Specimen of *Marpolia* showing microbial-induced iron oxide overgrowth (scale bar=100 μm). (b) Enlargement showing probable autolithified bacteria (scale bar=10 μm); Fine Pelitic Facies, Emigsville Member.

environments in which microbial consortia were able to reproduce.

## 5. Discussion

In this section, taphonomic data detailed above are compared with conditions expected for exceptional preservation under the influence of anoxia, salinity fluctuations, and obrution (see Babcock et al., 2001). This constitutes a test of whether one or more of these environmental factors are consistent with inferred depositional conditions associated with exceptional preservation in the Emigsville Member. In addition, the importance of predation, hypothesized to be the primary taphonomic filter (Babcock, 2003), is examined in the Emigsville Member.

### 5.1. Anoxia

Anoxia, or low-oxygen conditions, apparently played an important role in the exceptional preservation of organisms in the Emigsville Member. The magnitude of anoxic conditions, however, may have been limited to microenvironments, as there is little evidence present for widespread, long-term anoxia.

Anoxia has been proposed as prerequisite for the exceptional preservation of organisms in Cambrian strata (e.g., Conway Morris, 1985; Allison and Briggs, 1993; Butterfield, 1995; Capdevilla and Conway Morris, 1999). One hypothesis for Cambrian excep-

tional preservation involves multiple intervals of seafloor anoxia generated by regional or global oceanic cycling and the configuration of Cambrian continental shelves (Landing, 2000). Testing whether the anoxia scenario is applicable to a specific deposit requires determining whether evidence for anoxic bottom waters is present.

An anoxic depositional system is expected to have a distinctive sedimentary, biological, and chemical signature (Zangerl and Richardson, 1963; Savrda et al., 1984; Conway Morris, 1985). Fine, organic-rich pelitic sediments deposited under anoxic conditions typically contain abundant, small pyrite framboids (<18 μm) inferred to have formed in open, anoxic water (Wilkin and Arthur, 2001). Bioturbation should be absent or restricted to a few bedding-plane associations representing times of minor oxygenation (Crimes and Fedonkin, 1994; Droser et al., 2001, 2002; Schieber, 2003). Fossils preserved in black shales are expected to be largely allocthonous, derived from pelagic organisms in the overlying water column, from the predatory debris of nektonic organisms, or from organisms transported by currents (Zangerl and Richardson, 1963; Conway Morris, 1985; Kottachchi, 2001; Gaines and Droser, 2002).

The Emigsville Member of the Kinzers Formation was most likely deposited under exaerobic conditions (Savrda et al., 1984). Strata of the Fine Pelitic Facies of the Emigsville Member consist largely of fine-grained clay minerals; the matrix ranges from light to medium gray to tan, and chemical analysis rarely shows significant carbon content, suggesting that the majority of lithologies in the Fine Pelitic Facies is not carbon-enriched shales The Massive Pelitic Facies and the Impure Carbonate Facies contain silt grains and carbonates not typically present in anoxic deposits In Facies I, pyrite is largely restricted to large (0 1 mm or larger) crystals and spheroids, while true framboids are absent from analyzed samples Pyrite grains, associated with layers bearing more abundant organic material, are relatively abundant, suggesting the importance of reducing microbial consortia in pyrite distribution rather than anoxia Bioturbation is restricted to sparse, bedding-plane-parallel burrows in most of the deposit

Paleoecological information derived from the Emigsville Member also suggests that restrictive, low-oxygen sediment conditions were the norm Such conditions would have been required for the preservation of autolithified bacterial microenvironments Little evidence of buried mobile benthic organisms is present in most samples from the Fine Tempestite Taphofacies of the Fine Pelitic Facies, as remains consist largely of disarticulated trilobite sclerites, small-shelled organisms, and other inferred predatory debris The distribution of algal and cyanobacterial material, concentrated on some bedding planes and rare to absent on others, is consistent with an allochthonous, possibly pelagic, origin Differing biotic content in the Matground Taphofacies may be explained by the decreased sedimentation rates over this interval, which allowed for colonization by epifaunal taxa The presence of this community suggests that oxygen levels were sufficient to support an in situ restrictive metazoan community of moderate diversity, excluding the possibility of true anoxic conditions over this interval.

## 5 2 Salinity fluctuations

Fluctuations in salinity have been proposed as a mechanism to explain the abundance of exceptionally preserved organisms in the Chengjiang Biota and other Cambrian deposits of exceptional preservation (St John, 1999, Babcock et al, 2001) The model suggests that high frequency and high magnitude tidal fluctuations during the Cambrian caused fluctuations in sea water salinity in some shelf environments during the Cambrian This condition of fluctuating salinity would have temporarily excluded some biodegrading organisms, promoting exceptional preservation Similar modes of preservation may have been at work in some deposits of Late Paleozoic age (see Babcock et al, 2000)

Direct evidence indicating deposition in association with salinity fluctuations is problematic, since most chemical proxies for paleosalinity, such as boron content, are unstable in rocks exposed to metamorphism typical of Cambrian strata (Kloppmann et al, 2001) Instead, indirect evidence for salinity fluctuations is examined, in the form of tidal depositional patterns and presence or absence of taxonomic groups sensitive to such fluctuations (Babcock et al, 2001) Tidal depositional patterns are typically represented by rhythmic sedimentation with a periodic modulating pattern associated with changes in daily tide duration (Feldman et al, 1993, Babcock et al, 2001)

In strata of the Emigsville Member, the Massive Pelitic Facies shows the greatest potential of having been deposited under the influence of tidally influenced salinity modulation or salinity stressed conditions Inferred tidal laminae (see Feldman et al, 1993, Babcock et al, 2001) are present in the lowermost portions of the Massive Pelitic Facies Organisms indigenous to the Massive Pelitic Facies further suggest a salinity-stressed environment A low-diversity community, consisting of olenelloid trilobites, anomalocaridids, and nonmineralizing bivalved crustaceans, is interpreted to have been resistant to salinity fluctuations (Chulpac and Kordule, 2002) Notably absent are remains of echinoderms, mollusks, and most other sessile benthic organisms in these intervals

Although the biotic content and sedimentary characteristics of the Massive Pelitic Facies of the Emigsville Member suggest salinity fluctuations at various intervals during sedimentation, exceptionally preserved remains are significantly less abundant in this interval than in other lithologies of the Emigsville Member Thus, no direct comparison with the salinity-driven preservation in the Chengjiang Biota is present in the Emigsville Member, despite the probable presence of tidally generated sedimentation and restrictive depositional environments

*5 3 Obrution*

The third main model for exceptional preservation in the Cambrian is obrution, or the sudden blanketing and smothering of a benthic community (Seilacher et al , 1985, Robison, 1991, Liddell et al , 1997, Babcock et al , 2001) The sediment blanket must be of sufficient thickness to smother any non-mobile and non-burrowing organisms present The magnitude and frequency of obrution must be rare enough that the environment does not become inhospitable to organisms in the long term One cause of obrution is tempestite deposition generated by differential storm flows (Brett and Seilacher, 1991)

The application of obrution to exceptional preservation in Cambrian deposits is tied to the relatively poor ability of some Cambrian organisms to burrow through the sediments Reduced ability of organisms to burrow during the Cambrian, as compared to the later Phanerozoic, suggests that both aspects of obrution –smothering and protection of organic remains – would have been enhanced in Cambrian deposits (Droser and Bottjer, 1990)

In the Emigsville Member, the best candidates for obrution deposits are in the Impure Carbonate Facies, which contain articulated, in situ echinoderms The frequency of articulated echinoids throughout the unit suggests that burial of in situ organisms was a fairly regular occurrence and likely associated with both relatively shallow bathymetry, and proximity to an incipient carbonate buildup

*5 4 "Normal" pulsed sedimentation implications for Cambrian exceptional preservation*

Exceptional preservation in the Emigsville Member is largely limited to fragmentary remains in incrementally deposited, low oxygen sediments (Fine Tempestite Taphofacies and Matground Taphofacies). The scarcity of large, intact arthropods other than trilobites strongly suggests that most arthropod remains consist of predatory debris Small-shelled organisms such as *Pelagiella* and *Haplophrentis* were likely indigenous to this environment as detritus-feeders, but are also concentrated in coprolites. Other organisms are mainly restricted to fragmentary material

The presence of living organisms and debris left behind by predators in the Fine Tempestite Facies, in addition to the microenvironment-restricted distribution of pyrite grains and bedding-plane-parallel bioturbation, supports the presence of oxygen at the sediment–water interface This is consistent with an exaerobic depositional environment (Savrda et al , 1991, Powell et al , 2003) Organic remains preserved in this interval represent organisms that were sessile, slow-moving, or dead prior to obrution under normal marine conditions

In Cambrian fine-grained sediments, buried organic remains had a significantly higher chance of being exceptionally preserved, as clay laminae were "sealed" to one another, inhibiting the oxidation of organic material and promoting the development of isolated microenvironments and their microbial consortia (Gaines and Droser, 2002) In later deposits, exceptionally preserved fossils are largely allocthonous and are found only in highly restrictive environments with conditions inimical to biodegraders or other restrictive conditions (Conway Morris, 1985)

*5 5 Predation the primary taphonomic filter in Cambrian deposits*

The Emigsville Member demonstrates the profound effects of predation on Cambrian ecosystems (Babcock, 2003) The presence of macerated arthropod fragments, trilobites with healed injuries, coprolitic collections of *Pelagiella* and other organisms, and remains of the predators themselves is evidence for a complex trophic structure with a variety of modes of predation Emigsville Member lithologies show evidence of a mixture of anoxia, salinity fluctuations, and obrution, although none of these modes of deposition was pervasive While the absence of bioturbation in fine-grained strata was largely responsible for exceptional preservation of articulated organisms in Cambrian strata, predation intensity is inferred to be the primary taphonomic process responsible for the differences between the Emigsville Member and other DEPs containing abundant articulated organisms Predators are opportunistic yet sensitive to environmental change, mobile predators can be expected to avoid depositional environments with inimical conditions Thus, sedimentary anoxia, salinity fluctuations, and high sedimentation rates leading to sedimentary obrution would have controlled the presence and intensity of predation

Exceptional preservation of abundant articulated organisms may be interpreted as due largely to the absence both of carnivores and bioturbation Without the actions of predators and scavengers, complete, articulated exceptionally preserved organisms would have been the norm Instead, deposits like the Emigsville Member record the presence of exceptionally preserved organisms in depositional systems without obvious "special" depositional conditions

## 6. Conclusions

The Emigsville Member of the Cambrian Kinzers Formation contains abundant, fragmentary, exceptionally preserved remains in the Fine Pelitic Facies This facies contains evidence that anoxic sediment conditions, salinity fluctuations, and obrution through differential sedimentary input all had some role in exceptional preservation The presence of exaerobic conditions in the sediment is supported by the presence of autolithified bacteria involved in the growth of pyritic microenvironments The fragmentary nature of fossil evidence suggests the presence of an active benthic community with carnivores

The abundance of predation evidence in the Emigsville Member reinforces the idea that predation was the primary filter in Cambrian Taphonomic systems While more restrictive depositional conditions might have limited predators, allowing for preservation of whole articulated organisms, the Fine Pelitic Facies instead consists almost completely of debris derived from predation This suggests that exceptional preservation in the Fine Pelitic Facies of the Emigsville Member did not require exceptional depositional conditions to have exceptionally preserved remains

## Acknowledgements

I thank L E Babcock, S M Bergstrom, G Faure, M R Saltzman, L A Krissek, J Taylor, D U Wise, and E L Yochelson for helpful discussions Thanks to R D K. Thomas for arranging loans of specimens from the collection of Franklin and Marshall College Thanks to S Bhattiprolu and the MARC at Ohio State for helpful assistance with SEM–EDS Thanks to the Friends of Orton Hall Fund for help with funding my research This work was supported in part by grants from the National Science Foundation (EAR-0177229, EAR OPP-0229757) to L E Babcock

## References

Allison, P A , Briggs, D E G , 1991 The taphonomy of soft-bodied animals In Donovan, S K (Ed ), The Processes of Fossilization Columbia University Press, New York, pp 120–140

Allison, P A , Briggs, D E G , 1993 Burgess shale biotas, burrowed away? Lethaia 26, 225–229

Allison, P A , Brett, C E , Liddell, W D , Wright, E , 1995 Repetitive tapho, bio- and lithofacies, within the Middle Cambrian Swazey Limestone–Wheeler Shale transition of Utah Abstr Programs-Geol Soc Am 27 (6), 374

Ausich, W I , 2001 Echinoderm taphonomy In Jangoux, M , Lawrence, J M (Eds ), Echinoderm Studies, vol 6 A A Balkema, Rotterdam, pp 171–227

Babcock, L E , 1993 Trilobite malformations and the fossil record of behavioral asymmetry J Paleontol 67, 217–229

Babcock, L E , 1998 Experimental investigation of the processes of fossilization J Geosci Educ 46, 252–260

Babcock, L E , 2003 Trilobites in Paleozoic predator–prey systems, and their role in reorganization of early Paleozoic ecosystems In Kelley, P H , Kowalewski, M , Hansen, T H (Eds ), Predator–Prey Interactions in the Fossil Record Kluwer Academic/Plenum Publishers, New York, pp 55–92

Babcock, L E , Peel, J S , 2002 Anatomy, paleoecology, and taphonomy of the trilobite Buenellus from the Sirius Passet biota (Cambrian), North Greenland Abstr Programs-Geol Soc Am 34 (6), 171

Babcock, L E , Zhang, W , 1997 Comparative taphonomy of two nonmineralized arthropods Naraoia (Nektaspida, Early Cambrian, Chengjiang Biota, China) and Limulus (Xiphosurida, Holocene, Atlantic Ocean) Bull Natl Mus Nat Sci 10, 233–250

Babcock, L E , Merriam, D H , West, R R , 2000 Paleolimulus, an early limuline (Xiphosurida) from Pennsylvania–Permian Lagerstatten of Kansas and taphonomic comparison with modern Limulus Lethaia 33, 129–141

Babcock, L E , Zhang, W , Leslie, S A , 2001 The Chengjiang biota record of the Early Cambrian diversification of life and clues to exceptional preservation of fossils GSA Today 11 (2), 4–9

Barnaby, R J , Read, J F. 1990 Carbonate ramp to rimmed shelf evolution, lower to Middle Cambrian continental margin, Virginia Appalachians Geol Soc Amer Bull 102, 391–404

Blaker, M R , Peel, J S , 1997 Lower Cambrian trilobites from North Greenland Medd Gronl , Geosci 35, 1–145

Borkow, P S, Babcock, L E, 2003 Turning pyrite concretions outside-in role of biofilms in pyritization of fossils Sediment Record 1 (3), 4–7

Brasier, M D, Lindsay, J F, 2001 Did supercontinental amalgamation trigger the "Cambrian Explosion"? In Yu, A, Riding, R, Bottjer, D J, Zhuraviev, A Y (Eds), The Ecology of the Cambrian Radiation Columbia University Press, New York, pp 69–89

Brett, C E, Baird, G C, 1986 Comparative taphonomy a key to paleoenvironmental interpretation based on fossil preservation Palaios 1, 207–227

Brett, C E, Seilacher, A, 1991 Fossil lagerstaetten, a taphonomic consequence of event sedimentation In Einsele, G, Ricken, W, Seilacher, A (Eds), Cycles and Events in Stratigraphy Springer-Verlag, Berline, pp 283–297

Briggs, D E G, 1978 A new trilobite-like arthropod from the Lower Cambrian Kinzers Formation, Pennsylvania J Paleontol 52, 132–140

Briggs, D E G, Kear, A J, Martill, D M, Wilby, P R, 1993 Phosphatization of soft-tissue in experiments and fossils J Geol Soc London 150, 1035–1038

Butterfield, N J, 1995 Secular distribution of Burgess Shale-type preservation Lethaia 28, 1–13

Butterfield, N J, 2002 Leancholia guts and the interpretation of three-dimensional structures in Burgess Shale-type fossils Paleobiology 28, 155–171

Campbell, L D, 1969 Stratigraphy and paleontology of the Kinzers Formation, southeastern Pennsylvania MS thesis, Franklin & Marshall College, Lancaster, Pennsylvania, 51 pp

Campbell, L D, 1970 Occurrence of "Ogygopsis shale" fauna in southeastern Pennsylvania J Paleontol 45, 437–440

Campbell, L D, Kauffman, M E, 1969 Olenellus fauna of the Kinzers Formation, southeastern Pennsylvania Proc Pa Acad Sci 43, 172–176

Canfield, D E, Raiswell, R, 1991 Pyrite formation and fossil preservation In Allison, P A, Briggs, D E G (Eds), Taphonomy, Releasing the Data Locked in the Fossil Record, Topics in Geobiology, vol 9, pp 25–70

Capdevilla, D G, Conway Morris, S, 1999 New fossil worms from the Lower Cambrian of the Kinzers Formation, Pennsylvania, with some comments on Burgess Shale-type preservation J Paleontol 73, 394–402

Chen, J, Eidtmann, B D, Hou, X, 1989 New soft-bodied fossil fauna near the base of the Cambrian system at Chengjiang, eastern Yunnan, China In Zengquan, Li, Teichert, C, Shu, Sun (Eds), Developments in Geoscience Contributions to the 28th International Geological Congress Beijing Sci Press, pp 265–278

Chulpac I, Kordule, V, 2002 Arthropods of Burgess Shale type from the Middle Cambrian of Bohemia, Czech Republic Vestn Èes Geol Úst 77 (3), 167–182

Conway Morris, S, 1977 Fossil priapaulid worms Spec Pap Palaeontol 20, 1–95

Conway Morris, S, 1979 The burgess shale (Middle Cambrian) fauna Ann Rev Ecolog Syst 10, 327–349

Conway Morris, S, 1985 Cambrian lagerstatten their distribution and significance Philos Trans R Soc Lond, B Biol Sci 311, 49–65

Conway Morris, S, Robison, R A, 1986 Middle Cambrian priapulids and other soft-bodied fossils from Utah and Spain Univ Kans Paleontol Contrib. Pap 117, 1–22

Conway Morris, S, Robison, R A, 1988 More soft-bodied animals and algae from the Middle Cambrian of Utah and British Columbia Univ Kans Paleontol Contrib, Pap 122, 1–48

Conway Morris, S, Whittington, H B, 1985 Fossils of the burgess shale a national treasure in Yoho National Park, British Columbia Misc Rep -Geol Surv Can 43, 1–31

Conway Morris, S, Davis, N C, Higgins, A K, Peel, J S, Soper, N J, 1987 A burgess shale-like fauna from the Lower Cambrian of North Greenland Nature 326, 181–183

Crimes, T P, Fedonkin, M A, 1994 Evolution and dispersal of deep-sea traces Palaios 9, 74–83

Dean, W E, Anderson, R Y, Davies, G R, Graham, R, Loucks, R G, Robert, G, 1982 Continuous subaqueous deposition of the Permian Castile evaporates, Delaware basin, Texas and New Mexico SEPM Core Workshop, Special Issue Depositional and Diagenetic Spectra of Evaporites, Core Workshop, vol 3, pp 324–353

Dornbos, S Q, Bottjer, D J, 2001 Taphonomy and environmental distribution of helicoplacoid echinoderms Palaios 16, 197–204

Droser, M L, Bottjer, D J, 1990 Depth and extent of early Paleozoic bioturbation In Miller III, W (Ed), Paleocommunity Temporal Dynamics the Long-Term Development of Multispecies Assemblies. Paleontological Society Special Publication, vol 5, pp 153–165

Droser, M L, Jensen, S, Gehling, J G, 2001 Cambrian firm muddy substrates significance for the preservation of trace fossils Abstr Programs-Geol Soc Am 33 (6), 430

Droser M L, Jensen, S, Gehling, J G, Myrow, P M, Narbonne, G M, 2002 Lowermost Cambrian ichnofabrics from the Chapel Island Formation, Newfoundland implications of Cambrian substrates Palaios 17, 3–15

Dunbar, C O, 1925 Antennae in Olenellus getzi n sp Am J Sci, Fifth Series 5. 303–308

Durham, J W, 1996 Camptostroma, an Early Cambrian supposed scyphozoan, referable to echinodermata J Paleontol 40, 1216–1220

Erwin, D H, Collier, F J, Briggs, D E G, 1994 The Fossils of the Burgess Shale Smithsonian University Press, Washington, DC 238 pp

Feldman, H R, Archer, A W, Kvale, E P, Cunningham, C R, Maples, C G, West, R R, 1993 A tidal model of Carboniferous Konservat–Lagerstattten formation Palaios 8, 485–498

Fry, W L, 1983 An algal flora from the Upper Ordovician of the Lake Winnipeg region, Manitoba, Canada Rev Palaeobot Palynol 39, 313–341

Gaines, R R. 1999 Were Burgess Shale faunas burrowed away? Evidence from the Lower Cambrian Latham Shale, San Bernardino County, California PaleoBios 19 (1), 5–6 (Suppl)

Gaines, R R, Droser, M L, 2002 Bottom water oxygen content and proximal/distal control over paleontological assemblages in the Wheeler Shale, House Range, Utah PaleoBios 22 (1), 3–4 (Suppl)

Gaines, R R , Droser, M L , 2003 Paleoecology of the familiar trilobite *Elrathia kingii* an early exaerobic zone inhabitant Geology 31, 941–944

Ganis, G R , Hopkins, D , 1990 The West York block stratigraphic and structural setting In Scharnberger, C K (Ed ), Carbonates, Schists, and Geomorphology in the Vicinity of the Susquehanna River Guidebook for the 55th Annual Field Conference of Pennsylvania Geologists, pp 123–135

Gohn, G S , 1976 Sedimentology, stratigraphy, and paleogeography of Lower Paleozoic Carbonate Rocks, Conestoga Valley, Southeastern Pennsylvania Unpublished PhD thesis, University of Delaware, 315 pp

Hou, X , Bergstrom, J , Wang, H , Feng, X , Chen, A , 1999 The Chengjiang Fauna Exceptionally Well-Preserved Animals from 530 Million Years Ago Yunnan Science and Technology Press 170 pp

Janussen, D , Steiner, M , Zhu, M , 2001 New well-preserved scleritomes of Chancelloriidae from the Early Cambrian Yuanshan Formation (Chengjiang, China) and the Middle Cambrian Wheeler Shale (Utah, USA) and paleobiological implications J Paleontol 76, 596–606

Kloppmann, W , Casanova, J , Guerrot, C , Klinge, H , Negrel, P , Schelkes, K , 2001 Halite dissolution derived brines in the vicinity of a Permian salt dome (N German Basin), evidence from boron, strontium, oxygen, and hydrogen isotopes Geochim Cosmochim Acta 65, 4087–4101

Kottachchi, N , 2001 Fossils of the Middle Cambrian Burgess Shale, British Columbia, Canada, distribution and biostratinomic change through time Open File Rep -Geol Surv Can 4121, 1–143

Landing, E , 2000 Lower Paleozoic of the East Laurentian continental slope, eustatic/climate controls on macro- and microscale mudstone alternations Abstr Programs-Geol Soc Am 33 (1), 16

Landing, E , 2001 "Burgess biotas" and episodic slope and epeiric sea dysaerobia in the late Precambrian–Paleozoic Abstr Programs-Geol Soc Am 33 (6), 38

Liddell, W D , Wright, S H , Brett, C E , 1997 Sequence stratigraphy and paleoecology of the Middle Cambrian Spence Shale in northern Utah and southern Idaho Brigh Young Univ Geol Stud 42, 59–78

MacLachlan, D B , 1994 Some aspects of the lower Paleozoic Laurentian margin and slope in southeastern Pennsylvania In Faill, R T , Sevon, W D (Eds ), Various Aspects of Piedmont Geology in Lancaster and Chester Counties, Pennsylvania Guidebook for the 59th Annual Field Conference of Pennsylvania Geologists, pp 3–23

Martin, D , Briggs, D E G , Parkes, R J , 2003 Experimental mineralization of invertebrate eggs and the preservation of Neoproterozoic embryos Geology 31, 39–42

Meyers, J H , 1967 Clay mineralogy and illite polymorphism in the Lower Cambrian Kinzers Shale, Pennsylvania Piedmont BA thesis, Franklin and Marshall College, Lancaster, Pennsylvania, 76 pp

Mueller, K J , 1993 Palaeoscolecid worms from the Middle Cambrian of Australia Palaeontology 36, 549–592

Orr, P J , Kearns, S L , Briggs, D E G , 2002 Backscattered electron imaging of fossils exceptionally-preserved as organic compressions Palaios 17, 110–117

Petrovich R , 2001 Mechanisms of fossilization of the soft-bodied and lightly armored faunas of the Burgess Shale and of some other classical localities Am J Sci 301, 683–762

Powell, W G , Collom, C J , Johnston, P A , 2003 Geochemical evidence for oxygenated bottom waters during deposition of fossiliferous strata of the Burgess Shale Formation Palaeogeogr Palaeoclimatol Palaeoecol 201, 249–268

Pratt, B R , 1998 Probable predation on Upper Cambrian trilobites and its relevance for the extinction of soft-bodied Burgess Shale-type animals Lethaia 31, 73–88

Ressel, C E , Howell, B F , 1938 Lower Cambrian *Olenellus* zone of the Appalachians Geol Soc Amer Bull 49, 195–248

Rigby, J K , 1987 Early Cambrian sponges from Vermont and Pennsylvania, the only ones described from North America J Paleontol 61, 451–461

Robison, R A , 1991 Middle Cambrian biotic diversity examples from four Utah lagerstatten In Simonetta, A M , Conway Morris, S (Eds ), The Early Evolution of the Metazoa and the Significance of Problematic Taxa Cambridge University Press, Cambridge, pp 77–98

Satterthwait, D F , 1976 Paleobiology and paleoecology of Middle Cambrian algae from western North America Unpublished PhD thesis, University of California, Los Angeles, 134 pp

Savrda, C E , Bottjer, D J , Gorsline, D S , 1984 Development of a comprehensive oxygen-deficient marine biofacies model evidence from Santa Monica, San Pedro, and Santa Barbara Basins, California Continental Borderland Am Assoc Pet Geol Bull 68, 179–1192

Savrda, C E , Bottjer, D J , Seilacher, A , 1991 Redox-related benthic events In Einsele, G , Ricken, W , Seilacher, A (Eds ), Cycles and Events in Stratigraphy Springer-Verlag, pp 524–541 Chap 5 3

Schieber, J , 2003 Simple gifts and buried treasures—implications of finding bioturbation and erosion surfaces in black shales Sediment Record 1 (2), 4–8

Schieber, J , Baird, G , 2001 On the origin and significance of pyrite spheres in Devonian black shales of North America J Sediment Res 71, 155–166

Seilacher, A , Reif, W E , Westphal, F , 1985 Sedimentological, ecological, and temporal patterns of fossil Lagerstatten Philos Trans R Soc Lond , B 311, 5–23

Shu, D , Vannier, J , Luo, H , Chen, L , Zhang, X , Xu, S , 1999 Anatomy and lifestyle of *Kunmingella* (Arthropoda, Bradorrida) from the Chengjiang fossil Lagerstatte (Lower Cambrian, southwest China) Lethaia 32, 279–298

Skinner, E S , Babcock, L E , Peel, J S , 2002 Taphonomy of some exceptionally preserved organisms from the Lower Cambrian of Laurentia Abstr Programs-Geol Soc Am 34 (6), 171

Sprinkle, J , 1973 Morphology and evolution of blastozoan echinoderms Special Publication, Museum of Comparative Zoology Harvard University, Cambridge 284 pp

St John, J M , 1999 Abundance of Cambrian Lagerstatten, links among geophysical, astronomical, and sedimentological controls on exceptional preservation Abstr Programs-Geol Soc Am 31 (5), 73

Staub, J R , Among, H L , Gastaldo, R A , 2000 Seasonal sediment transport and deposition in the Rajang River delta, Sarawak, east Malaysia Sediment Geol 133, 249–264

Steiner, M , Fatka, O , 1996 Lower Cambrian tubular micro- to macrofossils from the Paseky Shale of Barrandian area (Czech Republic) Palaontol Z 70, 275–299

Stose, G W , Jonas A I , 1922 The Lower Paleozoic section of southeastern Pennsylvania J Wash Acad Sci 12, 358–366

Stose, G W , Stose, A J , 1944 Geology of the Hanover–York District, Pennsylvania U S Geol Surv Prof Pap 204, 1–84

Taylor, J F , Durika, N J , 1990 Lithofacies, trilobite faunas, and correlation of the Kinzers, Ledger, and Conestoga Formations in the Conestoga Valley In Scharnberger, C K (Ed ), Carbonates, Schists, and Geomorphology in the Vicinity of the Susquehanna River Guidebook for the 55th Annual Field Conference of Pennsylvania Geologists, pp 136–155

Vannier, J , Chen, J , 2002 Digestive system and feeding mode in Cambrian naraoiid arthropods Lethaia 35, 107–120

Whittington, H B , 1971 The Burgess Shale, history of research and preservation of fossils Atti-Istituto Veneto di Scienze, Lettere ed Arti Classe di Scienze Matematiche e Naturali 71 (1), 1170–1201

Whittington, H B , 1980 Exoskeleton, moult stage, appendage morphology, and habits of the Middle Cambrian trilobite *Olenoides serratus* J Paleontol 23, 171–204

Wilkin, R T , Arthur, M A , 2001 Variations in pyrite texture, sulfur isotope composition, and iron systematics in the Black Sea evidence for late Pleistocene to Holocene excursions of the $O_2$–$H_2S$ redox transition Geochim Cosmochim Acta 65, 1399–1416

Wilkin, R T , Barnes, H L , Brantley, S L , 1996 The size distribution of framboidal pyrite in modern sediments an indicator of redox conditions Geochim Cosmochim Acta 60, 3897–3912

Wood, E M , 1983 Corals of the World T F H Publications 256 pp

Zangerl, R , Richardson Jr , E S , 1963 The paleoecological history of two Devonian black shales Fieldiana, Geol Mem 4, 1–352

Zhang, W , 1987 Early Cambrian Chengjiang fauna with emphasis on the trilobites Acta Palaeontol Sin 26, 223–235

Zhang, X , Pratt, B R , 1996 Early Cambrian palaeoscolecid cuticles from Shaanxi, China J Paleontol 70, 275–279

Zhu, M , Babcock, L E , Steiner, M , 2004 Fossilization modes in the Chengjiang Lagerstätte (Cambrian of China) testing the roles of organic preservation and diagenetic alteration in exceptional preservation Palaeogeogr Palaeoclimatol Palaeoecol 220, 31–46

Available online at www.sciencedirect.com

Palaeogeography, Palaeoclimatology, Palaeoecology 220 (2005) 193–205

www.elsevier.com/locate/palaeo

# A new hypothesis for organic preservation of Burgess Shale taxa in the middle Cambrian Wheeler Formation, House Range, Utah

Robert R. Gaines[a,*], Martin J. Kennedy[b], Mary L. Droser[b]

[a]Geology Department, Pomona College, Claremont, CA 91711, USA
[b]Department of Earth Sciences-036, University of California, Riverside, CA 92521, USA

Received 15 March 2002, accepted 15 July 2004

## Abstract

Cambrian *konservat-lagerstatten* are the most significant fossil deposits for our understanding of the initiation of Phanerozoic life. Although many modes of preservation may occur, these deposits most frequently contain nonmineralized fossils preserved in the form of kerogenized carbon films, a rare yet important taphonomic pathway that has not previously been explained for any unit by a comprehensive model. The middle Cambrian Wheeler Formation of Utah, one of these *lagerstatten*, contains abundant kerogenized preservation of nonmineralized tissues, which occurs within a distinctive taphofacies that accumulated under the following conditions (1) domination of the siliciclastic fraction by clay-sized particles, (2) close proximity to a carbonate platform, which resulted in mixed carbonate-clay sediments, (3) a well-developed oxygen minimum precluding benthic colonization and burrowing, and (4) relative proximity to oxic bottom-waters, facilitating transport of organisms from a habitable environment to one that favored their preservation. We propose that preservation of nonmineralized tissues in the Wheeler Formation may have resulted from a combination of influences that reduced permeability and, thus, lowered oxidant flux, which in turn may have restricted microbial decomposition of some nonmineralized tissues. Those influences include near bottom anoxia, preventing sediment irrigation by restriction of bioturbation, reducing conditions near the sediment–water interface that may have acted to deflocculate aggregations of clay minerals, resulting in low permeability face-to-face contacts, early diagenetic pore occluding carbonate cements, and an absence of coarse grains such as silt, skeletonized microfossils, fecal pellets, or bioclasts. This model may be applicable to kerogenized preservation of macrofossils in other fossil *lagerstatten*.

*Keywords* Cambrian, *Lagerstatten*, Wheeler Formation, Burgess Shale, Taphonomy, Diagenesis

## 1. Introduction

Fossilization of nonmineralized tissues provides unparalleled anatomical and ecological information (Allison and Briggs, 1991) The celebrated fauna of

* Corresponding author
  *E-mail address* rgaines@pomona.edu (R.R. Gaines)

0031-0182/$ - see front matter © 2004 Elsevier B.V. All rights reserved
doi 10 1016/j palaeo 2004 07 034

the Burgess Shale contains more than 40 genera of nonmineralizing organisms, which were distributed globally during early and middle Cambrian time (Conway Morris, 1989) Although very uncommon, Cambrian deposits bearing preservation of nonmineralized tissues are more abundant than those of any other geologic period, even when normalized for outcrop area and time (Allison and Briggs, 1993a) Cambrian nonmineralized fossils are preserved in many taphonomic modes, the most significant of which is "Burgess Shale-type preservation", defined by Butterfield (1995) as the fossilization of non-mineralizing organisms as kerogenized organic carbon films under fully marine conditions While spectacular examples of preservation via mineralization of originally non-biomineralized tissues are known from Cambrian deposits (e g, Briggs and Nedin, 1997), Burgess Shale-type preservation is the most common mode of nonmineralized tissue preservation in lower and middle Cambrian *lagerstatten* worldwide (Butterfield, 1990) Despite considerable inquiry into the depositional environments of these deposits (e g, Allison and Brett, 1995, Babcock et al, 2001), the conditions that promoted Burgess Shale-type preservation remain poorly understood In this paper, we investigate the circumstances of this important taphonomic pathway in the middle Cambrian Wheeler Formation of Utah

Anoxia is frequently invoked as a causal agent in preservation of nonmineralized tissues However, although it is strongly correlated with preservation of nonmineralized tissues, anoxia alone does not inhibit decomposition (e g, Allison, 1988, Butterfield, 1990, 1995) Efforts to understand other physical factors that may have been, in part, responsible for the abundance of preservation of nonmineralized tissues in the Cambrian have focused on two related aspects of the problem specific taphonomic pathways that led to the preservation of nonmineralized fossils, and the secular trend in preservation of nonmineralized tissues over the Phanerozoic Butterfield (1995) argued that secular variation in dominant clay mineralogy resulted in temporal optima for the stabilization of organic material through absorption of decay-inducing enzymes onto the surfaces of particularly reactive clay minerals In this view, the early and middle Cambrian represented a fortuitous preservational window when conditions favoring nonmin-

eralized fossil preservation were maximized This idea was not supported by a metamorphic study of the Burgess Shale (Powell, 2003), which found that the original clay mineralogy of the Burgess Shale was neither unusual nor rich in highly reactive clay minerals Wollanke and Zimmerle (1990) also noted a correlation between preservation of nonmineralized tissues and clay-rich, fine-grained sediments, and suggested that burial in such sediments was an important prerequisite for preservation of nonmineralized tissues A recent conceptual model (Petrovich, 2001) suggested that under "sub-oxic" conditions, $Fe^{2+}$ bound to the surfaces of chitin and other organic biopolymers, inhibiting the ability of bacterial enzymes These hypotheses are difficult to test with available physical evidence

The most commonly cited hypothesis addressing the secular trend suggests that the likelihood of preservation of nonmineralized tissues was greatest before the advent of bioturbation in muddy substrates, implying that Burgess Shale and like faunas were "burrowed away" from the fossil record (Allison and Briggs, 1993a,b) The advent of significant bioturbation would have had a profound impact upon basic physical properties and early diagenetic processes in marine sediments, and has been held responsible for a number of secular changes including increased nutrient cycling (McIlroy and Logan, 1999) the taphonomic loss of Ediacaran biotas (Gehling, 1999), and the loss of thin event beds from the rock record, coincident with the development of the mixed layer (Droser et al, 2002) In a field test of the "burrowed away" hypothesis, Allison and Brett (1995) found that preservation of nonmineralized tissues and discrete trace fossils in the Burgess Shale were confined to mutually exclusive horizons, and concluded that bioturbation, as regulated by bottom-water oxygen content, was responsible for inhibiting preservation of nonmineralized tissues

In this paper, we develop an alternative, permeability-based taphonomic model for preservation of nonmineralized tissues of organic-walled macrofossils in the middle Cambrian Wheeler Formation. The Wheeler Formation of the House Range, Millard County, Utah (39°N, 113°W), is one of several Cambrian deposits traditionally considered *konservat-lagerstatten* (Conway Morris, 1998) and provides an excellent opportunity to address these questions

## 2. Geological setting

Middle Cambrian strata in the House Range and vicinity (Fig 1) were deposited in what has been interpreted as a fault-controlled trough, termed the

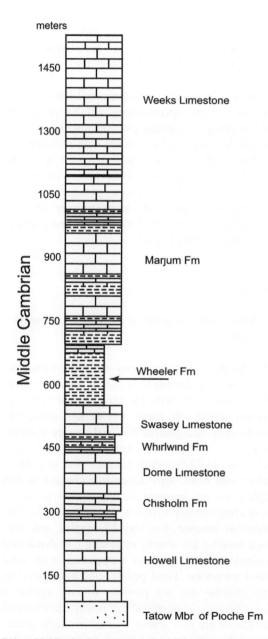

Fig 1 Middle Cambrian stratigraphy of the House Range, after Hintze and Robison (1975)

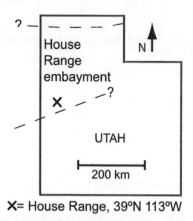

✗= House Range, 39°N 113°W

Fig 2 Location map of the study area, showing the areal extent of the House Range Embayment, as defined by Rees (1986)

House Range Embayment (Fig 2, Rees, 1986) Here, a geographically abrupt change in lithofacies juxtaposes platform carbonates found north and south of the embayment with fine-grained terrigenous and carbonate deposits within the embayment The lowest-energy deposits within the embayment, which include the Wheeler Formation, are characterized by a dominance of well-laminated strata, the presence of pyrite, and paucity of shelly benthic faunas, reflecting deposition in an oxygen-deficient environment (Rees, 1986) Wheeler Formation-age deposits in the House Range embayment occur across much of Western Utah (Fig 2), and are well known for dramatic lateral facies changes, including those that occur within the House Range (Rees, 1986) For this reason, a number of formation names have been applied (Rees, 1986) Units to which the name "Wheeler Shale" or "Wheeler Formation" have been applied include a variety of carbonate and mixed carbonate-siliciclastic lithofacies representing deposition in basinal, deep ramp, and shallow ramp deposits (Rees, 1986) In this study we have focused on well-known basinal localities in the Wheeler Amphitheater region of the House Range In these sections of Wheeler Formation, sediment was derived from two sources terrigenous mud delivered to the basin during times of increased continental runoff, and carbonate mud, most likely derived from the adjacent platform Typical Wheeler mudstone is dominated by clay-sized terrigenous material

Table 1

Weight percent of mineral phases present in six samples from the Wheeler Formation, as determined by Quantitative XRD using the method of Srodon et al (2001)

| | Quartz | Plagioclase | Calcite | Dolomite | 2 1 Al clay | 2 1 Fe clay | Chlorite | Total clay | Total CaCO$_3$ | Total |
|---|---|---|---|---|---|---|---|---|---|---|
| NTF 1 | 21 | 3 2 | 22 6 | 0 | 38 3 | 0 | 19 | 57 3 | 22 6 | 104 1 |
| NTF 2 | 17 3 | 3 8 | 36 3 | 0 | 19 7 | 3 3 | 12 5 | 35 5 | 36 3 | 92 9 |
| NTF 3 | 14 5 | 7 2 | 35 8 | 0 | 15 8 | 0 | 19 3 | 35 1 · | 35 8 | 92 6 |
| ETF 1 | 22 | 2 9 | 26 8 | 13 2 | 34 5 | 0 | 3 7 | 38 2 | 40 | 103 1 |
| ETF 2 | 19 7 | 2 8 | 12 0 | 8 5 | 39 1 | 0 | 9 9 | 49 9 | 20 5 | 92 0 |
| ETF 3 | 19 7 | 2 3 | 34 6 | 3 9 | 32 0 | 0 | 10 6 | 42 7 | 38 5 | 103 1 |

Samples NTF 1–3 come from the nonmineralized preservation taphofacies, samples ETF 1–2 come from unbioturbated intervals of the *Elrathia* taphofacies, sample ETF 3 comes from a bioturbated interval of the *Elrathia* taphofacies Totals indicate ≤8% total error in each case

(<5 μm) with 30–40% detrital micrite and authigenic carbonate (Table 1)

## 3. Methods

Mudstones commonly are very poorly exposed and appear featureless and massive where found in outcrop Sedimentary features such as grain-size, bedding, and ichnofabric are rarely observed even in the best outcrops, because they are obscured by weathering, cleavage, fracturing, and the generally small size of primary features For this reason, 1-to-5-m-thick sections in the Wheeler Formation were continuously sampled in duplicate after thorough outcrop study to capture maximum variation in sedimentologic and taphonomic attributes. Continuous sampling of mudstones is often impossible, but was facilitated in this case by significant carbonate content of Wheeler mudstones and aridity of the study area One set of samples was slabbed and logged on a millimeter scale Incorporating thin-section study and X-radiography, logs recorded features of bedding/lamination, ichnology (depth, extent, and style of bioturbation), and early diagenesis, particularly authigenic precipitation of pyrite, calcite, and dolomite The second set of samples was used for taphonomic and geochemical analyses Bulk mineralogy was assessed using X-ray diffraction, and weight percent carbonate was determined by coulometric analysis Quantitative X-ray diffraction also was performed on selected samples using the method of Srodon et al (2001) In this method, a zinc oxide standard with known peak intensity is mixed with the sample for analysis to determine weight percent of dominant minerals using integrated peak intensities Composi-

tion of organic fossils was determined using energy dispersive X-ray spectroscopy (EDX) on points selected using a scanning electron microscope For this analysis, samples were coated with platinum and analyzed with a 5-kV beam in order to avoid penetration of the beam into the matrix underlying the thin fossil Stable carbon isotopic ratios of authigenic and whole-rock carbonate were analyzed online after digestion with 100% phosphoric acid and are reported relative to PDB

## 4. Results

*4 1 Nature and composition of Wheeler Formation sediments*

Sedimentary features, grain-size, and carbonate/clay ratios of mudstones were highly consistent in all sections examined With the exception of very rare (aeolian) quartz silt grains (<1%) and authigenic crystals, only fine-grained clay and micrite particles are present Within each taphofacies, mudstones consist of 1-to-12-mm-thick couplets with gray bases grading into black tops Grain-size changes within couplets are not apparent in thin-section (although rare grading occurs) Couplets are dominated by thin continuous laminae Cut and fill, ripples, and convolute bedding are absent, indicating that deposition occurred dominantly from suspension with little current reworking Fecal pellets, clay aggregates, or wavy laminae are not present, arguing against a pelagic biogenic or microbial mat origin Skeletonized microfossils are also absent Fluorescent-light microscopy indicates that most laminae contain aggregates of micron-sized blocky calcite crystals which are morphologically consistent with microbially mediated calcite peloids described by Chafetz (1986) as common constituents of marine carbonate rocks, these indicate a detrital origin for carbonate Authigenic carbonate cements are the most prominent feature of Wheeler mudstone fabric (Fig 3) and occur as abundant dispersed, micron-sized crystals, as cone-in-cone calcite coatings on the ventral sides of trilobite carapaces, and in thin (<1 mm) continuous horizons

Quantitative X-ray diffraction indicates the prominent mineral phases are calcite (both micrite and authigenic), quartz, illite, chlorite, and minor plagio-

clase (in order of abundance, Table 1) During burial diagenesis, the original clay mineralogy was most likely altered to form the current illite, chlorite, quartz, and feldspar assemblage.

Cathodoluminescence indicates that carbonate diagenesis appears to have been restricted to a single early phase of authigenic carbonate cementation Thin sections show uniform orange luminescence of authigenic carbonates under cathodoluminescent light, whereas authigenic carbonates formed in multiple phases most commonly show distinctive luminescent color banding under cathodoluminescent light The cements show no evidence of recrystallization sub-

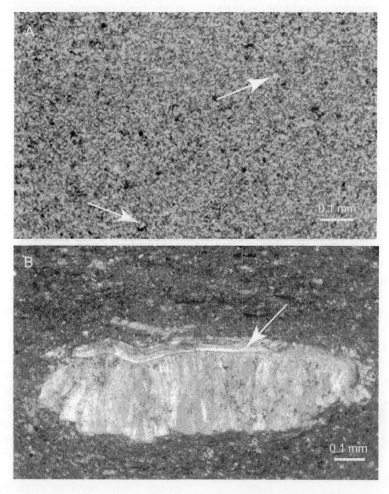

Fig 3 Polarized-light photomicrographs from the nonmineralized preservation taphofacies (A) and *Elrathia* taphofacies (B) In each, minute authigenic carbonate grains, which appear as bright spots, dominate the mudstone fabric, examples are indicated by arrows in A B additionally shows a recrystallized trilobite carapace (indicated by arrow) with cone-in-cone calcite cement coating the ventral side, and more prominent authigenic carbonate grains dispersed throughout the fabric

sequent to emplacement Deformation of sedimentary laminae around cement-lined trilobite nodules indicates cementation prior to burial compaction Additionally, the presence of authigenic calcite with the absence of authigenic dolomite in one taphofacies (Table 1) is noteworthy, the significance of this is discussed below $\delta^{13}C$ values from carbonate phases of whole-rock samples vary between $-1\,65\%o$ and $-0\,43\%o$ PDB ($n$=14) and are similar to authigenic phases. $-0\,47\%o$ to $-0\,31\%o$ PDB ($n$=4) All values fall within the range of $\delta^{13}C$ seawater values reported for this time interval of $-2\%o$ to $+0\,4\%o$ (Montañez et al , 2000)

Despite considerable continuity in sedimentary structures and bulk chemistry, mudstone sections examined exhibit considerable variation in body fossil content, presence/absence of in situ benthic faunas, ichnofabric, and taphonomic features In this paper, we describe two taphofacies within basinal Wheeler mudstones

### 4 2 Nonmineralized preservation taphofacies

The unbioturbated nonmineralized preservation taphofacies is the most commonly occurring taphofacies of the Wheeler Formation, and it is dominant in the lower portion of the unit This taphofacies contains abundant nonmineralized preservation of the macroscopic, non-calcified algae *Marpolia spissa* (Fig. 4), *Yuknessia simplex*, *Margaretia dorus* and *Morania fragmenta*, as well as unidentifiable worms, likely priapulids Other nonmineralized taxa which

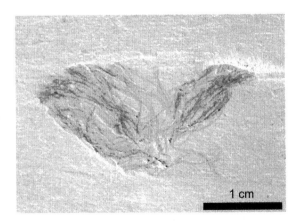

1 cm

Fig 4 Burgess Shale-type preservation of the non-calcified alga *Marpolia spissa* from the Wheeler Formation

were found to occur rarely in this taphofacies include the priapulid *Selkirkia*, the sponge *Choia*, and undetermined phyllocarid arthropods Additionally, at least 21 other genera of nonmineralized metazoa, dominantly arthropods, which were not recovered in this study, are known from the Wheeler Formation, most of which are common to the Burgess Shale (Robison, 1991) Nonmineralized fossils are uniformly black and reflective in appearance Energy dispersive X-ray spectroscopy of representative samples revealed that these fossils are composed of carbon in the form of thin, two-dimensional films (Fig 5) Nonmineralized taxa are found in association with abundant agnostid trilobites, presumably pelagic in life habit (Robison, 1972, Muller and Walossek, 1987), and rare examples of the "ptychoparid" trilobite *Elrathia kingii*

Examination of cut slabs as well as thin sections indicates that the nonmineralized preservation taphofacies is characterized by a near-total lack of bioturbation (Fig 6A) Very rare burrows are horizontal, commonly pyritized, and confined to individual bedding planes Authigenic pyrite is found within local aggregations dispersed throughout the sediment as well as within burrows Thin-section analysis reveals the common presence of micron-sized euhedral calcite crystals dispersed through the sediment Calcite crystals also line the ventral side of agnostid trilobite carapaces, imparting a three-dimensional relief uncommon to trilobite fossils in mudstones Dolomite is notably absent from this taphofacies (Table 1)

The absence of bioturbation and limited presence of authigenic pyrite suggest that the nonmineralized preservation taphofacies accumulated under dominantly anoxic conditions that excluded a benthic fauna Rare discrete, horizontal burrows were likely emplaced during short-lived variations in dissolved oxygen concentration Algae and priapulid worms, which are interpreted to have been infaunal (Conway Morris, 1998), represent an allochthonous fauna, as do rare occurrences of *Elrathia*. Agnostid trilobites inhabited overlying oxic waters, supplying molts and carcasses to the basin below Rare in situ occurrences of nonmineralized faunas (e g , the sponge *Choia*) occur on single bedding planes, and are interpreted to represent a "taphonomic window" In these cases, ichnologic evidence indicates that these single-horizon blooms occur at the transition from

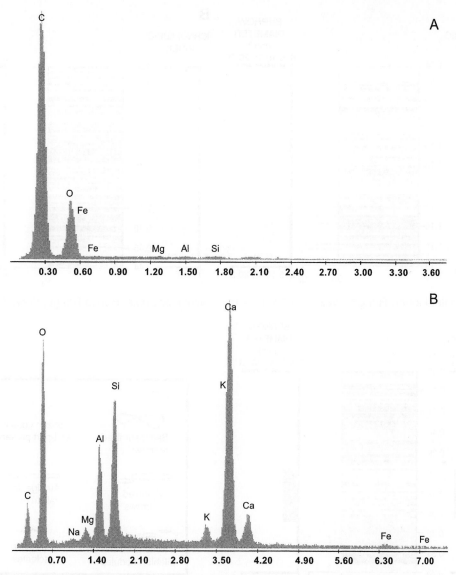

Fig 5 Energy dispersive X-ray spectra for an organic-walled vermiform metazoan (A) and adjacent matrix of the same specimen (B), using 5-kV beam  Note dominance of carbon in fossil (A) vs aluminosilicate and carbonate composition of matrix (B)

weakly bioturbated to laminated strata, indicating that short-lived habitable bottom-water conditions facilitated limited benthic colonization, but were followed by an abrupt shift to anoxic conditions, which encouraged preservation (see Discussion)

### 4 3 Elrathia taphofacies

The *Elrathia* taphofacies contains bioturbation in some horizons, and is the dominant taphofacies of the upper part of the Wheeler Formation  It is characterized by unusually abundant *Elrathia kingii* trilobites, preserved as nodules with the recrystallized exoskeleton forming a cap from which 0 5–3-mm calcite crystals radiate outward, from the ventral side  only *Elrathia kingii* is found in association with several species of other ptychoparid trilobites, which occur rarely, common agnostid trilobites, and rare algae and acrotretid brachiopods  Algae constitute the only preservation of nonmineralized tissues observed in this taphofacies

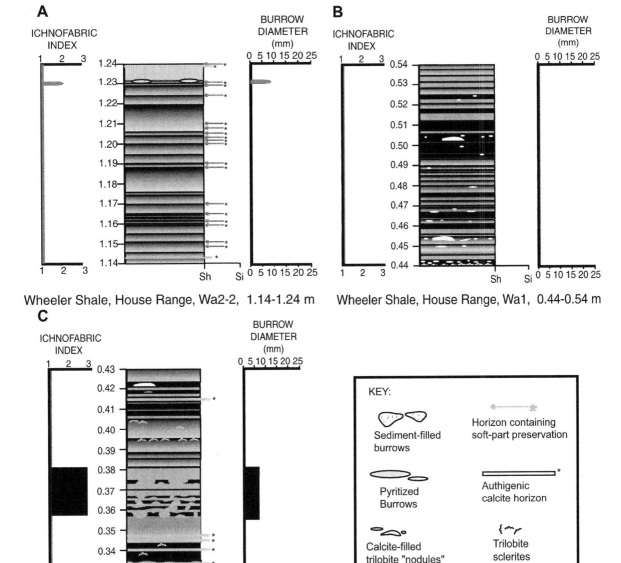

Fig 6 (A) Detailed log of 10-cm interval within the nonmineralized preservation taphofacies (B and C) Detailed logs of two 10-cm intervals within the *Elrathia* taphofacies, illustrating non-bioturbated (B) and bioturbated (C) intervals

Although dominantly unbioturbated (ii 1) (Fig 6B), 2-to-10-cm-thick bioturbated horizons are present (ii 2–3, Fig 6C), separated by 15-to-30-cm-thick intervals of unbioturbated strata (ii 1) Bioturbated intervals contain subhorizontal to vertical discrete burrows (10 mm max depth) including *Hormosiroidea* and/or indistinct mottling of mudstone fabric Authigenic pyrite occurs dispersed throughout the sediment and in local aggregations The shale fabric of the *Elrathia* taphofacies contains dispersed authigenic calcite and up to 13 2% (by weight) dolomite

Low levels of bioturbation in many horizons suggest that limited oxygenation of bottom-waters facilitated colonization of the seafloor by a benthic fauna, including *Elrathia*, but precluded significant infaunal activity Taphonomic, size-frequency, and orientation data indicate that *Elrathia* present are in situ even in unbioturbated intervals, suggesting at least minimal availability of bottom-water oxygen throughout the entire taphofacies Gradational increases in bioturbation suggest oscillations in bottom-water oxygen content that regulated benthic activity The restriction of dolomite to this taphofacies may imply a seawater source of $Mg^{+2}$ during diagenesis not present in the nonmineralized preservation taphofacies We interpret the presence of dolomite in this taphofacies to reflect more efficient pore–water exchange with bottom-waters, and/or more complete organic diagenesis than in the nonmineralized preservation taphofacies (see Discussion below)

## 5. Discussion

Any attempt to assign a causative mechanism to preservation of nonmineralized tissues in the Wheeler Formation must address the following observations (1) Nonmineralized fossils occur in laminated strata with no vertical bioturbation and rare horizontal bioturbation, while bioturbation is common in other lithologically similar facies (2) Nonmineralized fossils are housed in homogenous, clay-rich sediments containing abundant detrital and authigenic carbonate, and lacking in silt, fecal pellets and skeletonized microfossils (3) Taphofacies are sedimentologically and mineralogically similar, and are composed of laminated suspension deposits (4) Nonmineralized fossils are preserved as organic carbon films and are not replaced, mineralized, or preserved as molds

The most significant difference between the nonmineralized preservation taphofacies and the *Elrathia* taphofacies lies in the record of bioturbation and implied relative bottom-water oxygen content This, however, does not provide an adequate explanation for preservation of nonmineralized tissues Oxidation rates of many organic compounds are comparable under anoxic (given unlimited $SO_4^{2-}$) or oxic conditions (Henrichs and Reeburgh, 1987, Lee, 1992) Additionally, laboratory experiments have shown

rapid and complete decomposition of crustaceans buried in muddy sediments under anoxic conditions (Allison, 1988) Limited exposure to all oxidants (including $SO_4^{2-}$) is likely a more critical factor than the relatively minor metabolic efficiencies of one oxidant over another

Still, preservation of nonmineralized tissues in the Wheeler Formation shows a strong correlation with bottom-water anoxia We interpret this relation to be the indirect result of oxygen's influence on a primary control of preservation of nonmineralized tissues, permeability, which ultimately controls total oxidant exposure Oxygen influences permeability in two distinct ways First, anoxia reduces burrowing and the associated irrigation effects (Savrda et al , 1984) Burrowing directly disturbs only a limited amount of organic matter in the sediment, but churning and disruption of sediments promote a deeper redox boundary and open pathways enhancing pore–water exchange of oxidants, which sustain microbial decomposition of organic matter (Aller, 1982)

Second, anoxic conditions near the sediment–water interface may reduce permeability in clay-rich suspension deposits by deflocculating clays, allowing low permeability face-to-face contact of clay platelets (Moon and Hurst, 1984) Clay floccules (face-to-edge contacts forming aggregates) increase porosity and promote greater pore–water exchange Under reducing conditions in low turbulence environments, organic cations common in anoxic water such as amino and humic acids, coat charged clay surfaces, and may break surface-to-edge contacts and allow aggregates to disperse (Moon and Hurst, 1984) Laminated microfabrics formed by suspension settling of clay particles, such as that of the Wheeler Formation, have the lowest permeability of fine-grained sediments (Davies et al , 1991) The intervals of the Wheeler Formation bearing nonmineralized fossils are comprised of this laminated microfabric and do not contain fecal pellets, silt, mineralized microfossils, or other particles that otherwise might enhance permeability Biogenic particles, especially agnostid trilobite sclerites, are present here, but are volumetrically insufficient to impact permeability Differences in bottom-water oxygen content resulted in important permeability differences between the two taphofacies described here, as it restricted bioturbation, and also may have facilitated deflocculation of clay aggregates in the nonmineralized preservation taphofacies

Further reduction of permeability, limiting oxidant exchange with seawater also was driven by pore-occluding carbonate cementation Carbonate cementation is common in early diagenetic environments (Coleman, 1985) and is driven by sulfate reduction or dissolution and reprecipitation of highly reactive high-Mg calcite mineralogies typical of the extremely fine-grained micritic aggregates that comprise the detrital carbonate phase in the Wheeler. Evidence that carbonate cementation was early and derived predominantly from the detrital phase includes abundant dispersed authigenic carbonate, detrital-like carbon isotope values in authigenic cement, and evidence for early cementation from calcite-lined trilobite carapaces $\delta^{13}C$ values from carbonate phases in the Wheeler Formation support a detrital marine or marine origin of carbonate phases, rather than an origin from organic diagenesis Although organic diagenesis leads to the production of alkalinity, driving carbonate precipitation (e g, Mazzullo, 2000), several authors have pointed to the importance of a precursor detrital carbonate phase in the genesis of authigenic carbonates (Baker and Burns, 1985, Compton, 1988, Mazzullo, 2000, Mullins et al, 1985, Irwin, 1980) Sulfate reduction and coincident pyrite precipitation commonly lead to carbonate dissolution and subsequent re-precipitation (Coleman, 1985) Whole-rock values from the Wheeler Formation vary between $-1 65‰$ and $-0 43‰$ PDB ($n=14$) Authigenic phases range from $-0 47‰$ to $-0 31‰$ PDB ($n=4$) Uniformly, these values lie within the range of $\delta^{13}C$ seawater values reported for this time interval of $-2‰$ to $+0 4‰$ (Montañez et al, 2000) Any contribution from organic diagenesis must have been minor, as carbon derived as a result of these processes would have driven carbonate values toward lower $\delta^{13}C$ The minor presence of authigenic pyrite, however, indicates that limited sulfate reduction occurred at some stage of diagenesis Therefore, authigenic carbonate cements in the Wheeler Formation appear to have been derived from the dissolution and re-precipitation of a portion of the reactive detrital micrite phase

Low exchange of porewater with seawater and/or arrested organic diagenesis in intervals bearing preservation of nonmineralized tissues is suggested by the absence of dolomite as an authigenic phase in the nonmineralized preservation taphofacies, whereas dolomite is a significant component (up to 13 2% by weight) of the sedimentologically identical and interbedded *Elrathia* taphofacies Precipitation of dolomite implies a seawater source of $Mg^{+2}$, and, thus, may indicate seawater–porewater exchange in the Elrathia taphofacies that did not occur in the nonmineralized preservation taphofacies Alternately, if sufficient $Mg^{+2}$ to form dolomite was present in porewaters initially, dolomite absence from the non-mineralized preservation taphofacies may indicate incomplete sulfate reduction during carbonate cementation, since the presence of $SO_4^{2-}$ may inhibit dolomite formation (Baker and Kastner, 1981) Since authigenic dolomite typically forms only after sulfate reduction is complete (Coleman, 1985), dolomite absence from the nonmineralized preservation taphofacies would additionally imply an early cessation of organic diagenesis in this taphofacies, likely through occlusion of porosity Ongoing work will test these possibilities through isotopic analysis of sulfur to determine if the early diagenetic systems were open or closed to sea water sulfate

The combined effects of low original porosity, the porosity-reducing effects of anoxia, and early carbonate cementation may have resulted in porosity occlusion to an extent sufficient to restrict microbial activity and, thus, to facilitate Burgess Shale-type preservation in the Wheeler Formation (Fig 7) Original low porosity was facilitated by the inboard sequestration of all but the finest clastics and also by the absence of fecal pellets and skeletonized microfossils Skeletonized microfossils are not significant components of Cambrian sediments in general (Tucker, 1974), and the oldest known microfossil oozes do not occur until the late Cambrian (Tolmacheva et al, 2001) Within fine-grained sediments, the abundance of skeletonized microfossils is positively correlated with porosity and also with size of individual pore spaces (Kraemer et al, 2000). Therefore, the subsequent rise to abundance of skeletonized microfossils in marine sediments and likely increase in the abundance of fecal pellets may be important porosity-influencing secular trends

The hypothesis that a post-middle Cambrian increase in bioturbation may be responsible for the observed secular decline in preservation of nonmineralized tissues is not entirely supported in the

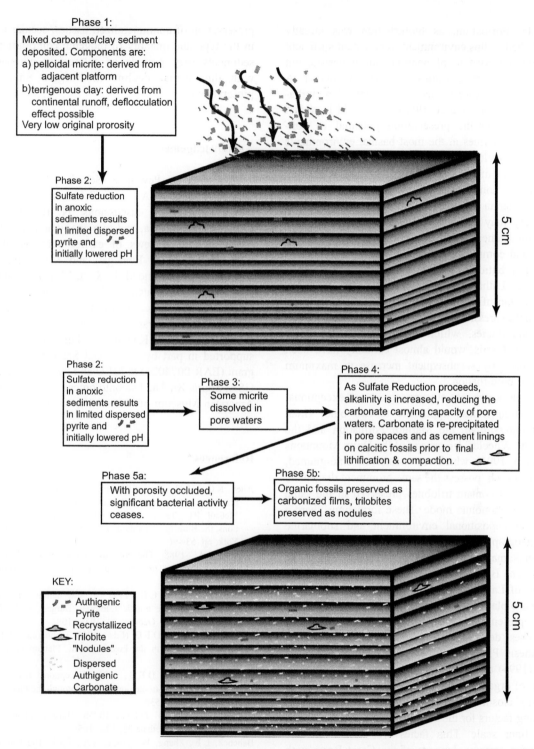

Fig 7  Schematic summary of porosity-occlusion model, showing early diagenetic processes proposed to have promoted Burgess Shale-type preservation in the nonmineralized preservation taphofacies of the Wheeler Formation  Trilobites shown in this summary are pelagic agnostids, however calcite coatings on the ventral sides of *Elrathia* found in the *Elrathia* taphofacies were emplaced via the same processes

Wheeler Formation, as bioturbation was already established in this environment to an extent sufficient to limit preservation of nonmineralized tissues, but was regulated by bottom-water oxygen content A subsequent increase in maximum depth of bioturbation (Droser and Bottjer, 1988) also would have had little effect on the preservation potential of nonmineralized tissues in the most basinal environments of the Wheeler Formation, as bioturbators were excluded for meters of preserved continuous section. Within the nonmineralized preservation taphofacies, however, strata deposited closer to the anoxic–dysoxic boundary may be interbedded at a decimeter scale with mudstones deposited under more oxic conditions, and commonly contain low levels of shallow-tier bioturbation, as also observed in the lower Cambrian Latham Shale (Gaines and Droser, 2002) Preservation of nonmineralized tissues in these more marginal expressions of the nonmineralized preservation taphofacies, which often contain the most metazoan fossils, would almost certainly have been jeopardized by a subsequent increase in maximum depth of bioturbation

The early porosity closure scenario of exceptional preservation is also applicable to two other Great Basin *lagerstatten*, the Marjum Formation and the "shallow Wheeler" *lagerstatte* of the Drum Mountains (Robison, 1991), which are exclusively fine-grained, carbonate-rich, possess the same prominent diagenetic fabrics, and contain trilobites preserved in the same unusual taphonomic mode These formations share a common depositional environment and submarine paleogeography proximal to a carbonate platform and, thus, may be considered expressions of the same facies The Burgess Shale and other Cambrian *lagerstatten* of Laurentia also occur near the shelf break/carbonate platform edge (Conway Morris, 1998), and may represent additional expressions of this taphofacies These deposits would provide a useful test of this hypothesis Physical mechanisms proposed by Butterfield (1995) and Petrovich (2001) for further stabilization of organic matter are not inconsistent with our porosity closure model and may be important contributing factors for the stabilization of organic tissues at a finer scale This facies was lost from late Cambrian cratonic deposits of the Great Basin with the widespread progradation of carbonate platforms If minimal porosity is indeed important to the organic

preservation of nonmineralized tissues, secular trends in the type and quantity of biogenic contributions to sediments may provide an interesting comparison with the secular decline in Burgess Shale-type preservation

## Acknowledgements

We thank D Pevear, J Scheiber, S Jensen, D Briggs, R Behl, T Algeo, S. Peters and S Hollingsworth for valuable discussions, N Butterfield, J Hagadorn, C Savrda, and two anonymous reviewers for thoughtful reviews of earlier versions of this manuscript, and S Finnegan and K Le for field assistance D Hammond, L Stott, M. Rincon and M. Prokopenko facilitated analyses conducted at the University of Southern California. M. Berke assisted with photomicrographs Special thanks to G H Johnson and M Kawabata This research was supported in part by a National Science Foundation grant (EAR-0074021) to MLD and by grants from the GSA, Sigma Xi, The Paleontological Society, and The American Museum of Natural History to RRG

## References

Aller, R C , 1982 The effects of macrobenthos on chemical properties of marine sediment and overlying water In McCall P L , Tevesz, M J S (Eds ), Animal–Sediment Relations, the Biogenic Alteration of Sediments Plenum Press, New York, pp 53–96

Allison, P A , 1988 The role of anoxia in the decay and mineralization of proteinaceous macro-fossils Paleobiology 14, 139–154

Allison, P A , Brett, C E , 1995 In situ benthos and paleo-oxygenation in the middle Cambrian Burgess Shale, British Columbia, Canada Geology 23, 1079–1082

Allison, P A , Briggs, D E G (Eds ), 1991 Taphonomy Releasing the Data Locked in the Fossil Record Plenum Press, New York, pp 25–90

Allison, P A , Briggs, D E G , 1993a Exceptional fossil record distribution of soft-tissue preservation through the Phanerozoic Geology 21, 527–530

Allison, P A Briggs D E G , 1993b Burgess Shale biotas burrowed away? Lethaia 26, 184–185

Babcock, L E , Zhang, W , Leslie, S A , 2001 The Chengjiang biota, record of the early Cambrian diversification of life and clues to exceptional preservation of fossils GSA Today 11, 4–9

Baker, P A , Burns, S J , 1985 Occurrence and formation of dolomite in organic-rich continental margin sediments AAPG Bull 69, 1917–1930

Baker, P A , Kastner, M , 1981 Constraints on the formation of sedimentary dolomite Science 213, 213–215

Briggs, D E G , Nedin, C , 1997 The taphonomy and affinities of the problematic fossil *Myoscolex* from the Lower Cambrian Emu Bay Shale of South Australia J Paleontol 71, 22–32

Butterfield, N J , 1990 Organic preservation of non-mineralizing organisms and the taphonomy of the Burgess Shale Paleobiology 16, 272–286

Butterfield, N J , 1995 Secular distribution of Burgess Shale-type preservation Lethaia 28, 1–13

Chafetz, H S , 1986 Marine peloids a product of bacterially induced precipitation of calcite J Sediment Petrol 56, 812–817

Coleman, M L , 1985 Geochemistry of diagenetic non-silicate minerals kinetic considerations Philos Trans R Soc Lond , A 315, 39–56

Compton, J S , 1988 Degree of supersaturation and precipitation of organogenic dolomite Geology 16, 318–321

Conway Morris, S , 1989 The persistence of Burgess Shale-type faunas implications for the evolution of deeper-water faunas Trans R Soc Edinb Earth Sci 80, 271–283

Conway Morris, S , 1998 The Crucible of Creation Oxford University Press, Oxford 242 pp

Davies, K D , Bryant, W R , Vessell, R K , Burkett, P J , 1991 Porosity, permeabilities and microfabrics of Devonian shales In Bennett, R H , Bryant, W R , Hulbert, M H (Eds ), Microstructure of Fine-grained Sediments Springer Verlag, New York, pp 109–119

Droser, M L , Bottjer, D J , 1988 Trends in depth and extent of bioturbation in Cambrian carbonate marine environments, western United States Geology 16, 233–236

Droser, M L , Jensen, S , Gehling, J G , 2002 Trace fossils and substrates of the terminal Proterozoic–Cambrian transition implications for the record of early bilaterians and sediment mixing Proc Natl Acad Sci 99, 12572–12576

Gaines, R R , Droser, M L , 2002 Depositional environments, ichnology, and rare soft-bodied preservation in the lower Cambrian Latham Shale, East Mojave SEPM Pac Sect Book 93, 153–164

Gehling, J G , 1999 Microbial mats in terminal Proterozoic siliciclastics Ediacaran death masks Palaios 14, 40–57

Henrichs, S M , Reeburgh, W S , 1987 Anaerobic mineralization of marine sediment organic matter rates and the role of anaerobic processes in the oceanic carbon economy Geomicrobiol J 5, 191–237

Hintze, L F , Robison, R A , 1975 Middle Cambrian stratigraphy of the House, Wah Wah, and adjacent ranges in Western Utah Geol Soc Amer Bull 86, 881–891

Irwin, H , 1980 Early diagenetic carbonate precipitation and pore fluid migration in the Kimmeridge Clay of Dorset, England Sedimentology 27, 577–591

Kraemer, L M , Owen, R M , Dickens, G R , 2000 Lithology of the upper gas hydrate zone, Blake Outer Ridge a link between diatoms, porosity, and gas hydrate Proc Ocean Drill Program Sci Results 164, 229–236

Lee, C , 1992 Controls on organic carbon preservation the use of stratified water bodies to compare intrinsic rates of decomposition in oxic and anoxic systems Geochim Cosmochim Acta 56, 3323–3355

Mazzullo, S J , 2000 Organogenic dolomitization in peritidal to deep-sea sediments J Sediment Res 70, 10–23

McIlroy, D , Logan, G A , 1999 The impact of bioturbation on infaunal ecology and evolution during the Proterozoic–Cambrian transition Palaios 14, 58–72

Montañez, I P . Osleger, D A , Banner, J L , Mack, L E , Musgrove, M L . 2000 Evolution of the Sr and C isotope composition of Cambrian oceans GSA Today 10, 1–7

Moon, C F , Hurst, C W , 1984 Fabric of mud and shales an overview In Stow, D A V , Piper, D J W (Eds ), Fine-grained Sediments, Geological Society Special Publication, vol 15, pp 579–593

Muller, K J , Walossek, D , 1987 Morphology, ontogeny, and lifehabit of *Agnostus pisiformis* from the Upper Cambrian of Sweden Fossils Strata 19, 1–124

Mullins, H T , Wise Jr , S W , Land, L S , Siegel, D I , Masters, P M , Hinchey, E J , Price, K R , 1985 Authigenic dolomite in Bahamian peri-platform slope sediment Geology 13, 292–295

Petrovich, R , 2001 Mechanisms of fossilization of the soft-bodied and lightly armored faunas of the Burgess Shale and of some other classical localities Am J Sci 3001, 683–726

Powell, W , 2003 Greenschist-facies metamorphism of the Burgess Shale and its implications for models of fossil formation and preservation Can J Earth Sci 40, 13–25

Rees, M N , 1986 A fault-controlled trough through a carbonate platform the middle Cambrian House Range embayment GSA Bull 97, 1054–1069

Robison, R A , 1972 Mode of life of agnostid trilobites 24th International Geological Congress, Montreal, Canada, Section 7 (Paleontology) pp 33–40

Robison, R A , 1991 Middle Cambrian biotic diversity, examples from four Utah *Lagerstatten* In Simonetta, A M , Conway Morris, S (Eds ), The Early Evolution of the Metazoa and the Significance of Problematic Taxa Cambridge University Press, Cambridge, pp 77–98

Savrda, C E , Bottjer, D J , Gorsline, D S , 1984 Development of a comprehensive oxygen-deficient marine biofacies model evidence from Santa Monica, San Pedro, and Santa Barbara Basins, California continental borderland AAPG Bull 68, 1179–1192

Srodon, J , Drits, V A , McCarty, D K , Hsieh, J C C , Eberl, D D , 2001 Quantitative X-ray diffraction analysis of clay-bearing rocks from random preparations Clays Clay Miner 49, 514–528

Tolmacheva, T J , Danelian, T , Popov, L E , 2001 Evidence for 15 m y of continuous deep sea biogenic siliceous sedimentation in early Paleozoic oceans Geology 29, 755–758

Tucker, M E , 1974 Sedimentology of Paleozoic pelagic limestones, the Devonian Griotte (southern France) and Cephalopodenkalk (Germany) Spec Publ Int Assoc Sedimentol 1, 71–92

Wollanke, G , Zimmerle, W , 1990 Petrographic and geochemical aspects of fossil embedding in exceptionally well preserved fossil deposits Mitt Geol -Palaontol Inst Univ Hamb 69, 57–74

Available online at www.sciencedirect.com

Palaeogeography, Palaeoclimatology, Palaeoecology 220 (2005) 207–225

www.elsevier.com/locate/palaeo

# Alpha, beta, or gamma: Numerical view on the Early Cambrian world

Andrey Yu. Zhuravlev[a,*], Elena B. Naimark[b]

[a]Área y Museo de Paleontología, faculdad de Ciences, Universidad de Zaragoza, C/ Pedro Cerbuna, 12, E-50009, Zaragoza, Spain
[b]Palaeontological Institute, Russian Academy of Sciences, ul. Profsoyuznaya 123, Moscow 117997, Russia

Received 28 March 2003, accepted 1 August 2004

## Abstract

Data on 220 sampling units from the Lower Cambrian section of the Siberian Platform (78), Mongolia (74), and some other regions (68) yielded qualitative and quantitative information on reef faunal composition and facies This data set served to find out the diversity components of the Early Cambrian radiation within reefal communities Additionally, dynamics of the development of faunal regions and of total archaeocyathan species diversity were computed The plotting of all these data sets indicated that the Early Cambrian global species diversity was mostly the factor of two variables Those variables were beta and gamma diversity, expressing taxonomic differentiation between communities and the overall taxonomic differentiation between geographic regions, respectively Alpha diversity showing the richness of taxa within a single community was significant during short episodes of the very initiation of Cambrian reef-building and of recovery that followed mass extinctions Thus, specialisation, which was the function of restriction of species to peculiar communities, and degree of endemicity were the principal factors of taxonomic diversity in the earliest Phanerozoic reefal palaeocommunities, while the significance of niche partitioning was restricted to peculiar intervals Endemicity itself was the factor of both geographic isolation and community differentiation leading to ecological specialisation The early evolution of Phanerozoic reefal communities was expansive and not intense
© 2005 Published by Elsevier B V

Keywords Diversity, Early Cambrian, Archaeocyath sponges, Reefal communities

## 1. Introduction

Alpha, beta, and gamma diversity are components of the global taxonomic diversity, and of these three categories, alpha diversity illustrates the richness of taxa at single locality or community levels Beta diversity indicates the taxonomic differentiation of fauna or flora between localities or communities usually within the limits of the same region, and gamma diversity expresses the overall taxonomic differentiation between geographic regions In general, alpha diversity reflects packing within a community, beta diversity serves as an approximation of habitat selection or specialization, while gamma diversity measures the degree of provinciality (Sepkoski, 1988)

\* Corresponding author
  E-mail address  ayzhur@mail ru (A Y Zhuravlev)

0031-0182/$ - see front matter © 2005 Published by Elsevier B V
doi 10 1016/j palaeo 2004 08 009

These factors are well-known among ecologists However, there are few palaeontological studies of these factors because of the difficulty of finding representative field collections Alpha, beta, and gamma diversity have been calculated for the entire Palaeozoic although reefs have been mostly avoided in the collection of data (Bambach, 1977; Sepkoski, 1988) These calculations revealed that alpha diversity was essentially constant for long intervals after a sufficient Early–Middle Ordovician rise, while it was low during the Cambrian Similarly, beta diversity was low in the Cambrian when compared with the later Palaeozoic and the significant increase during the Ordovician radiation

The present paper deals with Early Cambrian reefs and their alpha, beta, and gamma diversity as components of the famous Cambrian radiation and less known late Early Cambrian mass extinction Such a study is important for the understanding of hidden mechanisms of these phenomena Early Cambrian reefs yield a valuable information for an investigation of diversity pattern regularities They are relatively small in both geometric sizes and number of species contributing to them Publications on their ecology and species composition are plentiful For details, see recent reviews by Wood (1999), Pratt et al (2001), and Zhuravlev (2001) The Siberian data set is of special interest because during the Tommotian time archaeocyaths as well as some other Cambrian organisms and, thus, their communities developed within a very restricted area which was the Anabar–Sinsk belt of the Siberian Platform (Rozanov, 1980) As a result, the entire species pool can be scored with a high precision and problems related to an immigration of species are negligible The later phenomenon commonly overshadows pure evolutionary patterns making them unrecognizable in the fossil record At last but not at least, direct interactions of faunal and microbial components are observed in reefal sampling units under study which allow us to avoid the problem of time-averaging and mixing of different communitites

## 2. Material

The principal areas providing the data for the present work are the Siberian Platform (eastern part of Siberia) and Mongolia Therefore, Siberian stages and zones are selected here for the Early Cambrian time scale These are Nemakit–Daldynian (ca 550–540 Ma), Tommotian (ca 540–535 Ma), Atdabanian (ca 535–530 Ma), Botoman (ca 530–525 Ma), and Toyonian (ca 525–510 Ma) stages Cambrian reefal animals appeared during the second Tommotian stage, and further developed during the Atdabanian–Toyonian stages of the Early Cambrian epoch. The stages are subdivided into four zones each except the Toyonian Stage which includes three zones (Astashkin et al , 1991) *Nochoroicyathus sunnaginicus* (T1), lower (T2) and upper *Dokidocyathus regularis* (T3), *Dokidocyathus lenaicus–Tumuliolynthus primigenius* (T4), *Retecoscinus zegebarti* (A1), *Carinacyathus pinus* (A2), *Nochoroicyathus kokoulini* (A3), *Carinacyathus lermontovae* (A4), *Carinacyathus squamosus–Botomocyathus zelenovi* (B1), *Bergeroniellus gurarii* (B2), *Bergeroniellus asiaticus* (B3), *Bergeroniaspis ornata* (B4), *Bergeroniellus ketemensis* (TN1), *Lermontovia grandis* (TN2), and *Anabaraspis splendens* (TN3)

The Siberian data set includes 53 sampling units of the earliest Tommotian (T1) to latest Atdabanian (A4) age from the southeastern Siberian Platform (middle courses of the Aldan and Lena rivers) deployed in an ongoing study of community analysis. Here this data set is widened by the addition of 25 sampling units [10 of the late middle Tommotian age (T3), 2 of the late Atdabanian (A3 and A4), 10 of the earliest Botoman (B1), and 3 of the middle Toyonian (TN2)] from other southeastern localities as well as from the northern Siberian Platform (southern Anabar area) and the adjacent Kolyma Uplift Main features of Siberian Early Cambrian reefs are outlined by Zhuravleva (1972), Nikolaeva et al (1986), Kruse et al (1995), Riding and Zhuravlev (1995), and Debrenne and Zhuravlev (1996)

Another large data set [74 sampling units of the earliest Atdabanian (A1)–earliest Botoman (B1) age] represents several tectonic zones of western Mongolia which are the Khubsugul, Khan Khukhiy, Lake, Ider, and Tsagaan Olom zones in the southward direction In contrast to the relatively stable Siberian Platform, Mongolian tectonic zones delimit the Neoproterozoic–Cambrian terranes with a complicated geological history including microcontinents with predominantly carbonate sedimentation (Khubsugul and Tsagaan Olom terranes) as well as the former volcanic arcs

Table 1
Number of species per sampling unit for the Tommotian (T)–Atdabanian (A)–Botoman (B)–Toyonian (TN) stages of the early Cambrian (Siberian Platform sampling units in italics, Mongolian sampling units in bold)

| ZONE | T1 | T2 | T3 | T4 | A1 | A2 | A3 | A4 | B1 | B2 | B3 | B4 | TN1 | TN2 |
|---|---|---|---|---|---|---|---|---|---|---|---|---|---|---|
| | 2 | 10 | 18 | 18 | 11 | 10 | 16 | 11 | 14 | 1 | 3 | 7 | | 3 |
| | 7 | 10 | 6 | 12 | 9 | 5 | 10 | 8 | 8 | 6 | 1 | 27 | | 1 |
| | 5 | 10 | 7 | 10 | 16 | 8 | 11 | 11 | 6 | 6 | 5 | 7 | | 3 |
| | 8 | 10 | | | 1 | 7 | 12 | 12 | 14 | 4 | 8 | | | 1 |
| | 4 | 6 | | | 3 | 9 | 8 | 5 | 5 | 4 | 9 | | | 3 |
| | 9 | 18 | | | 4 | 10 | 11 | 9 | 4 | 13 | 8 | | | 7 |
| | 11 | 19 | | | 7 | 5 | 8 | 15 | 2 | 6 | | | | 2 |
| | 16 | 11 | | | 18 | 7 | 9 | 7 | 6 | 5 | | | | 1 |
| | 20 | 6 | | | 3 | 9 | 12 | 11 | 33 | 3 | | | | 1 |
| | 11 | 7 | | | 8 | 7 | 18 | 14 | 16 | 7 | | | | |
| | 15 | 10 | | | 10 | 17 | 7 | 15 | 20 | 3 | | | | |
| | 10 | | | | 10 | 11 | 25 | 11 | 17 | 3 | | | | |
| | 9 | | | | 4 | 10 | 18 | 10 | 15 | 4 | | | | |
| | | | | | 6 | 16 | 27 | 10 | 11 | | | | | |
| | | | | | 3 | 20 | 9 | | 5 | | | | | |
| | | | | | 5 | 12 | 11 | | 11 | | | | | |
| | | | | | 5 | 14 | 7 | | 13 | | | | | |
| | | | | | 9 | 21 | 10 | | 10 | | | | | |
| | | | | | 4 | 9 | 21 | | 8 | | | | | |
| | | | | | 7 | 11 | 26 | | 13 | | | | | |
| | | | | | 12 | 12 | 4 | | 13 | | | | | |
| | | | | | 8 | 5 | 6 | | 11 | | | | | |
| | | | | | 13 | 7 | 4 | | 15 | | | | | |
| | | | | | | 16 | 3 | | 11 | | | | | |
| | | | | | | 13 | 7 | | 15 | | | | | |
| | | | | | | 16 | 6 | | 10 | | | | | |
| | | | | | | 4 | 5 | | 14 | | | | | |
| | | | | | | 12 | 6 | | 12 | | | | | |
| | | | | | | | 8 | | 14 | | | | | |
| | | | | | | | | | 17 | | | | | |
| | | | | | | | | | 24 | | | | | |
| | | | | | | | | | 21 | | | | | |
| | | | | | | | | | 23 | | | | | |
| | | | | | | | | | 16 | | | | | |
| | | | | | | | | | 9 | | | | | |
| | | | | | | | | | 6 | | | | | |
| | | | | | | | | | 11 | | | | | |
| | | | | | | | | | 5 | | | | | |
| | | | | | | | | | 7 | | | | | |
| | | | | | | | | | 9 | | | | | |
| | | | | | | | | | 8 | | | | | |
| | | | | | | | | | 6 | | | | | |
| | | | | | | | | | 5 | | | | | |
| | | | | | | | | | 9 | | | | | |
| | | | | | | | | | 6 | | | | | |
| | | | | | | | | | 2 | | | | | |
| | | | | | | | | | 5 | | | | | |
| | | | | | | | | | 2 | | | | | |
| | | | | | | | | | 11 | | | | | |
| | | | | | | | | | 10 | | | | | |

Table 1 (continued)

| ZONE | T1 | T2 | T3 | T4 | A1 | A2 | A3 | A4 | B1 | B2 | B3 | B4 | TN1 | TN2 |
|---|---|---|---|---|---|---|---|---|---|---|---|---|---|---|
| | | | | | | | | | 13 | | | | | |
| | | | | | | | | | 7 | | | | | |
| | | | | | | | | | 11 | | | | | |
| | | | | | | | | | 16 | | | | | |
| | | | | | | | | | 13 | | | | | |
| | | | | | | | | | 13 | | | | | |
| | | | | | | | | | 10 | | | | | |
| | | | | | | | | | 17 | | | | | |
| | | | | | | | | | 7 | | | | | |
| | | | | | | | | | 3 | | | | | |
| | | | | | | | | | 11 | | | | | |

(Khan Khukhıy, Lake, and Ider terranes) (Kheraskova, 1995, Mossakovsky et al, 1996) It is not surprising that Mongolia yields numerical in terms of species composition and facies diversity reefs Only a minority of them were under scrutinised study since (Drosdova, 1980, Wood et al, 1993, Kruse et al, 1996; Zhuravlev, 2001)

Further data (68 sampling units) are obtained mostly from the literature on late middle Atdabanian (A3)–middle Toyonian (TN2) archaeocyathan reefs of Spain, Sardinia, France, Morocco, China, South Australia, and North America. However, the majority of localities supplied these data have been observed by the authors either in the field or in thin sections or both

Two sampling units are taken after Debrenne (1975) and Debrenne and Debrenne (1995) to represent early Atdabanian (A1 and A2) of Morocco Five (two of A2, two of B1, and one of TN2) sampling units are borrowed from Spanish data by Debrenne and Zamarreño (1970), Zamarreño and Debrenne (1977), Moreno-Eiris (1988), Perejón (1989), and Vennin et al (2001) A single (B1) sampling units is obtained from southern France (Debrenne, 1964, Courjault-Radé, 1988) Other two sampling units (both TN2) are from South China (Debrenne et al, 1991) A relatively large data set of 20 sampling units (1 of A3, 11 of A4, 5 of B1, 2 of B4, 1 of TN1) is Australian (Kruse and West, 1980, Gravestock, 1984, Debrenne and Gravestock, 1990, James and Gravestock, 1990, Kennard, 1991, Kruse, 1991, Zhuravlev and Gravestock, 1994) Well-known Sardinian sections provide 19 sampling units (9 of B1, 6 of B2, 3 of B3, and 1 of TN2) (Gandin and Debrenne, 1984, Debrenne and Gandin, 1985, Debrenne et al, 1989a, 1993; Perejón et al, 2000) The same

Table 2
Distribution and diversity of reefal communities during the Tommotian–middle Toyonian interval (Early Cambrian) A–Australia, C–South China, F–France, K–Kolyma Uplift, I–Iberian Peninsula, M–Mongolia, Mr–Morocco, P–Siberian Platform, S–Sardinia (sampling units yielding numerical information are in bold)

| COMMUNITY | T1 | T2 | T3 | T4 | A1 | A2 | A3 | A4 | B1 | B2 | B3 | B4 | TN1 | TN2 |
|---|---|---|---|---|---|---|---|---|---|---|---|---|---|---|
| *Cambrocyathellus/Archaeolynthus* framestone-P6 | P | | | | | | | | | | | | | |
| Ajacicyathid/*Renalcis*/hyolith grainstone-P3 | P | P | P | P | | | | | | | | | | |
| *Renalcis* mud mound-P15, M12 | | **P** | **P** | **P** | P | M | M | M | M | | | | | |
| *Cambrocyathellus* bafflestone-P20, M28 | | **P** | **P** | **P** | M | M | | | | | | | | |
| *Dictyocyathus* framestone-P10 | | **P** | **P** | **P** | P | I | I | S | S | | | | | |
| *Okulitchicyathus* bindstone-P17, M25 | | | **P** | **P** | M | | | | | | | | | |
| *Spinosocyathus* framestone-P16 | | **P** | **P** | **P** | Mr | I | | | | | | | | |
| Radiocyath bafflestone-P19, M27 | | | **P** | **P** | M | A | A | A | | | | | | |
| *Tumuliolynthus* bafflestone-P8 | | | **P** | | | | | | | | | | | |
| *Cambrocyathellus/Dictyocyathus* cementstone-P11, M20 | | **P** | **P** | **P** | M | **M** | **M** | | | | | | | |
| Ajacicyathid floatstone-P12, M7 | | | | **P** | **P** | P, K, **M** | **P, M** | **P, M** | **P, M** | | | | | |
| *Dictyosycon/Khasaktia* bindstone-P18, M18 | | | | | **P, M** | | | | | | | | | |
| *Batinevia* grainstone-P1, M4 | | | | | **P** | M | M | M | M, S | | | | | |
| *Renalcis/Epiphyton* mud mound-P9 | | | | | **P** | M | M | M | S, I | | | | | |
| *Tubomorphophyton* dendrolite-P13, M16 | | | | | **P** | **P, M** | **P, M** | **P, MA** | **P, MS** | | | | | P |
| Clotted ajacicyathid mound-P21 | | | | | | | | **P** | P | | | | | |
| Clotted ajacicyathid–hexactinellid mound-P22 | | | | | | | | **P, A** | P | | | | | |
| Ajacicyathid wackestone-P5 | | P | P | P | P | P, **Mr** | P | | P | | | | | |
| Ajacicyathid packstone-P2 | | P | P | P | P | P | P | P | **P, A** | | | | | |
| Ajacicyathid–oolite grainstone-P4, M2 | | | | | M | **M** | M | M | P | | | | | P, S |
| Capsulocyathid–thrombolite-P5 | | | | | | | | | K | | S | | | |
| *Archaeocyathus* framestone-P7 | | | | | | | | | L | L | L | L | | P, C, L |
| Clotted stromatolite-M1 | | | | | **M** | | | | | | | | | |
| *Proaulopora* grainstone-M3 | | | | | | | M | M | M | | | | | |
| *Tarthinia*/pelloidal mound-M5 | | | | | | M | | | | | | | | |
| Ajacicyathid/pelloidal mound-M6 | | | | | | **M** | M | M | M | | | | | |
| *Razumovskia/Tarthinia* mound-M8 | | | | | | **M** | M | M | M | | | | | |
| *Razumovskia/Gordonophyton* dendrolite-M9 | | | | | | **M** | M | M | M | | | | | |
| *Tubomorphophyton/Batinevia* dendrolite-M10 | | | | | | **M** | M | M | M | | | | | |
| *Renalcis*/archaeocyath framestone-M11 | | | | | | **M** | **M** | **M** | **M** | | | | | |
| *Renalcis/Epiphyton* dendrolite-M13 | | | | | | Mr | Mr | Mr | **M** | | | | | |
| *Renalcis/Gordonophyton* dendrolite-M14 | | | | | | | M | M | M, S | | | | | |
| Dense *Epiphyton* dendrolite-M15 | | | | | **M** | **M, Mr** | **M** | **M** | **M** | | | | | |
| *Archaeocyathus/Epiphyton* framestone-M17 | | | | | | | | | **M** | | | | | |
| Cribricyath framestone-M19 | | | | | | | | | **M** | | | | | |
| *Angusticellularia* cementstone-M21 | | | | | | **M** | | | | | | | | |
| *Angusticellularia/Renalcis*/hexactinellid mound-M22 | | | | | | | | | **M** | | | | | |
| *Mikhnocyathus* bindstone-M23 | | | | | | **M** | | | | | | | | |
| Echinoderm bafflestone-M26 | | | | | | | | | **M** | | | | | |
| *Sakhacyathus* bindstone-M24 | | | | | M | **M** | | | | | | | | |
| *Tubomorphophyton/Renalcis/Girvanella* bindstone | | | | | | | A | A | | | | | | |
| *Somphocyathus/Gravestockia* framestone | | | | | | | | A | | | | | | |
| *Anaptyctocyathus/Erismacoscinus* framestone | | | | | | | | | A | | | | | |
| *Agyrekocyathus* framestone | | | | | | | | | A | | | | | |
| *Archaeopharetra* framestone | | | | | | | | A | | | A | | | |
| *Gordonophyton*/archaeocyathid dendrolite | | | | | | | | | | | A | | | |
| Thrombolite | | | | | | I | I | I | I | | | | A | |
| *Archaeocyathus Razumovskia* bindstone | | | | | | | | | L | | S | | | C, I |
| *Anthomorpha* bindstone | | | | | | | | | F, S | | | | | |
| *Girvanella* silty grainstone | | | | | | | | | S | S | | | | |

Table 2 (continued)

| COMMUNITY | T1 | T2 | T3 | T4 | A1 | A2 | A3 | A4 | B1 | B2 | B3 | B4 | TN1 | TN2 |
|---|---|---|---|---|---|---|---|---|---|---|---|---|---|---|
| Girvanella dolostone | | | | | | | | | S | S | | | | |
| Protopharetra/Archaeosycon mound | | | | | | | | | L | | | | | |
| Mackenziecyathus/Spirocyathella boundstone | | | | | | | L | | L | | | | | |
| Loculicyathus/Spirocyathella framestone | | | | | | | | | L | | | | | |
| Fenestrocyathus/Tabulaconus floatstone | | | | | | | | | L | L | | | | |
| Claruscoscinus/Metaldetes oolitic grainstone | | | | | | | | | L | | | | | |
| Archaeocyathus bafflestone in clotted mud | | | | | | | | | | | L | | | |
| Archaeocyathus/Retilamina framestone | | | | | | | | | | | | L | | |
| Archaeocyathus bafflestone with stromatactis | | | | | | | | | | L | | | | |
| Archaeocyathus bafflestone in packstone | | | | | | | | | | | L | | | |

amount of data (7 of B1, 7 of B2, 3 of B3, 1 of B4, and 1 of TN2) represent North America (Debrenne and James, 1981, Voronova et al., 1987, Debrenne et al., 1989b, 1990, Mansy et al , 1993, McMenamin et al , 2000).

The entire list of 220 sampling units is given in Tables 1 and 2

## 3. Methods of calculation

### 3 1 Alpha diversity calculation

The alpha diversity is calculated as a mean number of species in sampling units representing the same palaeocommunity (Table 1). Early Cambrian pattern of alpha diversity is plotted in Fig 1

### 3 2 Community differentiation and beta diversity calculation

The indication of communities is the fundamental problem in the analysis of both alpha and beta diversity The very quality of such a study depends on a community identification This problem becomes even more influential when palaeocommunities are investigated because of an information loss through incomplete preservation and time-averaging during fossilization

Zhuravleva (1972) distinguished archaeocyathan reefal palaeocommunities of the middle Lena River area, which provided also the bulk of the information for the present study, by visual facies differences, relative archaeocyath content, reefal body shapes, and geochemical composition of the rocks containing

fossils Later, she (Zhuravleva in Nikolaeva et al, 1986) added data on principal reef-building organisms, microfacies, and mineral composition of reefal rocks As a result, nine Tommotian–Atdabanian archaeocyathan reefal palaeocommunities were outlined

In the present work, palaeocommunities are distinguished by their microfacies and general faunal and microbial differences as well as by an identification of hub species The latter are responsible for the organisation of a retinue of associated organisms (Miller, 1996) A unique combination of three factors, one of which is pure environmental and other two are influential for the guild structure, provide an approximation of the overall disparity of communities Besides, we have recognised as palaeocommunities only those assemblages of fossils where direct interactions of their faunal and microbial components can be observed These interactions are intergrowing and encrustation with features of mutual responses such as devastation, dwarfing, malformations, and contact avoiding The aforementioned features are preserved in mineralised parts of organisms under study

Hub species are defined here as either the principal reef-building agents or as the most significant in terms of either biovolume or abundance species The hub species were mostly archaeocyaths or renalcids Their role was sometimes taken by other extinct reef-building creatures, which were radiocyaths, coralomorphs, or cribricyaths Archaeocyaths were calcified sponges and probably demosponge relatives Renalcids (Renalcis, Tarthinia, Angusticellularia, Epiphyton, Gordonophyton, Tubomorphophyton, Proaulopora, Batinevia, Ra-

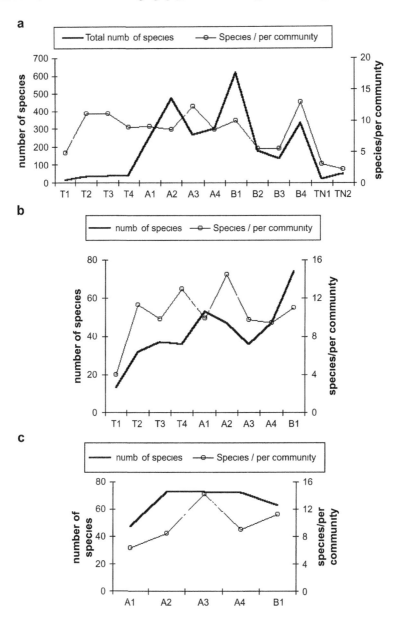

Fig 1 Archaeocyath species diversity and alpha diversity expressed as mean number of species per sampling unit through the Tommotian (T)–Atdabanian (A)–Botoman (B)–Toyonian (TN) stages of the Early Cambrian (a) Total species diversity against global alpha diversity after Table 1 (b) Species diversity against alpha diversity in reefal palaeocommunities of the Siberian Platform (c) Species diversity against alpha diversity in reefal palaeocommunities of Mongolia

zumovskia, Girvanella) were possibly calcified cyanobacteria while coralomorphs (Cysticyathus, Khasaktia, Rackovskia, and Hydroconus) were probable cnidarians, radiocyaths could be another group of problematic sponges, and cribricyaths were tiny enigmatic animals (See Debrenne and Reitner, 2001, Riding, 2001, for details of palaeobiology and palaeoecology of these groups )

Despite different approaches to the classification of reefal palaeocommunities, even in unmatched microfacies definitions between previous works (Zhuravleva, 1972, Nikolaeva et al, 1986) and our own study, the results were basically the same The differences in the number of palaeocommunities are related to the addition of Botoman and Toyonian sites (P4, P7, P14), new finds of clotted microbial microfacies (P21, P22), and to scrutinized subdivision of archaeocyathan-cyanobacterial mounds of Zhuravleva (Nikolaeva et al, 1986) Here, such mounds are further subdivided into 10 palaeocomunities due to different composition of these overall assemblages and hub-species as well as more precise data on microfacies in some cases (P6, P8–P11, P15–P18, and P20)

Mongolian early Atdabanian–early Botoman buildups were diverse Those of the Tsagaan Olom and Khubsugul terrains developed in conditions of pure carbonate sedimentation with a minor siliciclastic and volcanic supply in Tsagaan Olom sites In the Khan Khukhiy and Lake terrains, volcanic arcs provided relatively calm conditions for settlement of archaeocyaths in shallow water and mixed archaeocyath-hexactinellid assemblage in a deeper environment All reef facies of the Ider terrane represent a separate kind of reef which developed under the harsh condition of a high volcanic ash input As a result, endemics comprise half of the species restricted to this terrain

Despite a difference in tectonic regimes between the Siberian Platform and Mongolia, a number of communities occur in both regions, such as *Batinevia* grainstone (P1 and M4), ajacicyathid floatstone (P12 and M7), *Renalcis* mound (P15 and M12), *Gordonophyton* or *Tubomorphophyton* dendrolite (P13 and M16), *Gordonophyton/Khasaktia* bindstone (P18 and M18), cementstone (P11 and M20), *Okulitchicyathus* bindstone (P17 and M26), radiocyath bafflestone (P19 and M28), and *Cambrocyathellus* bafflestone (P20 and M25) These rarely developed concurrently, but contained the same hub and similar accessory species and occupied the same environments Similar community structures are observed in other regions Some localities contain relative species which succeeded ancestor species, creating similar communities For instance, an early Atdabanian *Agastrocyathus* framestone of Spain and Morocco is organised by a genus related to Tommotian *Spinosocyathus* of the Siberian

Platform (P16) Such palaeocommunities are classified together

In terms of microfacies, general taxonomic composition, and hub species, the principal Early Cambrian reefal palaeocommunities recognized on the Siberian Platform and in Mongolia are listed from the shallowest to deepest in Appendices 1 and 2 Table 2 contains the data on the community differentiation plotted against the Early Cambrian time scale.

Beta diversity itself is calculated here judging by data on a number of concurrent communities and is expressed as $\beta = 1 - {}_mS_J$ where ${}_mS_J$ is mean Jaccard similarity coefficient (Fig 2b and c) The latter is computed for each pair of sampling units $i$ and $j$ representing different palaeocommunities as follows

$$S_J = a/a + b + c$$

In this formula, $a$ represents the number of species occurring in sampling units $i$ and $j$ (joint occurrence), $b$ is the number of species present in sampling unit $i$ but not in $j$, $c$ is the number of species present in sampling unit $j$ but not in $i$ This coefficient is less sensitive to variation in sample size, has both upper and lower limits, and is independent of negative matches, all of which makes this formula highly influential in the analysis of palaeocommunities, and it increases linearly with any increase of the number of shared taxa (Sepkoski, 1988, Shi, 1993, Naimark, 2001)

The higher is the similarity in species composition of palaeocommunities, the lower is beta diversity and vice versa This value ($\beta$) is calculated for the early Tommotian–early Botoman (T1-B1) palaeocommunities of the Siberian Platform and for the early Atdabanian–early Botoman (A1-B1) ones of Mongolia (Fig 2b and c)

*3 3 Estimation of the number of provinces and gamma diversity*

Despite multiple publications treating the palaeogeography and palaeobiogeography of the Cambrian world (see Brock et al, 2000, Smith, 2001, for review), the resulting picture is too controversial to

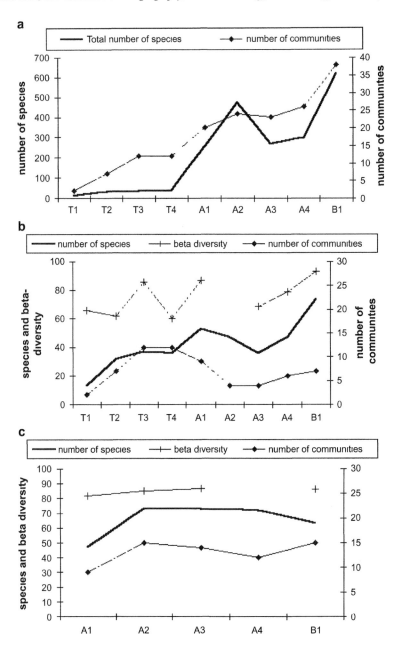

Fig 2 Archaeocyath species diversity and beta diversity expressed as $\beta = 1 - S_J$ and as number of palaeocommunities through the Tommotian (T)–Atdabanian (A)–Botoman (B) stages of the Early Cambrian (a) Total species diversity against total number of palaeocommunities after Table 2 (b) Species diversity against beta diversity in reefal palaeocommunities of the Siberian Platform (c) Species diversity against beta diversity in reefal palaeocommunities of Mongolia

provide a firm framework for the estimation of provincial differentiation within Cambrian marine basins

Thus, the simplest approach has been used in order to get the data on provincial differentiation Here, the faunal basins possessing the same endemic species are scoured to estimate the pattern of provinciality during the Early Cambrian These Early Cambrian faunal basins are

(1) The Siberian Platform including the Kolyma Uplift and the bordering part of the Russian Far East (Yudoma Maya Depression and Shevli terrane) and Transbaikalia (Kylar Graben) (T1-B1, B4, TN2),
(2) The Altay Sayan Foldbelt together with Mongolia, and the remainder of Transbaikalia and of the Russian Far East (A1-TN3),
(3) The South Urals (B1),
(4) Kazakhstan (A2-A3, B4),
(5) Central-East Asia (Tajikistan, Uzbekistan, and Tarim of China) (A4; TN2);
(6) Morocco (A1-B1),
(7) Spain and Germany (A2-B2, B4, TN2),
(8) France together with Sardinia (A2-B4, TN2),
(9) South China (B1, B4; TN2),
(10) Laurentian part of present North America and Koryakia (A4-B4; TN2),
(11) Australia and Antarctica with South African allochthonous clasts (A2-B1, B3-TN2)

In spite of the empirical character of our identification of faunal basins, the resulting pattern closely matches the palaeogeographic interpretation by Kruse and Shi (in Brock et al, 2000) based on the calculation of the mutual phenetic similarities between 15 regions for archaeocyathan genera The later authors using the Jaccard and Ochiai similarity indices to calculate a cophenetic correlation came to the following interpretation of the Early Cambrian archaeocyathan provincial differentiation

(1) Siberia–Mongolia (including the Siberian Platform–Kolyma Uplift, Altay Sayan Foldbelt, Mongolia, Transbaikalia, and the Russian Far East-Hinggan),
(2) Central-East Asia (South Urals, Kazakhstan, Tajikistan, Uzbekistan, Kyrgyzstan, Tarim, and South China),
(3) Europe–Morocco,
(4) Laurentian part of present North America and Koryakia.
(5) Australia and Antarctica with South African authochthonous localities

Some differences are derived from the principal data set (species vs genera) and from the simplification of subdivision that was used by Kruse and Shi For instance, they treated the Russian Far East as a single region, though it is an artificial unit consisting of parts of the Siberian Platform separated by modern administrative boundaries and different ancient terrains which possess a common history with the Altay Sayan Foldbelt Recent monographic analysis of Mongolian reefal species as well as new data on Tuva and Altay Sayan Foldbelt revealed a single species pool existing during the Early Cambrian in these regions as well as in Transbaikalia (Zhuravleva et al, 1997a,b, Osadchaya and Kotel'nikov, 1998, Zhuravlev, 1998) The data on Serbia, Kyrgyzstan, and Hinggan (North China) have been omitted here due to the absence of a representative species set from these regions Finally, the South Urals and Kazakhstan occur within Central-East Asia branch of Kruse and Shi's dendrograms due to negative matches rather than the presence of any species in common The subdivision of the Europe–Morocco cluster into three separate basins (Morocco, Spain and Germany, France and Sardinia) is due to different species pools (over 90 % of endemic species) confined to each of them (Debrenne, 1964, Elicki and Debrenne, 1993, Perejón, 1994, Debrenne and Debrenne, 1995)

The application of the Jaccard coefficient of similarity to a matrix containing data on all Early Cambrian skeletal genera also revealed extremely low values of the index for each pair of 11 provinces listed above (Debrenne et al, 1999) Another numerical approach has been demonstrated by Naimark and Rozanov (1997, Naimark 2001), who calculated the percentage of endemics per region per time slice as well as the average geographic distribution index This index is calculated with the following method An appearance of a genus in a single region is accepted arbitrary as 1 unit, an appearance of a genus in several regions of the same province is scored as 5 units, a global distribution is scored for 10 units Markov and Naimark (1998) showed that the change of unit values does not influence the general pattern of the geographic distribution Both calculations determined an increase of archaeocyathan provinciality from the middle Atdabanian to late Botoman time

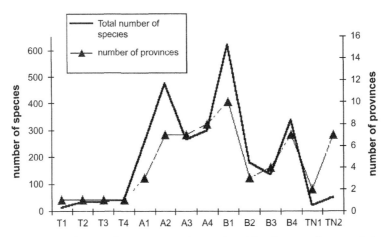

Fig 3 Archaeocyath species diversity and gamma diversity expressed as number of faunal regions ("provinces") through the Tommotian (T)–Atdabanian (A)–Botoman (B)–Toyonian (TN) stages of the Early Cambrian

In accordance, 11 faunal basins are preferable here over five provinces  Also, such a subdivision is due to the absence of firm data on the Cambrian biogeographic hierarchy, allowing the justification of biogeographic provinces  However, as it would follow below, this difference is not influential for the subject of the paper

Gamma diversity is calculated here as a number of concurrent faunal basins per Early Cambrian zone of a total number of 11  Alternately, the Dice (Sørenson) coefficient of similarity is computed for four of six archaeocyathan orders (Monocyathida, Ajacicyathida, Tabulacyathida, and Capsulocyathida)

$$S_D = 2c/a + b$$

In this formula, $a$ represents the number of species occurring in sampling units $i$ and $j$ (joint occurrence); $b$ the number of species present in sampling unit $i$ but not in $j$, $c$ is the number of species present in sampling unit $j$ but not in $i$  This index is calculated for each pair of faunal basins per each early Cambrian zone  Then, the results are averaged for all pairs  The final data on gamma diversity are plotted in Fig 3

## 4. Patterns of diversity

### 4 1  Alpha diversity

The Early Cambrian reefal species are mostly archaeocyaths with minor admixture of cribricyaths,

coralomorphs, and radiocyaths  Some Mongolian sampling units show that the number of cribricyath species approaches closely the number of archaeocyath species  For instance, in the earliest Atdabanian (A1) sampling unit of *Okulitchicyathus* bindstone from Zuune Arts, cribricyaths (9 species) together with a radiocyath (1 species) and a coralomorph (1 species) comprise the bulk of the species diversity (11 of total 16 species)  However, archaeocyaths dominate still in number of individuals as well as volumetrically  In terms of biovolume, cribricyaths are significant in a single cribricyath framestone sampling unit from the Lake terrane of Mongolia (B1)  However, such a high diversity of non-archaeocyath species is a rare exclusion within the entire set of Early Cambrian reefal sampling units

The average number of species per Early Cambrian reefal community is relatively constant and slightly varies abutting 10 (9 1–13 7) species per sampling unit during the middle Tommotian (T2)–the earliest Botoman (B1) and the latest Botoman (B4) intervals (Fig 1a)

Individual sampling units within the limits of these intervals contain from 1 (some sampling units of Mongolia) to 27–33 species (some earliest Botoman localities of Mongolia and latest Botoman sites of South Australia)  The lowermost species diversity is typical of sampling units which possess features of harsh for reefal faunal milieus in stromatolites, dense renalcid dendrolites built by *Gordonophyton* or *Tubomorphophyton*, under volcanic ash falls, increased salinity and under extremely shallow highly agitated conditions If unfavourable physical conditions suppressed the species richness, conditions did not influence the abundance of individuals On the contrary, microbial (stromatolites) and some calcimicrobial (renalcid dendrolites) neighbourhood were depressive for reefal animals in the aspects of species richness and abundance

The largest palaeocommunities, in terms of species number, developed under nearly stable conditions where renalcids were represented by tiny botryoidal forms of mostly pendant habit (*Renalcis, Angusticellularia*) Archaeocyath and coralomorph species were almost equally distributed in terms of both abundance and biovolume within such palaeocommunities As a result, the palaeocommunities lacked any distinct hub species among animals

During the earliest Tommotian (T1), middle Botoman (B2, B3), and Toyonian (TN1, TN2) intervals, the average species number per sampling unit was relatively low, from 2 4 to 5 7 (Fig 1a). The earliest Tommotian low value (4.6) reflects a small size of available species pool (13 archaeocyath species only) Middle Botoman (5 0–5 7) and, more particularly, the Toyonian (2 4) falls in the average number of species per palaeocommunity that coincide with intervals following severe mass extinctions of Early Cambrian marine faunas (Zhuravlev and Wood, 1996) These events were especially devastating for reefal communities Again, the shrinking of the available species pool probably led to the lowering of species number per community.

The alpha diversity pattern was calculated separately for Mongolia and the Siberian Platform, independent of the growing dynamics of the total species diversity (Fig 1b and c) Even at the beginning of the Botoman Stage, when the total species pool drastically enlarged, the alpha diversity did not increase in concert and was held at the same stable level

In summary, with the exception of the earliest Tommotian time of the reefal biota appearance and both middle Botoman and Toyonian intervals of its near disappearance, the plot of values of average species number per sampling unit and the total species diversity do not fluctuate in concert (Fig 1) In contrast to the persistent growth in overall species diversity during the middle Tommotian–early Botoman, alpha diversity does not increase through the same interval and the local species richness stays nearly constant

## 4 2 Beta diversity

The data set on palaeocommunity diversity includes a large array of various sampling units for the earliest Tommotian–middle Toyonian interval (Table 2) A number of palaeocommunities scored per each time slice is plotted against total species diversity (Fig 2a) Fig 2b and c shows the dynamics of beta diversity calculated as both a rough number of concurrent communities and $\beta$ ($\beta = 1 - {}_mS_J$) for the Siberian Platform and Mongolia where sizable and the most evenly sampled with respect to time and environment data matrices are available

Both methods of beta diversity estimation resulted in merely similar plots However, the plot of $\beta$ calculated as the function of Jaccard similarity coefficient reflects a regional species diversity dynamics better This is not surprising because Jaccard similarity coefficient depends to a certain extent on the total species diversity if the number of species per community is almost persistent In order to avoid circularity, it would be cautious to use different and less dependent methods of beta diversity calculation

Equally, large scale (Fig 2a) and a smaller scale (Fig 2b and c) patterns indicate satisfactory matches between diversity of palaeocommunities and total species number despite of an impossibility to score a real number of palaeocommunities This figure should be larger as it follows from an ordinary multiplication of the number of species by the number of communities Nonetheless, the increases and decreases of the species number follow the course of the plot expressing the community number indicating a direct positive dependence of the total diversity on beta diversity

A similar conclusion follows from a simple comparison of Jaccard similarity coefficient values obtained for selected Siberian Platform sampling units The selected sampling units represent

(1) different facies of the same time slice,
(2) different facies of sequential time slices,
(3) same facies of the same time slice, and
(4) same facies of sequential time slices

Table 3
Mean values of beta diversity expressed as Jaccard similarity coefficient between two types of Tommotian–Atdabanian reefal palaeocommunities of the Siberian Platform

| Sampling units | Mud mounds | Dendrolites |
|---|---|---|
| Different time slices and facies | 0 26 | 0 22 |
| Same time slice, different facies | 0 23 | 0 25 |
| Different time slices same facies | 0 63 | 0 55 |
| Same facies and time slice | 0 42 | 0 46 |

Higher deviations are observed in species composition between different facies rather than within the same facies, with time again clearly demonstrating the influence of average beta diversity onto the total species diversity (Table 3) A close statement has been inferred by Zhuravleva (1972) who based on qualitative observations of middle Tommotian–early Botoman archaeocyathan assemblages in the middle courses of the Lena River where a facies pattern varies widely

The palaeocommunities developed within the same facies demonstrate the highest value of Jaccard similarity coefficient at 0 55–0 63 Such values are indicative for the biggest proportion of species shared between palaeocommunities and, therefore, the lowest beta diversity When applying to a wider scale and total number of sample units, this trend is also well expressed (Table 4) The increasing of average beta diversity with time reveals a multiplication of facies variations and of disparity in palaeocommunities

This pattern is particularly well expressed during the late Tommotian–early Atdabanian interval On the Siberian Platform, a transitional phase of especially rapid turnovers within reefal paleocommunities was restricted to this interval The rapidity of reorganisation itself was related to the dispersion of reefal biota into a variety of environments from extremely shallow agitated waters to relatively deep calm conditions below fair-weather wave base

It is noteworthy that mean values of Jaccard similarity coefficient for much more environmentally heterogeneous Mongolian palaeocommunities are lower than those for less diverse Siberian ones Mongolian basins covered a high number of different volcanic arcs, accretionary wedges, microcontinents, and sea mounts while on the Siberian Platform, reefal communities were restricted to a relatively simple ramp-like margins of the epeiric platform (Kheraskova, 1995, Sukhov, 1997) The comparison of data on community number plotting for the Siberian Platform and Mongolia that reveals the community diversity (and, as a result, the species diversity) is the factor of seascape heterogeneity which was very wide in Early Cambrian marine basins of Mongolia This observation is in accordance with a general recognition of an environmental heterogeneity with strong influence on diversity (Zherikhin, 1992, Huston, 1994)

Beta diversity is also commonly expressed as a direct function of turnover (Waide et al , 1999, Bell, 2001) The turnover itself is merely a rate of community composition decay with distance The analysis of our data set with Jaccard similarity coefficient calculation does not indicate such a simple correlation Mean Jaccard similarity coefficient values for both middle Tommotian mud mounds (P15) and late Atdabanian dendrolites (P13) on the Siberian Platform vary independently on the distance between palaeocommunities (Table 5) Even at a larger scale, a late Tommotian *Spinosocyathus* framestone palaeocommunity (P16) of the southern Siberian Platform is much closer to any of the similar but very remote (700 km away) palaeocommunities of the northern

Table 4
Average beta diversity expressed as $\beta=1-S_J$ for total reefal palaeocommunities of the Siberian Platform (Tommotian–early Botoman) and Mongolia (Atdabanian–early Botoman) (number of cell pairs in parentheses)

| Region | T1 | T2 | T3 | T4 | A1 | A2 | A3–4 | B1 |
|---|---|---|---|---|---|---|---|---|
| Siberian Platform | 0 61 [3] | 0 66 [78] | 0 85 [42] | | 0 77 [45] | | 0 76 [6] | 0 95 [15] |
| Western Mongolia | | | | | 0 82 [21] | 0 85 [197] | 0 87 [138] | 0 86 [317] |

Table 5
Variations in mean values of beta diversity expressed as Jaccard similarity coefficient along the same marker bed exemplified by Tommotian (T) and Atdabanian (A) reefal palaeocommunities of the Siberian Platform

| Distance (km) | 0 | 3 5 | 10 | 20 | 24 | 31 | 46 |
|---|---|---|---|---|---|---|---|
| Middle Tommotian (T2) | 0 37 | | 0 34 | | | | 0 38 |
| Late Atdabanian (A4) | 0 25 | 0 23 | 0 08 | 0 31 | 0 33 | 0 18 | |

Siberian Platform ($_mS_J$= 0 19, $N$=4 cell pairs) rather than any of northern palaeocommunities to each other ($_mS_J$=0 16; $N$=6 cell pairs), although the latter are restricted to an area of about 30 km$^2$ Similarly, Sepkoski (1988) did not found any decay of Jaccard similarity coefficient values for Cambrian palaeocommunities along environmental gradient

Mean values of the Jaccard similarity coefficient between adjacent environmental zones calculated for the Cambrian by Sepkoski (1988) are 0 31 It is higher than that for later Palaeozoic periods (0 19–0 22) Comparable index computed for the generic composition of Siberian palaeocommunities comprises values of 0 5 (T1), 0 49 (T2), 0 5 (T4), 0 22 (A1), 0 31 (A3), 0 29 (A4), and 0 15 (B1) Being averaged, this index approaches closely Sepkoski's mean Cambrian value −0 34 However, such a high value is due to extremely high values of the Jaccard similarity coefficient obtained for Tommotian palaeocommunities where each genus includes several species It decreases rapidly in the Atdabanian when almost each genus is represented by a single species Perhaps this phenomenon is related to specialisation This is the very time when obligate cryptobionts appeared among archaeocyaths as well as rheophile species In general, 0 34 value is rather unusual for the Early Cambrian and 0 25 value calculated as the averaged Jaccard similarity coefficient for Atdabanian–Botoman palaeocommunities, probably, better characterises Cambrian beta diversity

### 4 3 Gamma diversity

Gamma diversity is measured here by plotting of number of archaeocyath regions (Fig 3) Independently of a number of separate provinces under consideration (11 or 5), the plot is qualitatively the same but flatter in the last case

Similar to the plot of community numbers, it matches closely the plot of total species diversity for Early Cambrian reefal biota with the earliest and latest Botoman (B1, B4) and middle Toyonian (TN2) peaks as well as with the middle Botoman (B2, B3) and earliest Toyonian (TN1) depressions Interestingly, the rapid growth in the number of separated faunal regions during the first half of the Atdabanian Stage (from 1 to 7) is accompanied by a

similar temps increase in community diversity (nearly 4 times) and in total species number (over 10 times)

This pattern is not surprising Empirically, the mutual dependence of overall species number and number of isolated provinces was postulated customarily since Valentine and Moores (1972, Valentine, 1973) However, such a mirror stroke of both plots is evident for a significant input of gamma diversity into overall species diversity It is not lower than the input of beta diversity

The explanation of this phenomenon adjusted either to purely tectonic events or to some intrinsic biotic features such as higher temps of extinction among non-specialised forms over those of specialists. Because as a rule, specialists were characterised by a restricted distribution, this pattern led to a higher endemicity and, thus, increasing provinciality (Naimark, 2001)

## 5. Summary

The precise and volumetric data on Early Cambrian reefal palaeocommunities allow us to estimate the significance of alpha, beta, and gamma diversity components for the overall species diversity Although our approaches to the distinguishing of palaeocommunities and faunal basins are arbitrary to a certain extent, their consistent application to the entire set of sampling units provides some useful conclusions at least

In general, palaeoecologists have not found a significant influence of alpha diversity onto the global taxonomic diversity (Bambach, 1977, 1985, Sepkoski et al, 1981, Bottjer et al, 1996, DiMichele and Phillips, 1996, Ivany, 1996) Although opposite data are not uncommon (Johansen, 1996, Flynn et al, 1991, Zherikhin, 1992), they are intimately tied with the Mesozoic–Cenozoic "pull of the Recent" Our data on the earliest Phanerozoic reefal palaeocommunities concur with the previous conclusions on the independence of the total species diversity on alpha diversity The number of species per reefal community, probably, was constant in the Early Cambrian, and alpha diversity did not play a significant role in the shaping of species diversity dynamics Only in the earliest Tommotian, when archaeocyathan reefal communities were appearing, and during short intervals of recovery followed early Botoman and early Toyonian mass extinctions, the overall diversity started to increase by alpha diversity addition

The total species diversity is mostly the factor of two variables which are beta and gamma diversity, expressing taxonomic differentiation between communities and the overall taxonomic differentiation between geographic regions, respectively These two components fluctuate in accordance with the total species diversity at the global scale as well as at regional scales Thus, specialisation, which is expressed in the restriction of species to peculiar communities, and degree of endemicity are the principal factors of taxonomic diversity, while the significance of niche partitioning is negligible Endemicity itself is the function of both geographic isolation and community differentiation leading to ecological specialisation (Naimark and Rozanov, 1997) The early evolution of reefal communities was mostly expansive and not intense

Sepkoski (1988) estimated the growth of both alpha and beta diversity from the Cambrian to the Ordovician as 1 5 times (from 0 31 to 0 19 for beta diversity) Such a growth would explain a twofold increase of total diversity in the Early–Middle Ordovician but hardly a four-fold increase However, according to his own data, the decay of Jaccard similarity coefficient due to environmental separation grew from 0 04 (Cambrian) to 0 19 (Ordovician) which would be the source of a four-fold increase Thus, a sufficient increase of total diversity in the Ordovician was mostly the factor of beta diversity growth related to an appearance of new communities and their further separation

Almost constant alpha diversity during most of the existence of archaeocyathan reefal communities is also interesting and, probably, is not merely an artefact of uneven preservation and sampling Kirchner (2002), using spectral analysis of the fossil record, has inferred that some intrinsic factors exist that maintain diversification rates independently on external shocks Although the very nature of such factors is unknown, they can be responsible for alpha diversity persistence Probably, a relatively firm structure of communities prevented an increase in alpha diversity Species were replaced by their

own functional copies recruited from their close relatives New species had to wedge into the habitat gradient at points between overlap of two or more species, thus creating their own, separate communities rather than sharing a common resource with the existing species As a result, beta and gamma diversity grew

## Acknowledgements

The authors thank Loren Babcock for the kind invitation to participate in a special issue of *Palaeogeography, Palaeoclimatology, Palaeoecology*. The bulk of the Mongolian material that served as our study have been sampled by N A Drosdova, V D Fonin, T A Sayutina, Yu I Voronin, and E A Zhegallo from the Palaeontological Institute of the Russian Academy of Sciences The present work was supported in part by the Russian Foundation for Basic Research, project no 00-04-49182

## Appendix A. Early Cambrian reefal palaeocommunities of the Siberian Platform listed from the shallowest to the deepest (their stratigraphic distribution is given in brackets)

(P1) *Batinevia* grainstone representing the shallowest, above fair-weather base facies in the area accumulated in highly agitated conditions (A1)

(P2) ajacicyathid packstone of numerous fragmented archaeocyathan cups formed in highly agitated conditions (T3-B1)

(P3) ajacicyathid–*Renalcis*–hyolith mostly glauconitic fine-grained grainstone, agitated conditions (T1–T4)

(P4) ajacicyathid/oolite grainstone where large archaeocyathan fragments and ooids dominate, agitated conditions (B1, TN2)

(P5) ajacicyathid wackestone consisting mostly of large ajacicyathid and capsulocyathid cups, agitated conditions (T3-B1)

(P6) *Cambrocyathellus/Archaeolynthus* framestone built under shallow subtidal agitated conditions by branching modular *Archaeolynthus* and *Cambrocyathellus* strengthened by encrusting *Renalcis* and minor fibrous synsedimentary cementation (T1)

(P7) *Archaeocyathus* framestone built in shallow subtidal environment by large branching modular *Archaeocyathus*, agitated conditions (TN2)

(P8) *Tumuliolynthus* bafflestone built under shallow subtidal agitated conditions by tiny branching modular *Tumuliolynthus* encrusting by *Renalcis* (T3)

(P9) *Renalcis/Epiphyton* mud mound developed under shallow subtidal conditions with archaeocyaths and coralomorphs (*Khasaktia* and *Hydroconus*) restricted to primary cavities, medium energy (A1)

(P10) *Dictyocyathus* framestone built in shallow subtidal environment by branching modular *Dictyocyathus*, medium energy (T2-A1)

(P11) *Cambrocyathellus/Dictyocyathus* cementstone formed in shallow subtidal conditions by thick synsedimentary cement which encrusted and, thus strengthened archaeocyathan and *Cysticyathus* cups, medium energy (T2–T4)

(P12) ajacicyathid floatstone composing of relatively large, archaeocyathan fragments densely packed in a red argillaceous mudstone where dominance of solitary mushroom-shape cups is indicative of archaeocyathan ability to inhabit a soft substrate, medium energy (T4-B1)

(P13) *Tubomorphophyton* dendrolite in which branching tubular renalcid was almost the exclusive framework-builder that dominated relatively deeper subtidal waters, medium energy (A1-B1; TN2)

(P14) capsulocyathid thrombolite composed of relatively large, chambered archaeocyaths (*Clathricoscinus* and *Tylocyathus*) encrusted by filamentous bacteria, medium energy (B1)

(P15) *Renalcis* mud mound with a participation of archaeocyaths and coralomorphs (*Cysticyathus* and *Khasaktia*) as frame-builders, medium energy (T2-A1)

(P16) *Spinosocyathus* framestone built in subtidal environment by modular *Spinosocyathus* and strengthened by its exoskeletal calcified tissue, contains a diverse archaeocyathan assemblage, medium energy (T2–T4)

(P17) *Okulitchicyathus* bindstone built in subtidal environment by plate-like modular *Okulitchicyathus* and deployed by diverse cavity dwellers, low energy (T3–T4)

(P18) *Dictyosycon/Khasaktia* bindstone built under subtidal conditions by plate-like modular *Khasaktia* and large stick-like *Dictyosycon*, providing together a heterogenous space for numerous inhabitants, low energy (A1)

(P19) radiocyath bafflestone built in subtidal environment by large branching modular radiocyaths, low energy (T3–T4)

(P20) *Cambrocyathellus* bafflestone built in subtidal environment almost exclusively by tiny branching modular *Cambrocyathellus*, low energy (T2–T4)

(P21) red clotted ajacicyathid mud mound bearing a few archaeocyathan fragmented skeletons, low energy (A4-B1)

(P22) clotted ajacicyathid mound of bacterial microclots and mostly hexactinellid spicules, while archaeocyaths composed the third significant component, developed below fair-weather base, low energy (A4-B1)

## Appendix B. Early Cambrian reefal palaeocommunities of Mongolia listed from the shallowest to the deepest (their stratigraphic distribution is given in brackets)

(M1) clotted stromatolite with rare dwarfed *Cambrocyathellus* only of the Tsagaan Olom terrane (A1)

(M2) ooid shoal with abundant rheophile *Ajacicyathus* and *Robustocyathellus* of the Tsagaan Olom and Ider terranes (A1–A4)

(M3) *Proaulopora* grainstone with rheophile *Ajacicyathus–Robustocyathellus–Plicocyathus* assemblage of the Lake terrane (A3-B1)

(M4) *Batinevia* grainstone with either *Cambrocyathellus foraminosus* or *Ardrossacyathus* depending on age of the Khubsugul and Lake terranes (A2-B1)

(M5) *Tarthinia*/pelloidal mound with *Nochoroicyathus* spp , *Alataucyathus*, and *Rackovskia*, of the Lake terrane (A3)

(M6) ajacicyathid-pelloidal mound with *Leptosocyathus–Plicocyathus* surrounded by grainstone of the Lake terrane (A2-B1)

(M7) ajacicyathid floatstone with either *Nochoroicyathus-Rotundocyathus* or *Thalamocyathus–Pycnoidocyathus* of the Lake, Ider and Khubsugul terranes (A2-B1)

(M8) *Razumovskia/Tarthinia* mound with *Orbicyathellus–Loculicyathus–Sakhacyathus* embraced by grainstone of the Lake terrane (A2-B1)

(M9) *Razumovskia/Gordonophyton* dendrolite with *Archaeopharetra* and *Rackovskia* of the Ider and Lake terranes (A2-B1)

(M10) *Tubomorphophyton/Batinevia* dendrolite with *Cambrocyathellus, Khasaktia*, and *Hydroconus* of the Tsagaan Olom and Lake terranes (A2-B1)

(M11) *Renalcis*/archaeocyath framestone with either *Usloncyathus–Khasaktia* or *Irinaecyathus–Clathricoscinus* in floatsone of the Lake terrane (A2-B1)

(M12) *Renalcis* mound with *Archaeopharetra, Alataucyathus*, and *Hydroconus* of the Lake terrane (A2-B1)

(M13) *Renalcis/Epiphyton* dendrolite with *Irinaecyathus–Archaeocyathus* and cryptic capsulocyathids in grainstone of the Lake terrane (B1)

(M14) *Renalcis/Gordonophyton* dendrolite with *Nochoroicyathus* spp , *Erismacoscinus*, and *Cambrocyathellus minutus* of the Lake and Khan Khukhiy terranes (A3-B1)

(M15) dense *Epiphyton* dendrolite with radiocyaths and cryptic capsulocyathids of the Tsagaan Olom and Khubsugul terranes (A1-B1)

(M16) *Gordonophyton* dendrolite mostly with cryptic capsulocyathids of the Tsagaan Olom, Lake, and Khubsugul terranes (A1-B1)

(M17) *Archaeocyathus/Epiphyton* framestone with cryptic capsulocyathids in grainstone of the Lake terrane (B1)

(M18) *Gordonophyton/Khasaktia* bindstone of the Tsagaan Olom terranes (A1)

(M19) cribricyath framestone with *Tollicyathus* of the Lake terrane (B1)

(M20) cementstone with juvenile *Archaeopharetra* and *Usloncyathus* of the Tsagaan Olom and Lake terranes (A1-A3)

(M21) *Angusticellularia* cementstone with *Nochoroicyathus–Archaeopharetra* in floatstone of the Khan Khukhiy terrane (A2)

(M22) *Angusticellularia/Renalcis*–pharetronid/hexactinellid sponge mound of the Lake terrane (B1)

(M23) *Mikhnocyathus* bindstone with radiocyaths and *Khasaktia* of the Ider and Khubsugul terranes (A2)

(M24) *Sakhacyathus* bindstone of the Khubsugul terrane (A1–A2)

(M25) *Cambrocyathellus* bafflestone with cryptic *Hydroconus* and *Khasaktia* of the Tsagaan Olom and Ider terranes (A1–A2)

(M26) *Okulitchicyathus* bindstone of the Tsagaan Olom terrane (A1)

(M27) echinoderm (?eocrinoid) bafflestone with archaeocyaths possessing extremely compound skeletons of the Lake terrane (B1)

(M28) radiocyath bafflestone of the Tsagaan Olom terrane (A1)

## References

Astashkin, V A Pegel'. T V. Repina, L N , Rozanov, A Yu , Shabanov, Yu Ya , Zhuravlev, A Yu , Sukhov S S , Sundukov, V M . 1991 The Cambrian system on the Siberian platform Int Union Geol Sci Publ 27. 1–133

Bambach, R K , 1977 Species richness in marine benthic habitats through the Phanerozoic Paleobiology 3, 152–167

Bambach, R K . 1985 Classes and adaptive variety the ecology of diversification in marine faunas through the Phanerozoic In Valentine, J W (Ed ), Phanerozoic Diversity Patterns Profiles in Macroevolution Princeton Univ Press, Princeton, NJ, pp 191–253

Bell, G , 2001 Neutral macroecology Science 293, 2413–2418

Bottjer. D J. Schubert, J K , Droser, M L , 1996 Comparative evolutionary palaeoecology assessing the changing ecology of the pas In Hart, M B (Ed ), Biotic Recovery from Mass Extinction Events, Geol Soc Spec Publ vol 102 The Geological Society, London, pp 1–13

Brock, G A . Engelbretsen, M J , Jago. J B . Kruse, P D , Laurie, J R , Shergold, J H , Shi, G R , Sorauf, J E , 2000 Palaeobiogeographic affinities of Australian Cambrian faunas Assoc Australas Palaeontol Mem 23. 1–61

Courjault-Radé, P, 1988 Analyse sédimentologique de la formation de l'Orbiel ("alternaces gréso-calcaires" auct, Cambrien inférieur) Evolution tectono-sédimentaire et climatique (versant sud de la Montagne Noire, Massif Central, France) Bull Soc Géol Fr 8e Sér 4, 1003–1013

Debrenne, F, 1964 Archaeocyatha Contribution à l'étude des faunas cambriennes du Maroc, de Sardaigne et de France Notes Mém Serv Géol Maroc 179, 1–265

Debrenne, F, 1975 Formations organogènes du Cambrien inférieur du Maroc In Sokolov, B S (Ed), Iskopaemye Cnidaria, T 2 (Fossil Cnidaria, V 2), Tr Inst geol geofiz Sibirsk otd vol 202 Akad nauk SSSR, pp 19–24

Debrenne, F, Debrenne, M, 1995 Archaeocyaths of the Lower Cambrian of Morocco Beringeria, Spec Issue 2, 121–145

Debrenne, F, Gandin, A, 1985 La formation de Gonnesa (Cambrien, SW Sardaigne) biostratigraphie, paléogéographie, paléoécologie des Archéocyathes Bull Soc Géol Fr, 8e Sér 1 (4), 531–540

Debrenne, F, Gravestock, D I, 1990 Archaeocyatha from the Sellick Hill Formation and Fork Tree limestone on Fleurieu Peninsula, South Australia In Jago, J B, Moore, P S (Eds), The Evolution of a Late Precambrian–Early Palaeozoic rift complex the Adelaide geosyncline, Geol Soc S Austral Spec Publ vol 16, pp 290–309

Debrenne, F, James, N P, 1981 Reef-associated archaeocyathans from the Lower Cambrian of Labrador and Newfoundland Palaeontology 24, 343–378

Debrenne, F, Reitner, J, 2001 Sponges, cnidarians, and ctenophores In Zhuravlev, A Yu, Riding, R (Eds), Ecology of the Cambrian Radiation Columbia Univ Press, New York, pp 301–325

Debrenne, F, Zamarreño, I, 1970 Sur la découverte d'archéocyathes dans le Cambrien du NW de l'Espagne Breviora Geol Asturica 14 (1), 1–11

Debrenne, F, Zhuravlev, A Yu, 1996 Archaeocyatha, palaeoecology a Cambrian sessile fauna Bol Soc Paleontol It, Spec Vol 3, 77–85

Debrenne, F, Gandin, A, Pillola, G L, 1989a Biostratigraphy and depositional setting of Punta Manna Member type-section (Nebida Formation, Lower Cambrian, SW Sardinia, Italy) Riv It Paleontol Stratigr 94, 483–514

Debrenne, F, Gandin, A, Rowland, S M, 1989b Lower Cambrian bioconstructions in northwestern Mexico (Sonora) Depositional setting, paleoecology and systematics of archaeocyaths Géobios 22, 137–195

Debrenne, F, Gandin, A, Gangloff, R A, 1990 Analyse sédimentologique et paléontologie de calcaires organogènes du Cambrien inférieur de Battle Mountain (Nevada, U S A) Annal Paléontol (Vert -Invert) 76, 73–119

Debrenne, F, Gandin, A, Zhuravlev, A, 1991 Palaeoecological and sedimentological remarks on some Lower Cambrian sediments of the Yangtze platform (China) Bull Soc Géol Fr 162, 575–584

Debrenne, F, Gandin, A, Debrenne, M, 1993 Calcaires à archeocyathes du Membre de la Vallée de Matoppa (Formation de Nebida), Cambrien inférieur du sud-ouest de la Sardaigne (Italie) Annal Paléontol (Vert -Invert) 79, 77–118

Debrenne, F, Maidanskaya, I D, Zhuravlev, A Yu, 1999 Faunal migrations of archaeocyaths and Early Cambrian plate dynamics Bull Soc Géol Fr 170, 189–194

DiMichele, W A, Phillips, T L, 1996 Clades, ecological amplitudes, and ecomorphs phylogenetic effects and persistence of primitive plant communities in the Pennsylvanian-age tropical wetlands Palaeogeogr, Palaeoclimat, Palaeoecol 127, 83–105

Drosdova, N A, 1980 Vodorosli v organogennykh postroykakh nizhnego kembriya Zapadnoy Mongolii (Algae in Lower Cambrian organogenous buildups of Western Mongolia) Tr, Sovm Sovet -Mongol Paleontol Eksp 10, 1–140 (in Russian)

Elicki, O, Debrenne, F, 1993 The Archaeocyatha of Germany, Freiberg Forsch h C450 Palaontol, Stratigr, Fazies 1, 3–41

Flynn, L J, Tedford, R H, Zhanxiang, Qui, 1991 Enrichment and stability in the Pliocene mammalian fauna of North China Paleobiology 17, 246–265

Gandin, A, Debrenne, F, 1984 Lower Cambrian bioconstructions in southwestern Sardinia (Italy) Géobios, Mém Spécial 8, 231–240

Gravestock, D I, 1984 Archaeocyatha from the lower part of the Lower Cambrian carbonate sequence in South Australia Assoc Australas Palaeontol Mem 2, 1–139

Huston, M A, 1994 Biological Diversity The Coexistence of Species in Changing Landscapes Cambridge Univ Press, Cambridge, UK

Ivany, L C, 1996 Coordinated stasis or coordinated turnover? Exploring intrinsic vs extrinsic controls on pattern Palaeogeogr, Palaeoclimat, Palaeoecol 127, 239–256

James, N P, Gravestock, D I, 1990 Lower Cambrian shelf and shelf margin buildups, Flinders Ranges, South Australia Sedimentology 37, 455–480

Johansen, M B, 1996 Adaptive radiation, survival and extinction of brachiopods in the northwest European Upper Cretaceous–Lower Paleocene chalk Palaeogeogr, Palaeoclimatol, Palaeoecol 74, 147–204

Kennard, J M, 1991 Lower Cambrian archaeocyathan buildups, Todd River Dolomite, northeast Amadeus Basin, Central Australia sedimentology and diagenesis In Korsch, R J, Kennard, J M (Eds), Geological and Geophysical Studies in the Amadeus Basin, Central Australia, Bull Bur Miner Res vol 236, pp 195–223

Kheraskova, T N, 1995 Paleogeography of Central Asia paleoocean in Vendian and Cambrian In Rodriguez Alonso, M D, Gonzalo Corral, J C (Eds), XIII Reunión de Geología del Oeste Península Caracterización y evolución de la cuenca Neoproterozoico–Cámbrico en la Península Ibérica, 19–30 de Septiembre de 1995 Salamanca-Coimbia, pp 77–80

Kirchner, J W, 2002 Evolutionary speed limits inferred from the fossil record Nature 415, 65–67

Kruse, P D, 1991 Cyanobacterial–archaeocyathan–radiocyathan bioherms in the Wirrealpa Limestone of South Australia Can J Earth Sci 28, 601–615

Kruse, P D, West, P W, 1980 Archaeocyatha of the Amadeus and Georgina Basin J Austral Geol Geophys 5, 165–181

Kruse, P D, Zhuravlev, A Yu James, N P, 1995 Primordial metazoan-calcimicrobial reefs Tommotian (Early Cambrian) of the Siberian Platform Palaios 10, 291–321

Kruse P D, Gandin, A, Debrenne, F, Wood, R, 1996 Early Cambrian bioconstructions in the Zavkhan Basin of western Mongolia Geol Mag 133, 429–444

Mansy, J L, Debrenne, F, Zhuravlev, A Yu, 1993 Calcaires a archéocyathes du Cambrien inférieur du Nord de la Colombie britannique (Canada) Implications paléogéographiques et précisions sur l'extension du continent Américano-Koryakien Géobios 26, 643–683

Markov, A V, Naimark, E B, 1998 Kolichestvennye zakonomernosti makroevolyutsii opyt primeneniya sistemnogo podkhoda k analizu razvitiya nadvidovykh taksonov (Quantitative regularities of macroevolution an attempt of systematic analysis of the development of taxa above species rank) GEOS, Moscow (in Russian with English summary 318 pp )

McMenamin, M A S, Debrenne, F, Zhuravlev, A Yu, 2000 Early Cambrian Appalachian archaeocyaths further age constraints from the fauna of New Jersey and Virginia, U S A Géobios 33, 693–708

Miller III, W, 1996 Ecology of coordinated stasis Palaeogeogr, Palaeoclimatol, Palaeoecol 127, 177–190

Moreno-Eiris, E 1988 Los montéculos arrecifales de Algas y Arqueociatos del Cémbrico Inferior de Sierra Morena III Microfacies y diagénesis Bol Geol Min 98, 591–621

Mossakovsky, A A, Pushcharovsky, Yu M, Ruzhentsev, S V, 1996 Prostranstvenno-vremennye sootnosheniya struktur Tikhookeanskogo i Indo-Atlanticheskogo tipov v pozdnem dokembrii i vende (Relationships of Pacific- and Indian–Atlantic-type structures in the Late Precambrian and Vendian) Dokl Akad Nauk 350, 799–802 (in Russian)

Naimark, E B, 2001 Evolyutsionnay biogeografiya iskopaemykh morskikh bespozvonochnykh model' i primery (Evolutionary Biogeography of Fossil Marine Invertebrates Model and Examples), Avtoref Diss D Biol Sci Palaeontol Inst, Russian Acad Sci, Moscow, 64 pp (in Russian)

Naimark E B, Rozanov, A Yu, 1997 Regularities in the development of regional faunas of archaeocyaths Stratigr, Geol Korrelyatsiya 5, 67–78

Nikolaeva, I V, Zhuravleva, I T, Borodaevskaya, Z V, Repina, L N Kosukhina, I G, Meshkova, N P, Pelman, Yu L, Mandrikova, N T, Perozio, G N, Kameneva, M Yu, Kozlov, G N, 1986 Nizhniy kembriy yugo-vostoka Sibirskoy platformy (litologiya, fatsii, paleoekologiya) (Lower Cambrian of the south-eastern Siberian Platform (lithology, facies, palaeoecology)) Tr Inst, geol geofiz, Sibirsk otdel vol 659 Akad Nauk SSSR, pp 1–230 (in Russian)

Osadchaya, D V, Kotel'nikov, D V, 1998 Archaeocyaths from the Atdabanian (Lower Cambrian) of the Altay–Sayan Foldbelt, Russia Geodiversitas 20, 5–18

Perejón, A, 1989 Arqueociatos del Ovetiense el la sección del Arroyo Pedroche, Sierra de Córdoba, España Bol R Soc Española Hist Nat (Secc Geol ) 84, 143–247

Perejón A, 1994 Palaeogeographic and biostratigraphic distribution of Archaeocyatha in Spain Cour Forsch -Inst Senkenberg 172, 341–354

Perejón, A, Fréhler, M, Bechstédt, T, Moreno-Eiris, E, Boni, M, 2000 Archaeocyathan assemblages from the Gonnesa Group, Lower Cambrian (Sardinia, Italy) and their sedimentological context Bol Soc Paleontol It 39, 257–291

Pratt, B R, Spincer, B R, Wood, R A, Zhuravlev, A Yu, 2001 Ecology and evolution of Cambrian reefs In Zhuravlev, A Yu, Riding, R (Eds ), Ecology of the Cambrian Radiation Columbia Univ Press, New York, pp 254–274

Riding, R, 2001 Calcified algae and bacteria In Zhuravlev, A Yu, Riding, R (Eds ), Ecology of the Cambrian Radiation Columbia Univ Press, New York, pp 445–473

Riding, R, Zhuravlev, A Yu, 1995 Structure and diversity of oldest sponge-microbe reefs lower Cambrian, Aldan River, Siberia Geology 23, 649–652

Rozanov, A Yu, 1980 Centers of origin of Cambrian fauna 26th Ses Congr géol Int, Paris, Résumés, vol 1, pp 181

Sepkoski Jr, J J, 1988 Alpha, beta, or gamma where does all the diversity go? Paleobiology 14, 221–234

Sepkoski Jr, J J, Bambach, R K, Raup, D M, Valentine, W 1981 Phanerozoic marine diversity and the fossil record Nature 293, 435–437

Shi, G R, 1993 Multivariate data analysis in palaeoecology and palaeobiogeography—a review Palaeogeogr, Palaeoclimatol, Palaeoecol 105, 199–234

Smith, A G, 2001 Paleomagnetically and tectonically based global maps for Vendian to mid-Ordovician time In Zhuravlev, A Yu, Riding, R (Eds ), Ecology of the Cambrian Radiation Columbia Univ Press, New York, pp 11–46

Sukhov, S S 1997 Cambrian depositional history of the Siberian craton evolution of the carbonate platforms and basins Sediment Facies Palaeogeogr 17, 27–39

Valentine, J W, 1973 Evolutionary Paleoecology of the Marine Biosphere Prentice-Hall, Englewood Cliffs, NJ

Valentine, J W, Moores, E M, 1972 Global tectonics and the fossil record J Geol 80, 167–184

Vennin, E, Moreno-Eiris, E, Perejón, A Álvaro, J J, 2001 Fracturación sinsedimentaria y diagénesis precoz en las bioconstrucciones del Cámbrico Inferior de Alconera (Ossa-Morena) Rev Soc Geol España 14, 75–88

Voronova, L G, Drosdova, N A, Esakova, N V, Zhegallo, E A, Zhuravlev, A Yu, Rozanov, A Yu, Sayutina, T A Ushatinskaya, G T, 1987 Nizhnekembriyskie iskopaemye gor Makkenzi (Kanada) (Lower Cambrian fossils of the Mackenzie Mountains [Canada]) Tr, Paleontol Inst vol 224 USSR Acad Sci, pp 1–88 (in Russian)

Waide, R B, Willig, M R, Steiner, C F, Mittelbach G, Gough, L, Dodson, S I, Juday, G P, Parmenter R, 1999 The relationship between productivity and species richness Annu Rev Ecol Syst 30, 257–300

Wood, R, 1999 Reef Evolution Oxford Univ Press, Oxford

Wood, R, Zhuravlev, A Yu, Chimed Tseren, A, 1993 The ecology of Lower Cambrian buildups from Zuune Arts, Mongolia implications for early metazoan reef evolution Sedimentology 40, 829–858

Zamarreño, I, Debrenne, F, 1977 Sédimentologie et biologie des constructions organogènes du Cambrien inférieur du Sud de l'Espagne B R G M Mém 89, 49–61

Zherikhin, V V, 1992 Istoricheskie izmeneniya raznoobraziya nasekomykh (Historical changes in the insect diversity) In Yurtsev, B A (Ed ), Biologicheskoe Raznoobrazie Podkhody k Izucheniyu i Sokhraneniyu (Biological Diversity Approaches to the Study and Preservation) Zool Inst Russian Acad Sci, St Petersburg, pp 53–65 (in Russian)

Zhuravlev, A Yu, 1998 Early Cambrian archaeocyathan assemblages of Mongolia IV Field Conference Cambrian Stage Subdivision Working Group, Int Subcom Cambrian Stratigr, Sweden, 24–31 August 1998, Abstracts, pp 24–25

Zhuravlev, A Yu, 2001 Paleoecology of Cambrian reef ecosystems In Stanley Jr, J D (Ed ), The History and Sedimentology of Ancient Reef Systems Plenum Press, New York, pp 121–157

Zhuravlev, A Yu, Gravestock, D I, 1994 Archaeocyaths from Yorke Peninsula, South Australia and archaeocyathan Early Cambrian zonation Alcheringa 18, 1–54

Zhuravlev, A Yu, Wood, R A, 1996 Anoxia as the cause of the mid-Early Cambrian (Botomian) extinction event Geology 24, 311–314

Zhuravleva, I T, 1972 Rannekembriyskie fatsial'nye kompleksy arkheotsiat (r Lena, srednee techenie) (Early Cambrian facies assemblages of archaeocyats (Lena River, middle cources)) In Zhuravleva, I T (Ed ), Problemy Stratigrafii i Paleontologii Nizhnego Kembriya Sibiri (Problems in the Lower Cambrian Stratigraphy and Palaeontology of SiberiaNauka, Novosibirsk, pp 31–109 (in Russian)

Zhuravleva, I T, Boyarinov, A S, Konyaeva, I A, Osadchaya, D V, 1997a Biostratigraphy of the Kiya River section Early Cambrian archaeocyaths from the Kiya River section (Kuznetsk Alatau) Ann Paléontol (Vert -Invert ) 83, 3–92

Zhuravleva, I T, Konyaeva, I A, Osadchaya, D V, Boyarinov, A S, 1997b Biostratigraphy of the Kiya River section Early Cambrian archaeocyaths and spicular sponges from the Kiya River section (Kuznetsk Alatau) Ann Paléontol (Vert -Invert ) 83, 115–200

Printed and bound by CPI Group (UK) Ltd, Croydon, CR0 4YY

08/05/2025

01864934-0002